Introduction to Avionics Systems

R. P. G. Collinson

Introduction to Avionics Systems

Fourth Edition

 Springer

R. P. G. Collinson
(deceased)
Maidstone, Kent, UK

ISBN 978-3-031-29217-0   ISBN 978-3-031-29215-6   (eBook)
https://doi.org/10.1007/978-3-031-29215-6

This Springer imprint is published by the registered company Springer Nature Switzerland AG
The registered company address is: Gewerbestrasse 11, 6330 Cham, Switzerland

Foreword

This book 'Introduction to Avionics Systems' is now into its fourth edition. This reflects the fact that the book has over the years become a standard for students and practising engineers in the field of avionics. The book is available in a number of languages other than English.

Inevitably since the third edition was published in 2011, both component and system technology have continued to advance. Turning the telescope of history around, avionics has its origins in analogue technology and each function had a unique 'black box' requiring extensive cabling between the units. The steady march of miniaturization of components and the transfer to digital technology could be predicted leading to multi-function boxes connected by digital data busses. The reliability of modern systems has shown an equally remarkable improvement. Some advances were not seen, and this is most notable in display systems where after decades of only minor changes we have seen the arrival of optical waveguide technology and solid-state displays.

The extensive revisions include, on displays, updates for the latest developments in helmet-mounted displays (HMDs), use of helmet-mounted rate gyros for helmet tracking, updates on HUD/HMD optical waveguide system technology and the latest advances with replacing CRTs with solid-state displays in HUDs. On controls and fly-by-wire, the section on civil aircraft has been updated to cover the Airbus A350 and the advances in its flight control system over the Airbus A380. The opportunity has also been taken to add a new section on automatic flight control of vectored thrust aircraft covering both the BAE Systems Harrier and the Lockheed Martin F-35B Lightning 2 Joint Strike Fighter. For the F-35B, the flight control systems for vertical landing is covered in some detail.

Dick Collinson's career spanned the whole of this period of avionics development both during his time working in the industry and in retirement as an enthusiast for technology.

It is a credit to his knowledge that the fundamental principles described in this book are just as fresh and comprehensible as ever; it is only the mechanisation of those fundamentals that has changed. This fourth edition was largely completed just before Dick Collinson died in February 2022 with his brain as sharp as ever. His two sons (Matthew and Shawn) have picked up the challenge of tidying up the work to produce a textbook which will continue to be a reference and inspiration for many years to come.

Formerly Chief Technologist, BAE Systems
Rochester, Kent, UK

Christopher Bartlett

Preface

My aims in writing this book were to explain the basic principles underlying the key, or core, avionic systems in modern civil and military aircraft and their implementation using modern technology. Technology is continually advancing, and this fourth edition incorporates the advances being made since the third edition was finished 12 years ago. The opportunity has been taken to extend the coverage of particular areas, where relevant, and helicopter flight control has been added.

The core systems covered comprise pilot's display systems and man-machine interaction, fly-by-wire flight control systems, inertial sensor systems, navigation systems, air data systems, autopilots and flight management systems, and avionic system integration. Unmanned air vehicles (UAVs) are briefly discussed.

The systems are analysed mathematically (where appropriate) from the physical laws governing their behaviour so that the system design and response can be understood and the performance analysed. Worked examples are included to show how the theory can be applied to a representative system. Physical explanations are set out of the system behaviour and the text is structured so that readers can 'fast forward' through the maths and just accept the results if they so wish.

The systems covered are all 'flight safety critical'. Their implementation using modern digital technology to meet the very high safety and integrity requirements is explained together with the overall integration of these systems.

A particular aim, based on my experience over many years, is to meet the needs of graduates (or equivalent) entering the avionics industry who have been educated in a wide variety of disciplines, for example, electronic engineering, computer science, mathematics, physics, mechanical and aeronautical engineering. The book also aims to meet the needs of engineers at all levels working in particular areas of avionics who require an appreciation and understanding of other areas and disciplines.

A further objective is to show the very wide range of disciplines which are involved in avionic systems, as this makes the subject an interesting and challenging field. Apart from the interest inherent in aircraft, the range of disciplines and technologies which are exploited covers aerodynamics and aircraft control, satellite navigation, optical gyroscopes, man-machine interaction, speech recognition, advanced display systems, holographic optics, intelligent knowledge-based systems, closed-loop control systems, high-integrity failure survival systems, high-integrity software, integrated circuit design and data bus systems.

    Personally, I have found avionics to be a very interesting and challenging area to work in, and I hope this book will help the reader to share this interest.

    Note: The illustration shows the visual impact of the avionic systems in a modern aircraft.

Maidstone, Kent, UK   Dick Collinson
February 2022

Acknowledgements

I would like to thank the management of BAE SYSTEMS Avionics Group, Rochester for their assistance in writing this book and their permission to use a number of illustrations showing BAE SYSTEMS equipment and concepts.

I would also like to express my appreciation to former senior managers within the company who have given their support and encouragement over the years in producing the first and second editions of this book, and who have now retired. In particular, Sue Wood, Brian Tucker, Robin Sleight and Ron Howard.

My thanks and appreciation to my former colleagues in the company, Chris Bartlett, Ted Lewis and Paul Wisely for their help and advice on display technology and John Corney for his help and advice on all aspects of flight control, particularly helicopters. The 'teach-ins' over a pint of real ale at the Tiger Moth pub have been a pleasant part of writing this book. Paul refers to them as 'literary lunches'.

I would like to express my appreciation and thanks to Derek Jackson, former Managing Director of Smiths Aerospace, Cheltenham for his helpful comments and for writing the foreword to the third edition.

I would also like to thank my former colleagues Gordon Belcher, Andrew Gibson, Derek Hamlin, Robin Heaps and Dave Jibb for their help in obtaining information, checking the draft chapters and providing helpful and constructive comments in the earlier editions of this book; most of the theoretical coverage remains unchanged.

I would also like to thank Professor David Allerton, of Cranfield University and Professor John Roulston, Technical Director, BAE SYSTEMS, Avionics Group, for their help and support in producing the first and second editions.

Grateful acknowledgement is made to the following companies and organisations for permission to use their illustrations and material.

AIRBUS
BAE SYSTEMS
BEI Systron Donner Inertial Division
Honeywell
Royal Aeronautical Society Library

SavetheRoyalNavy.org
Schlumberger Industries,
Smiths Industries.
The Boeing Company

Finally, I would like to thank my wife and family for their whole-hearted support and encouragement in writing this book.

Post note: Sadly, just before Richard Collinson completed the fourth edition, he passed away on the 16th February 2022 at the age of 95. The fourth edition has been completed by his sons Matthew Collinson and Shawn Collinson with the invaluable support of Chris Bartlett, retired former Chief Technologist at BAe Systems, Rochester, Kent who has reviewed and suggested amendments to the Edition 4 changes.

Contents

Introduction

1.1 Importance and Role of Avionics

'Avionics' is a word derived from the combination of aviation and electronics. It was first used in the USA in the early 1950s and has since gained wide-scale usage and acceptance although it must be said that it may still be necessary to explain what it means to the lay person on occasions.

The term 'avionic system' or 'avionic sub-system' is used in this book to mean any system in the aircraft that is dependent on electronics for its operation, although the system may contain electro-mechanical elements. For example, a fly-by-wire (FBW) flight control system depends on electronic digital computers for its effective operation, but there are also other equally essential elements in the system. These include solid slate rate gyroscopes and accelerometers to measure the angular and linear motion of the aircraft and air data sensors to measure the height, airspeed and incidence. There are also the pilot's control stick and rudder sensor assemblies and electro-hydraulic servo actuators to control the angular positions of the control surfaces.

The avionics industry is a major multi-billion dollar industry worldwide and the avionics equipment on a modern military or civil aircraft can account for around 30% of the total cost of the aircraft. This figure for the avionics content is more like 40% in the case of a maritime patrol/anti-submarine aircraft (or helicopter) and can be over 75% of the total cost in the case of an airborne warning and control system (AWACS) aircraft.

Modern general aviation aircraft also have significant avionics content. For example, colour head down displays (HDDs), global positioning system (GPS) satellite navigation systems, radio communications equipment. Avionics can account for 10% of their total cost.

The avionic systems are essential to enable the flight crew to carry out the aircraft mission safely and efficiently, whether the mission is carrying passengers to their destination in the case of a civil airliner, or, in the military case, intercepting a hostile aircraft, attacking a ground target, reconnaissance or maritime patrol.

A major driver in the development and introduction of avionic systems has been the need to meet the mission requirements with the minimum flight crew. In the case of a modern civil airliner, this means a crew of two only, namely the First Pilot (or Captain) and the Second Pilot. This is only made possible by reducing the crew workload by automating the tasks that used to be carried out by the Navigator and Flight Engineer. The achievement of safe two-crew operation has very considerable economic benefits for the airline in a highly competitive market with the consequent saving of crew salaries, expenses and training costs. The reduction in weight is also significant and can be translated into more passengers or longer range on less fuel. This is because unnecessary weight is geared up ten to one, as will be explained later. In the military case, a single-seat fighter or strike (attack) aircraft is lighter and costs less than an equivalent two-seat version. The elimination of the second crew member (Navigator/Observer/Radar Operator) has also significant economic benefits in terms of reduction in training costs. (The cost of training and selection of aircrew for fast jet operation is very high.)

Other very important drivers for avionic systems are increased safety; air traffic control (ATC) requirements; all weather operation; reduction in fuel consumption; improved aircraft performance, control and handling; and reduction in maintenance costs.

Military avionic systems are also being driven by a continuing increase in the threats posed by the defensive and offensive capabilities of potential aggressors.

The role played by the avionic systems in a modern aircraft in enabling the crew to carry out the aircraft mission can be explained in terms of a hierarchical structure comprising layers of specific tasks and avionic system functions as shown in Fig. 1.1. This shows the prime, or 'core', functions that are mainly common to both military and civil aircraft. It must be pointed out, however, that some avionic systems

R. P. G. Collinson, *Introduction to Avionics Systems*, https://doi.org/10.1007/978-3-031-29215-6_1

Fig. 1.1 Core avionic systems

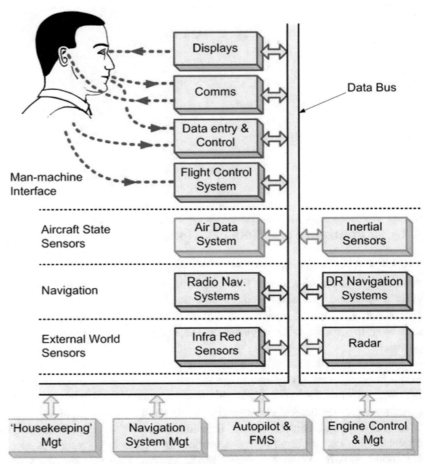

have been left off this diagram for clarity. For example, the Air Traffic Control (ATC) transponder system, the Ground Proximity Warning System (GPWS) and the Threat Alert/ Collision Avoidance System (TCAS), all of which are mandatory equipment for civil airliners.

GPWS provides warning by means of a visual display and audio signal ('Pull up, Pull up ... ') that the aircraft is on a flight path that will result in flying into the terrain, and that action must be taken to change the flight path.

TCAS provides an alerting and warning display of other aircraft in the vicinity in terms of their range, course and altitude together with advisory collision avoidance commands.

Referring to Fig. 1.1, it can be seen that the main avionic sub-systems have been grouped into five layers according to their role and function. These are briefly summarised below in order to provide an overall picture of the roles and functions of the avionic systems in an aircraft.

It should be noted that unmanned aircraft (UMAs) are totally dependent on the avionic systems. These are briefly discussed in Chap. 10.

1.1.1 Systems that Interface Directly with the Pilot

These comprise displays, communications, data entry and control and flight control.

The *Display Systems* provide the visual interface between the pilot and the aircraft systems and comprise head up displays (HUDs), helmet mounted displays (HMDs) and head down displays (HDDs). Most combat aircraft are now equipped with an HUD. A small but growing number of civil aircraft have HUDs installed. The HMD is also an essential system in modern combat aircraft and helicopters. The prime advantages of the HUD and HMD are that they project the display information into the pilot's field of view so that the pilot can be head up and can concentrate on the outside world.

The HUD now provides the primary display for presenting the essential flight information to the pilot and in military aircraft has transformed weapon-aiming accuracy. The HUD can also display a forward-looking infrared (FLIR) video picture one to one with the outside world from a fixed FLIR imaging sensor installed in the aircraft. The infrared (IR) picture merges naturally with the visual scene operations to be carried out at night or in conditions of poor visibility clouds. The HMD enables the pilot to be presented with information while looking in any direction, as opposed to the limited forward field of view of the HUD. An essential element in the overall HMD system is the Helmet Tracker system to derive the direction of the pilot's sight line relative to the aircraft axes.

This enables the pilot to designate a target to the aircraft's missiles. It also enables the pilot to be cued to look in the direction of a threat(s) detected by the aircraft's Defensive Aids system. The HMD can also form part of an indirect viewing system by driving a gimballed infrared imaging sensor to follow the pilot's line of sight. Image intensification devices can also be integrated into the HMD. These provide a complementary night-vision capability enabling the aircraft (or helicopter) to operate at night or in poor visibility.

Colour head down displays have revolutionised the civil flight-deck with multi-function displays eliminating the inflexible and cluttered characteristics of 1970s' generation flight-decks with their numerous dial-type instrument displays dedicated to displaying one specific quantity only.

The multi-function colour displays provide the primary flight displays (PFDs) of height; airspeed; Mach number; vertical speed; artificial horizon; pitch angle, bank angle and heading; and velocity vector. They provide the navigation displays, or horizontal situation indicator (HSI) displays, which show the aircraft position and track relative to the destination or waypoints together with the navigational information and distance and time to go. The weather radar display can also be superimposed on the HSI display. Engine data are presented on multi-function colour displays so that the health of the engines can easily be monitored and differences from the norm highlighted. The aircraft systems, for example, electrical power supply system, hydraulic power supply system, cabin pressurisation system and fuel management system, can be shown in easy-to-understand line diagram format on the multi-function displays. The multi-function displays can also be reconfigured in the event of a failure in a particular display.

The *Communications Systems* play a vital role: the need for reliable two-way communication between the ground bases and the aircraft or between aircraft is self-evident and is essential for air traffic control. A radio transmitter and receiver equipment was in fact the first avionic system to be installed in an aircraft and goes back as far as 1909 (Marconi Company). The communications radio suite on modern aircraft is a very comprehensive one and covers several operating frequency bands. Long-range communication is provided by high-frequency (HF) radios operating in the band 2–30 MHz. Near- to medium-range communication is provided in civil aircraft by very-high-frequency (VHF)

radios operating in the band 30–100 MHz and in military aircraft by ultra-high-frequency (UHF) radios operating in 250–400 MHz (VHF and UHF are line of sight propagation systems). Equipment is usually at duplex level of redundancy; the VHF radios are generally at triplex level on a modern airliner. Satellite communications (SATCOM) systems are also installed in many modern aircraft and these are able to provide very reliable worldwide communication.

The *Data Entry and Control Systems* are essential for the crew to interact with the avionic systems. Such systems range from keyboards and touch panels to the use of direct voice input (DVI) control, exploiting speech recognition technology, and voice warning systems exploiting speech synthesisers.

The *Flight Control Systems* exploit electronic system technology in two areas, namely auto-stabilisation (or stability augmentation) systems and FBW flight control systems. Most swept wing jet aircraft exhibit a lightly damped short-period oscillatory motion about the yaw and roll axes at certain height and speed conditions, known as 'Dutch roll', and require at least a yaw auto-stabiliser system to damp and suppress this motion; a roll auto-stabiliser system may also be required. The short-period motion about the pitch axis can also be insufficiently damped and a pitch auto-stabiliser system is necessary. Most combat aircraft and many civil aircraft in fact require three axis auto-stabilisation systems to achieve acceptable control and handling characteristics across the flight envelope.

FBW flight control enables a lighter, higher-performance aircraft to be produced compared with an equivalent conventional design by allowing the aircraft to be designed with a reduced or even negative natural aerodynamic stability. It does this by providing continuous automatic stabilisation of the aircraft by computer control of the control surfaces from appropriate motion sensors. The system can be designed to give the pilot a manoeuvre command control that provides excellent control and handling characteristics across the flight envelope. 'Carefree manoeuvring' characteristics can also be achieved by automatically limiting the pilot's commands according to the aircraft's slate. A very high integrity, failure survival system is of course essential for FBW flight control.

1.1.2 Aircraft State Sensor Systems

These comprise the air data systems and the inertial sensor systems.

The *Air Data Systems* provide accurate information on the air data quantities, that is, the altitude, calibrated airspeed, vertical speed, true airspeed, Mach number and airstream incidence angle. This information is essential for the control and navigation of the aircraft. The air data computing system computes these quantities from the outputs of very accurate sensors that measure the static pressure, total pressure and the outside air temperature. The airstream incidence angle is derived from airstream incidence sensors.

The *Inertial Sensor Systems* provide the information on aircraft attitude and the direction in which it is heading, which is essential information for the pilot in executing a manoeuvre or flying in conditions of poor visibility, flying in clouds or at night. Accurate attitude and heading information are also required by a number of avionic sub-systems, which are essential for the aircraft's mission – for example, the autopilot and the navigation system and weapon aiming in the case of a military aircraft.

The attitude and heading information is provided by the inertial sensor system(s). These comprise a set of gyros and accelerometers that measure the aircraft's angular and linear motion about the aircraft axes, together with a computing system that derives the aircraft's attitude and heading from the gyro and accelerometer outputs. Modern attitude and heading reference systems (AHRS) use a strapped down (or body mounted) configuration of gyros and accelerometers as opposed to the earlier gimballed systems.

The use of very-high-accuracy gyros and accelerometers to measure the aircraft's motion enables an inertial navigation system (INS) to be mechanised, which provides very accurate attitude and heading information together with the aircraft's velocity and position data (ground speed, track angle and latitude/longitude co-ordinates). The INS in conjunction with the air data system also provides the aircraft velocity vector information. The INS is thus a very important aircraft slate sensor system – it is also completely self-contained and does not require any access to the outside world.

1.1.3 Navigation Systems

Accurate navigation information, that is, the aircraft's position, ground speed and track angle (direction of motion of the aircraft relative to true North), is clearly essential for the aircraft's mission, whether civil or military. Navigation systems can be divided into dead reckoning (DR) systems and position fixing systems; both types are required in the aircraft.

The *Dead Reckoning Navigation Systems* derive the vehicle's present position by estimating the distance travelled from a known position from knowledge of the speed and direction of motion of the vehicle. They have the major advantages of being completely self-contained and independent of external systems.

The main types of DR navigation systems used in aircraft are as follows:

(a) Inertial navigation systems: The most accurate and widely used systems.

(b) Doppler/heading reference systems: These are widely used in helicopters.

(c) Air data/heading reference systems: These systems are mainly used as a reversionary navigation system being of lower accuracy than (a) or (b).

A characteristic of all DR navigation systems is that the position error builds up with time and it is therefore necessary to correct the DR position error and update the system from position fixes derived from a suitable position fixing system.

The *Position Fixing Systems* used are now mainly radio navigation systems based on satellite or ground-based transmitters. A suitable receiver in the aircraft with a supporting computer is then used to derive the aircraft's position from the signals received from the transmitters.

The prime position fixing system is without doubt GPS – global positioning system. This is a satellite navigation system of outstanding accuracy that has provided a revolutionary advance in navigation capability since the system started to come into full operation in 1989.

There are also radio navigation aids such as VOR/DME and TACAN that provide the range and bearing (R/Θ) of the aircraft from ground beacon transmitters located to provide coverage of the main air routes.

Approach guidance to the airfield/airport in conditions of poor visibility is provided by the instrument landing system (ILS), or by the later microwave landing system (MLS).

A full navigation suite on an aircraft is hence a very comprehensive one and can include INS, GPS, VOR/DME, ILS and MLS. Many of these systems are at duplex level and some may be at triplex level.

1.1.4 Outside World Sensor Systems

These systems, which comprise both radar and infrared sensor systems, enable all weather and night time operation and transform the operational capability of the aircraft (or helicopter).

A very brief description of the roles of these systems is given below.

The *Radar Systems* installed in civil airliners and many general aviation aircraft provide weather warning. The radar looks ahead of the aircraft and is optimised to detect water droplets and provide warning of storms, cloud turbulence and severe precipitation so that the aircraft can alter course and avoid such conditions, if possible. It should be noted that in severe turbulence, the violence of the vertical gusts can subject the aircraft structure to very high loads and stresses. These radars can also generally operate in ground mapping and terrain avoidance modes.

Modern fighter aircraft generally have a ground attack role as well as the prime interception role and carry very sophisticated multi-mode radars to enable them to fulfil these dual roles. In the airborne interception (AI) mode, the radar must be able to detect aircraft up to 100 miles away and track while scanning and keeping tabs on several aircraft simultaneously (typically at least 12 aircraft). The radar must also have a 'look down' capability and be able to track low-flying aircraft below it.

In the ground attack or mapping mode, the radar system is able to generate a map-type display from the radar returns from the ground, enabling specific terrain features to be identified for position fixing and target acquisition.

The *Infrared Sensor Systems* have the major advantage of being entirely passive systems. Infrared (IR) sensor systems can be used to provide a video picture of the thermal image scene of the outside world either using a fixed FLIR sensor or, alternatively, a gimballed IR imaging sensor. The thermal image picture at night looks very much like the visual picture in daytime, but highlights heat sources, such as vehicle engines, enabling real targets to be discriminated from camouflaged decoys. An IR system can also be used in a search and track mode; the passive detection and tracking of targets from their IR emissions is of high operational value as it confers an all-important element of surprise.

FLIR systems can also be installed in civil aircraft to provide enhanced vision in poor visibility conditions in conjunction with an HUD.

1.1.5 Task Automation Systems

These comprise the systems that reduce the crew workload and enable minimum crew operation by automating and managing as many tasks as appropriate so that the crew role is a supervisory management one. The tasks and roles of these are very briefly summarised below.

Navigation Management comprises the operation of all the radio navigation aid systems and the combination of the data from all the navigation sources, such as GPS and the INS systems, to provide the best possible estimate of the aircraft position, ground speed and track. The system then derives the steering commands for the autopilot so that the aircraft automatically follows the planned navigation route, including any changes in heading as particular waypoints are reached along the route to the destination. It should be noted that this function is carried out by the flight management system (FMS) (if installed).

The *Autopilots and Flight Management Systems* have been grouped together because of the very close degree of integration between these systems on modern civil aircraft. It should be noted, however, that the Autopilot is a 'stand-alone' system and not all aircraft are equipped with an FMS.

The autopilot relieves the pilot of the need to fly the aircraft continually with the consequent tedium and fatigue and so enables the pilot to concentrate on other tasks associated with the mission. Apart from basic modes, such as height hold and heading hold, a suitably designed high-integrity autopilot system can also provide a very precise control of the aircraft flight path for such applications as automatic landing in poor or even zero visibility conditions. In military applications, the autopilot system in conjunction with a suitable guidance system can provide automatic terrain, following, or terrain avoidance. This enables the aircraft to fly automatically at high speed at very low altitudes (100–200 ft) so that the aircraft can take advantage of terrain screening and stay below the radar horizon of enemy radars.

Sophisticated FMS has come into wide-scale use on civil aircraft since the early 1980s and has enabled two-crew operations of the largest, long-range civil airliners. The tasks carried out by the FMS include the following:

- Flight planning.
- Navigation management.
- Engine control to maintain the planned speed or Mach number.
- Control of the aircraft flight path to follow the optimised planned route.
- Control of the vertical flight profile.
- Ensuring the aircraft is at the planned three-dimensional (3D) position at the planned time slot – often referred to as four-dimensional (4D) navigation. This is very important for air traffic control.
- Flight envelope monitoring.
- Minimising fuel consumption.

The *Engine Control and Management Systems* carry out the task of control and the efficient management and monitoring of the engines. The electronic equipment involved in a modern jet engine is very considerable: it forms an integral part of the engine and is essential for its operation. In many cases, some of the engine control electronics are physically mounted on the engine. Many modern jet engines have a full authority digital engine control system (FADEC). This automatically controls the flow of fuel to the engine combustion chambers by the fuel control unit so as to provide a closed-loop control of engine thrust in response to the throttle command. The control system ensures that the engine limits in terms of temperatures, engine speeds and accelerations are not exceeded, and the engine responds in an optimum manner to the throttle command. The system has what is known as full authority in terms of the control it can exercise on the engine and a high-integrity failure survival control system is essential. Otherwise, a failure in the system could seriously damage the engine and hazard the safety of the aircraft. An FADEC engine control system is thus similar in many ways to an FBW flight control system.

Fig. 1.2 The Eurofighter Typhoon. The aircraft behind the Typhoon is a development prototype of the new BAE Systems Tempest, the next-generation strike fighter. (Courtesy of BAE Systems)

Other very important engine avionic systems include engine health monitoring systems that measure, process and record a very wide range of parameters associated with the performance and health of the engines. These give early warning of engine performance deterioration, excessive wear, fatigue damage, high vibration levels, excessive temperature levels, etc.

Housekeeping Management is the term used to cover the automation of the background tasks that are essential for the aircraft's safe and efficient operation. Such tasks include the following:

- Fuel management. This embraces fuel flow and fuel quantity measurement and control of fuel transfer from the appropriate fuel tanks to minimise changes in the aircraft trim.
- Electrical power supply system management.
- Hydraulic power supply system management.
- Cabin/cockpit pressurisation systems.
- Environmental control system.
- Warning systems.
- Maintenance and monitoring systems. These comprise monitoring and recording systems that are integrated into an on-board maintenance computer system. This provides the information to enable speedy diagnosis and rectification of equipment and system failures by pin-pointing faulty units and providing all the information, such as part numbers, for replacement units down to module level in some cases.

The above brief summaries cover the roles and importance of the avionic systems shown in Fig. 1.1. It should be pointed out, however, that there are several major systems, particularly on military aircraft, which have not been mentioned.

Space constraints limit the coverage of the systems shown in Fig. 1.1 to Displays, Data Entry and Control, Flight Control, Inertial Sensor Systems, Navigation Systems, Air Data Systems, Autopilots and Flight Management Systems, Data Buses and avionic systems integration.

It is not possible to cover Communications, Radar, Infra-red Systems, Housekeeping Management and Engine Control, except for a very limited description of the vertical compressor fan and its control.

1.1.6   Illustrative Examples of Impact on Modern (2021) Avionic Systems

The impact of the avionic systems on a modern 2021 aircraft can be seen in the following illustrations.

Figure 1.2 shows the Eurofighter Typhoon fighter aircraft in service with the air forces of the UK, Germany, Italy and Spain. The Typhoon is a highly agile single-seat fighter/strike aircraft with outstanding performance.

The fore-plane and elevons provide positive trim lift. It is highly unstable in pitch at subsonic speeds (time to double pitch amplitude following a disturbance is around 0.2 s). It is also unstable in yaw at supersonic speeds because of the reduced size. A high-integrity FBW flight control system compensates for the lack of natural stability and provides excellent control and handling across the flight envelope and under all loading conditions.

Figure 1.3 shows the Typhoon Cockpit, which is designed to deliver optimum levels of tactical and functional information without overloading the pilot.

The wide field of view holographic HUD and the HDDs, including the centrally located video colour moving map display, can be seen together with the small centrally

Fig. 1.3 Eurofighter cockpit showing displays and pilot with the new BAE Systems Striker® II Binocular HMD. (Courtesy of BAE Systems)

Fig. 1.4 'Striker® II' HMD night-vision display. (Courtesy of BAE Systems)

mounted pilot's control stick. The unobstructed view of the display resulting from the use of a small control stick is apparent.

A major attribute of the Typhoon cockpit is the new BAE Systems/'Striker® II' binocular helmet mounted display, which the pilot is shown wearing in Fig. 1.4. The 'Striker® II' Binocular HMD offers major improvements in reliability and performance over the previous version. The 0.5″ diameter cathode ray tubes (CRTs) have been replaced with high-resolution, matrix-addressable colour displays, eliminating the 13,000 volt supply that was required for the CRTs to

give a major improvement in reliability. The solid-state display enables important features to be enhanced by the use of colour to assist the pilot's situation awareness. The two night-vision cameras mounted on the sides of the 'Striker® II' helmet have been eliminated and replaced with a single new-technology high-performance vision sensor mounted in the front of the helmet (in what has been called a 'Cyclops' configuration after the Greek mythological one-eyed monster), which achieves significant weight savings. A new high-performance hybrid opto-inertial technology has been developed, which creates an unlimited 'head motion box' and minimises any delay in determining where the pilot is looking. The tracking system immediately calculates the pilot's exact head position and angle; this means that no matter where the pilot is looking, the HMD displays accurate targeting information and symbology with near-zero latency. The system can position symbology onto the visor, even if optical tracking is lost.

Figure 1.4 illustrates the quality of the Striker II HMD night-vision display.

The F-35 Lightning design would not be possible without modern precision avionics. The design exploits vectored thrust control to achieve vertical landing and the number of controls that need to be operated to achieve this are too many for the pilot to operate, and a high-integrity Automatic Flight Control System is used to manoeuvre the aircraft. The pilot commands the direction and the speed of flight path velocity vector operating the pilot's control stick and throttle, and the automatic flight control system follows the command manoeuvre. Excellent control and handling characteristics have been achieved. Figure 1.5 shows the Lockheed Martin F-35 Lightning 2 carrying out a backwards landing on HMS Queen Elizabeth II, the Royal Navy's new aircraft carrier.

The F-35 cockpit design has eliminated the head up display and the pilot relies on a helmet mounted display together with an enormous head down display totally replacing the conventional cockpit instruments. The large display panel is programmable into functional areas, as required.

The Boeing Chinook Helicopter first flew around 1960 and has been in continual service and production. It has been continually updated throughout its life and remains a highly competitive helicopter by making full use of the latest digital avionic systems. Figure 1.6 shows the Boeing Chinook Mk6 and a photo of the Chinook Mk6 Helicopter cockpit, showing the pilot cockpit displays and control and throttle sticks, and motion view. It features a fly-by-wire flight control system with a mechanical reversionary control system. It incorporates computer controlled 'Force Feel' pilot's control stick and throttle.

The 'Force Feel' enables the pilot to exert fingertip precision control manoeuvring.

The advantages of the computer controlled 'Force Feel' active inceptor are as follows:

1. Increased pilot awareness of flight mode and conditions through tactile feedback.
2. Improved handling in hover, low-speed flight and degraded visual conditions.
3. Electronic linking of two inceptors across the cockpit between pilot and co-pilot avoids issues inherent with non-linked passive inceptors.
4. Greater mission success: Reduced exceedances of airframes, gearbox and engine limits extend maintenance intervals, which reduces costs through increased task accuracy and precision.

Fig. 1.5 Lockheed Martin F-35 Lightning 2 carrying out a backwards landing on HMS Queen Elizabeth II, the Royal Navy's new aircraft carrier. (Courtesy of http://savetheroyalnavy.org/wp)

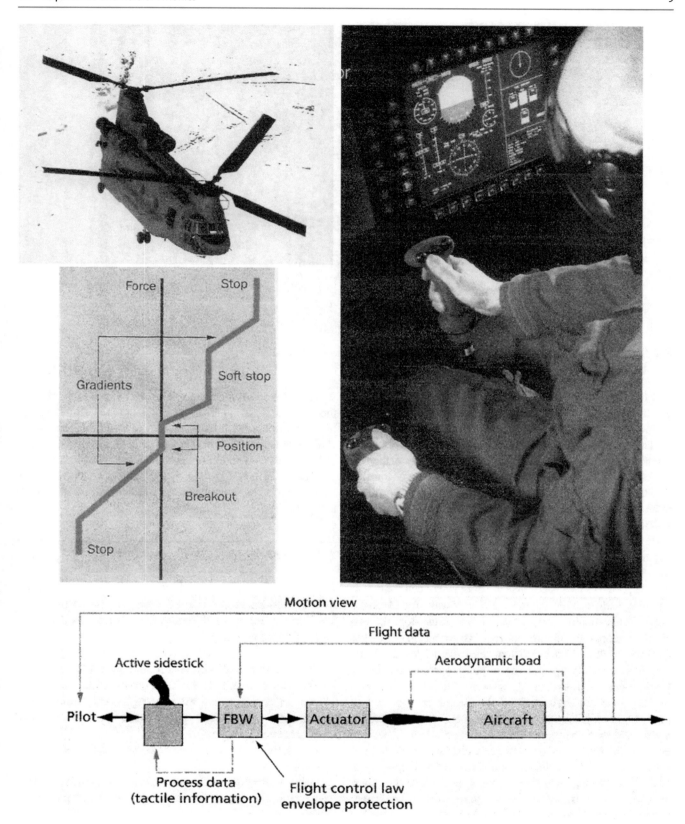

Fig. 1.6 Boeing Chinook Mk6 and a photo of Chinook Mk6 Helicopter cockpit, showing the pilot cockpit displays and control and throttle sticks, and motion view. (Courtesy of BAE Systems)

Fig. 1.7 Photograph of the new Airbus A-350 long-range airliner. (Courtesy of Airbus)

Modern avionic systems are just as important in civil aviation and Fig. 1.7 is a photograph of the new Airbus A-350 long-range airliner (by courtesy of Airbus Limited).

The A-350 is capable of carrying over 300 passengers over a 6000–8000 nautical mile (NM) range. It makes extensive use of carbon composite structures throughout the airframe.

The A-350 has achieved a significant structural weight reduction thereby improving the fuel economy. The Airbus A-350 competes directly with the Boeing 787.

It should be noted that two high-thrust, high by-pass jet engines to propel the aircraft are more fuel efficient than the four high by-pass engines required on the Boeing 747's and Airbus A-380's. This is because the drag is reduced (only two engines) and the total engine weight reduced for the same reason. This enables a lower total engine thrust from the two engines compared with the four required for the 747's and the A-380's. It should be noted that the orders for the Airbus A-380 have dried up and the aircraft is no longer being produced although existing A-380's are still being operated.

The advanced flight-deck of the A-350 can be seen in Fig. 1.8 and features six widescreen, high-resolution, matrix-addressable organic light-emitting diode (OLED) display panels. Each pilot has two wide display screens able to show a primary flight display and a navigation display. The central console has two large screens, one above the other, which normally display engine and systems data. The display screens are identical and fully interchangeable and have the same part number for reduced spares requirements, while the advanced design and mature technology reduce maintenance costs by up to 80%.

The pilot's side stick controllers can be seen at the sides of the flight-deck; the FBW flight control system eliminates the bulky control column between the pilots and instruments of earlier-generation aircraft. This ensures an unobstructed view of the displays. The multipurpose control and display units, in addition to accessing the flight management system, are also used to give systems maintenance data in the air and on the ground.

1.2　The Avionic Environment

Avionic systems equipment is very different in many ways from ground-based equipment carrying out similar functions. The reasons for these differences are briefly explained in view of their importance.

1. The importance of achieving minimum weight.
2. The adverse operating environment in military aircraft in terms of operating temperature range, acceleration, shock, vibration, humidity range and electro-magnetic interference.
3. The importance of very high reliability, safety and integrity.
4. Space constraints particularly in military aircraft requiring an emphasis on miniaturisation and high packaging densities.

The effects on the design of avionic equipment to meet these requirements can result in the equipment costing up to

Fig. 1.8 Airbus A-350 flight-deck. (Courtesy of Airbus)

ten times as much as equivalent ground-based electronic equipment. The aircraft environment requirements are briefly discussed below.

1.2.1 Minimum Weight

There is a gearing effect on unnecessary weight, which is of the order of 10:1. For example, a weight saving of 10 kg enables an increase in the payload capability of the order of 100 kg. The process of the effect of additional weight is a vicious circle. An increase in the aircraft weight due to, say, an increase in the weight of the avionics equipment requires the aircraft structure to be increased in strength, and therefore made heavier, in order to withstand the increased loads during manoeuvres (assuming the same maximum normal acceleration, or g, and the same safety margins on maximum stress levels are maintained). This increase in aircraft weight means that more lift is required from the wing and the accompanying drag is thus increased. An increase in engine thrust is therefore required to counter the increase in drag, and the fuel consumption is thus increased. For the same range, it is thus necessary to carry more fuel and the payload has to be correspondingly reduced, or, if the payload is kept the same, the range is reduced. For these reasons, tremendous efforts are made to reduce the equipment weight to a minimum and weight penalties can be imposed if equipment exceeds the specified weight.

1.2.2 Environmental Requirements

The environment in which avionic equipment has to operate can be a very severe and adverse one in military aircraft; the civil aircraft environment is generally more benign but is still an exacting one.

Considering just the military cockpit environment, such as that experienced by the HUD and HDD, the operating temperature range is usually specified from −40 °C to +70 °C. Clearly, the pilot will not survive at these extremes but if the aircraft is left out in the Arctic cold or soaking in the Middle-East sun, for example, the equipment may well reach such temperatures. A typical specification can demand full performance at 20,000 ft. within 2 minutes of take-off at any temperature within the range.

Vibration is usually quite severe and, in particular, airframe manufacturers tend to locate the gun right under the displays. Power spectral energy levels of 0.04 g^2 per Hz are encountered in aircraft designed in the 1970s and levels of 0.7 g^2 per Hz at very low frequencies are anticipated in future installations. It is worth noting that driving over cobblestones will give you about 0.001 g^2 per Hz.

The equipment must also operate under the maximum acceleration or g to which the aircraft is subjected during manoeuvres. This can be 9 g in a modern fighter aircraft and the specification for the equipment would call up at least 20 g.

The electro-magnetic compatibility (EMC) requirements are also very demanding. The equipment must not exceed the specified emission levels for a very wide range of radio frequencies (RFs) and must not be susceptible to external sources of very high levels of RF energy over a very wide frequency band.

The equipment must also be able to withstand lightning strikes and the very high electro-magnetic pulses (EMP) that can be encountered during such strikes.

Design of electronic equipment to meet the EMC requirements is in fact a very exacting discipline and requires very careful attention to detail design.

1.2.3 Reliability

The overriding importance of avionic equipment reliability can be appreciated in view of the essential roles of this equipment in the operation of the aircraft. It is clearly not possible to repair equipment in the operation of the aircraft. It is clearly not possible to repair equipment in flight, so equipment failure can mean aborting the mission or significant loss of performance or effectiveness in carrying out the mission. The cost of equipment failures in airline operation can be

Fig. 1.9 Young Isaac Newton about to experience a force of 1 Newton

very high – interrupted schedules, loss of income during 'aircraft on the ground' situations, etc. In military operations, aircraft availability is lowered and operation capability lost.

Every possible care is taken in the design of avionic equipment to achieve maximum reliability. The quality assurance (QA) aspects are very stringent during the manufacturing processes and also very frequently call for what is referred to as 'reliability shake-down testing', or RST, before the equipment is accepted for delivery. RST is intended to duplicate the most severe environmental conditions to which the equipment could be subjected, in order to try to eliminate the early failure phase of the equipment life cycle. A typical RST cycle requires the equipment to operate satisfactorily through the cycle described below.

- Soaking in an environmental chamber at a temperature of +70 °C for a given period.
- Rapidly cooling the equipment to −55 °C in 20 minutes and soaking at that temperature for a given period.
- Subjecting the equipment to vibration, for example 0.5 g amplitude at 50 Hz, for periods during the hot and cold soaking phases.

A typical specification would call for 20 RST cycles without a failure before acceptance of the equipment. If a failure should occur at the nth cycle, the failure must be rectified and the remaining $(20 - n)$ cycles repeated. All failures in services (and in testing) are investigated by the QA team and remedial action taken, if necessary.

The overall cost differential in meeting all the factors imposed by the avionic equipment environment can thus be appreciated.

1.3 Choice of Units

It is appropriate at this point to explain the units that are used in this book. All quantities in this book are defined in terms of the SI system of units – *Le Systèm Internationale d'Unités.*

The SI unit of mass is the kilogram (kg), the unit of length is the metre (m) and the unit of time is the second (s). The SI unit of temperature is the Kelvin (K). Zero degree Kelvin corresponds to the absolutely lowest temperature possible at which almost all molecular translational motion theoretically stops. To convert temperature in degree Celsius (°C) to degree Kelvin, it is necessary to add 273.15. Thus, 0 °C = 273.15°K and 30 °C = 303.15°K, and conversely 0° K = −273.15 °C.

The SI unit of force is the Newton (N), one Newton being the force required to accelerate a mass of one kilogram by one metre per second. Converting to pound, foot, second units, one Newton is approximately equal to a force of 0.22 pound (lb) weight. The apocryphal story of 'an apple falling on Newton's head triggering off his theory of gravitation' gives an appropriate twist to the unit of the Newton, as an average apple weighs about 0.22 lb.! Figure 1.9 illustrates the event!

Pressure in the SI system is measured in N/m^2 or Pascals (Pa). The millibar (mb), which is equal to 1/1000 bar, is also widely used, the bar being derived from BARometric. One bar corresponds to the atmospheric pressure under standard sea level conditions and is equal to 100,000 N/m^2 or 100 kPa, and 1 millibar is equal to 100 N/m^2 or 100 Pa.

Altitude is quoted in feet as this is the unit used by the Air Traffic Control authorities, and altimeters in the USA and the UK are calibrated in feet.

Speed is also quoted in knots as this unit is widely used in navigation. One knot is one nautical mile (NM) per hour and one nautical mile is equal to the length of the arc on the Earth's surface subtended by an angle of 1 minute of arc measured from the Earth's centre, which can be related directly to latitude and longitude.

1 NM = 6076.1155 ft. (or 1852 m exactly). The conversion from knots to metres/second is given by 1 knot = 0.5144 m/s and conversely 1 m/s = 1.9438 knots. A useful approximate conversion is 1 knot = 0.5 m/s, or 1 m/s = 2 knots.

Displays and Man–Machine Interaction

2

2.1 Introduction

The cockpit display systems provide a visual presentation of the information and data from the aircraft sensors and systems to the pilot (and crew) to enable the pilot to fly the aircraft safely and carry out the mission. They are thus vital to the operation of any aircraft as they provide the pilot, whether civil or military, with:

- Primary flight information.
- Navigation information.
- Engine data.
- Airframe data.
- Warning information.

The military pilot has also a wide array of additional information to view, such as:

- Infrared (IR) imaging sensors.
- Radar.
- Tactical mission data.
- Weapon aiming.
- Threat warnings.

The pilot is able to rapidly absorb and process substantial amounts of visual information but it is clear that the information must be displayed in a way which can be readily assimilated, and unnecessary information must be eliminated to ease the pilot's task in high workload situations. A number of developments have taken place to improve the pilot–display interaction and this is a continuing activity as new technology and components become available. Examples of these developments are:

- Head up displays (HUDs)
- Helmet mounted displays (HMDs)
- Multi-function colour displays
- Digitally generated colour moving map displays
- Synthetic pictorial imagery
- Displays management using intelligent knowledge-based system (IKBS) technology
- Improved understanding of human factors and involvement of human factor specialists from the initial cockpit design stage

Equally important and complementary to the cockpit display systems in the 'man–machine interaction' (MMI) are the means provided for the pilot to control the operation of the avionic systems and to enter data. Again, this is a field where continual development is taking place. Multi-function keyboards and multi-function touch panel displays are now widely used. Speech recognition technology has now reached sufficient maturity for 'direct voice input' (DVI) control to be installed in the new generation of military aircraft. Audio warning systems are now well established in both military and civil aircrafts. The integration and management of all the display surfaces by audio/tactile inputs enables a very significant reduction in the pilot's workload to be achieved in the new generation of single seat fighter/strike aircraft. Other methods of data entry which are being evaluated include the use of eye trackers.

It is not possible in the space of one chapter to cover all aspects of this subject which can readily fill several books. Attention has, therefore, been concentrated on providing an overview and explanation of the basic principles involved in the following topics:

- Head up displays (Sect. 2.2)
- Helmet mounted displays (Sect. 2.3)
- Computer aided optical design (Sect. 2.4)
- Discussion of HUDs vs HMDs (Sect. 2.5)
- Head down displays (Sect. 2.6).=
- Data fusion (Sect. 2.7)
- Intelligent displays management (Sect. 2.8)
- Display technology (Sect. 2.9)
- Control and data entry (Sect. 2.10)

© The Author(s), under exclusive license to Springer Nature Switzerland AG 2023
R. P. G. Collinson, *Introduction to Avionics Systems*, Methods in Molecular Biology 2687,
https://doi.org/10.1007/978-3-031-29215-6_2

2.2 Head up Displays

2.2.1 Introduction

Without doubt the most important advance to date in the visual presentation of data to the pilot has been the introduction and progressive development of the head up display or HUD (The first production HUDs, in fact, went into service in 1962 in the Buccaneer strike aircraft in the UK).

The HUD has enabled a major improvement in man–machine interaction (MMI) to be achieved as the pilot is able to view and assimilate the essential flight data generated by the sensors and systems in the aircraft whilst head up and maintaining full visual concentration on the outside world scene.

The basic concept of a head up display is shown in Fig. 2.1. An HUD basically projects a collimated display in the pilot's head up forward line of sight (LOS) so that he can view both the display information and the outside world scene at the same time. The fundamental importance of collimating the display cannot be overemphasised and will be explained in more detail later. Because the display is collimated, that is, focused at infinity (or a long distance ahead), the pilot's gaze angle of the display symbology does not change with head movement so that the overlaid symbology remains conformal, or stabilised, with the outside display data at the same time without having to change the direction of gaze or refocus the eyes. There are no parallax errors so aiming symbols for either a flight path director, or for weapon aiming in the case of a combat aircraft, remain overlaid on a distant 'target' irrespective of the pilot's head movement. (Try sighting on a landmark using a mark on a window. The aiming mark moves off the 'target' if the head is moved sideways – this effect is parallax.)

The advantages of head up presentation of essential flight data such as the artificial horizon, pitch angle, bank angle, flight path vector (FPV), height, airspeed and heading can be seen in Fig. 2.2, which shows a typical head up display as viewed by the pilot during the landing phase.

It should be noted that the location of the basic flight data on the HUD display follows that of the basic 'T' of the cockpit instrument panel, viz.:

Indicated Airspeed   Artificial Horizon   Altitude
Compass Heading

This is because all pilots learn to fly with the basic instruments located in these positions; an internationally accepted convention.

Figure 2.2 shows a presentation of primary flight information on an HUD. The display shows the artificial horizon with the aircraft making a 3.6^0 descent, on a heading of 00^0. The left-hand scale shows an airspeed of 137 knots and the right-hand scale an altitude of 880 ft. The flight path vector symbol shows where the aircraft's Centre of Gravity (CG) is moving relative to the horizon – a conventional blind flying panel only shows where the aircraft is pointing relative to the horizon; if the aircraft flight path was maintained, it would be the impact point with the ground.

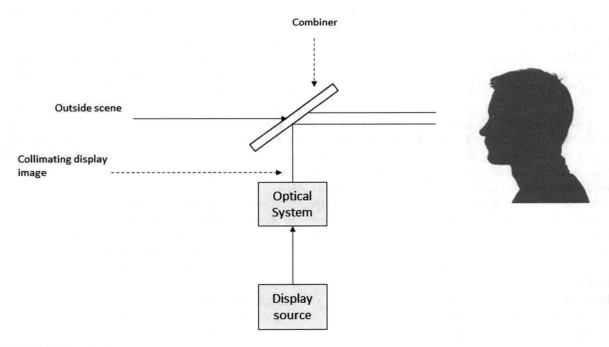

Fig. 2.1 Basic head up display concept

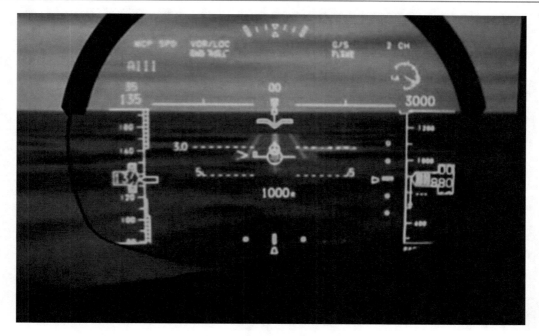

Fig. 2.2 Head up presentation of primary flight information. (Courtesy of BAE Systems)

The pilot is thus free to concentrate on the outside world during manoeuvres and does not need to look down at the cockpit instruments or head down displays (HDDs). It should be noted that there is a transition time of one second or more to refocus the eyes from viewing distant objects to viewing near objects a metre or less away, such as the cockpit instruments and display, and adapt to the cockpit light environment. In combat situations, it is essential for survival that the pilot is head up and scanning for possible threats from any direction. The very high accuracy which can be achieved by an HUD and computerised weapon aiming system together with the ability to remain head up in combat have made the HUD an essential system on all modern combat aircraft.

Figure 2.3 illustrates a typical weapon aiming display. The illustration shows a weapon aiming display for a 'Dive Attack' using the 'Continuously Computed Impact Point' (CCIP) to determine when to release the bombs. The Impact Point is where a bomb would fall if released at that instant, and this is continuously computed and displayed as the 'Bomb Fall Line'.

The pilot manoeuvres the aircraft so that the Bomb Fall Line goes through the Target Marker symbol, making any necessary correction to maintain this alignment. When the 'Bomb Impact Point' marker symbol reaches coincidence with the 'Target Marker' symbol, the pilot releases the bombs.

The illustration depicts a dive attack with a pitch attitude of minus 30 degrees at an altitude of 3000 feet (at that instant) and an indicated airspeed of 450 knots.

2.2.2 Basic Principles

The basic configuration of an HUD is shown schematically in Fig. 2.4. The pilot views the outside world through the HUD combiner glass (and windscreen). The combiner glass is effectively a 'see-through' mirror with a high optical transmission efficiency so that there is little loss of visibility looking through the combiner and windscreen. It is called a combiner as it optically combines the collimated display symbology with the outside world scene viewed through it. Referring to Fig. 2.4, the display symbology generated from the aircraft sensors and systems (such as the inertial navigation system [INS] and air data system) is displayed on the surface of a cathode ray tube (CRT). The display images are then relayed through a relay lens system which magnifies the display and corrects for some of the optical errors which are otherwise present in the system. The relayed display images are then reflected through an angle of near 90° by the fold mirror and thence to the collimating lens which collimates the display images which are then reflected from the combiner glass into the pilot's forward field of view (FOV). The virtual images of the display symbology appear to the pilot to be at infinity and overlay the distant world scene, as they are collimated. The function of the fold mirror is to enable a compact optical configuration to be achieved so that the HUD occupies the minimum possible space in the cockpit.

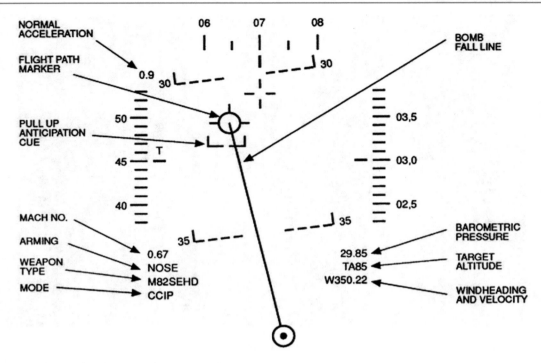

Fig. 2.3 Typical weapon aiming display. (Courtesy of BAE Systems)

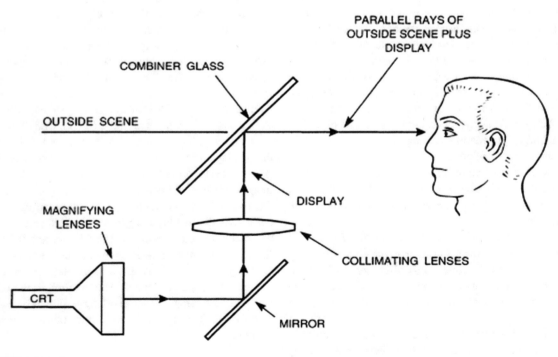

Fig. 2.4 HUD schematic

The fundamental importance of collimation to any HUD system merits further explanation for the benefit of readers whose knowledge of optics needs refreshing.

A collimator is defined as an optical system of finite focal length with an image source at the focal plane. Rays of light emanating from a particular point on the focal plane exit from the collimating system as a parallel bunch of rays, as if they came from a source at infinity.

Figure 2.5a, b shows a simple collimating lens system with the rays traced from a source at the centre, O, and a

point, D, on the focal plane respectively. A ray from a point on the focal plane which goes through the centre of the lens is not refracted and is referred to as the 'non-deviated ray'. The other rays emanating from the point are all parallel to the non-deviated ray after exiting the collimator.

It should be noted that the collimating lens, in practice, would be made of several elements to minimise the unacceptable shortcomings of a single element.

It can be seen from Fig. 2.5a that an observer looking parallel to the optical axis will see point O at eye positions A, B and C, and the angle of gaze to view O is independent of the displacement of the eye from the optical axis.

Similarly, it can be seen from Fig. 2.5b that the angle of gaze from the observer to see point D is the same for eye positions A, B and C and is independent of translation.

The refractive HUD optical system, in fact, is basically similar to the simple optical collimation system shown in Fig. 2.6. The rays are traced for the observer to see points D, O and E on the display with eye positions at points A, B and C. It can be seen that the angles of gaze to see points D, O and E are the same from points A, B or C. The appearance of the collimated display is thus independent of the position (or translation) of the eye and is only dependent on the angle of gaze. Also because of collimation, the display appears to be at infinity as the rays emanating from any point on the display are all parallel after exiting the collimating system.

It should be noted that the display images must be correctly collimated. De-collimation is the term used when the light from a given point on the display does not leave the optical system parallel over the entire lens surface. The light can converge, diverge or otherwise 'misbehave' resulting in 'swimming' of the display images when the pilot's head moves. Sometimes this creates discomfort and in the case of convergence can even cause nausea.

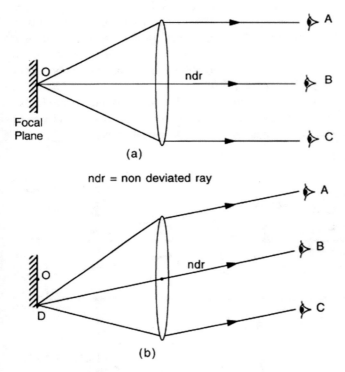

ndr = non deviated ray

Fig. 2.5 Simple optical collimator

Fig. 2.6 Simple optical collimator ray trace

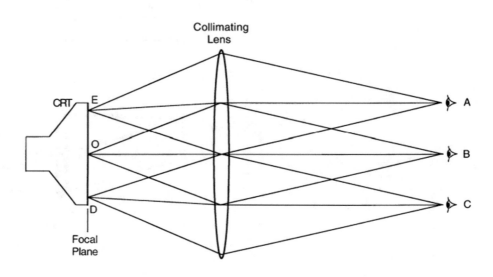

A very important parameter with any HUD is the field of view (FOV), which should be as large as possible within the severe space constraints imposed by the cockpit geometry. A large horizontal FOV is particularly important to enable the pilot to 'look into turns' when the HUD forms part of a night vision system and the only visibility the pilot has of the outside world is the forward-looking infrared (FLIR) image displayed on the HUD.

It is important to distinguish between the instantaneous field of view (IFOV) and the total field of view (TFOV) of a HUD as the two are not the same in the case of the refractive type of HUD.

The instantaneous field of view is the angular coverage of the imagery which can be seen by the observer at any specific instant and is shown in the simplified diagram in Fig. 2.7. It is determined by the diameter of the collimating lens, D, and the distance, L, of the observer's eyes from the collimating lens.

$$\text{IFOV} = 2\tan^{-1}D/2\,L$$

The total field of view is the total angular coverage of the CRT imagery which can be seen by moving the observer's eye position around. TFOV is determined by the diameter of the display, A, and effective focal length of the collimating lens, F.

$$\text{TFOV} = 2\tan^{-1}A/2\,F$$

Reducing the value of L increases the IFOV as can be seen in Fig. 2.7b, which shows the observer's normal eye position brought sufficiently close to the collimating lens for the IFOV to equal the TFOV. However, this is not practical with the conventional type of HUD using refractive optics. This is because of the cockpit geometry constraints on the pilot's eye position and the space constraints on the diameter of the collimating lens. The IFOV is generally only about two-thirds of the TFOV.

It can be seen from Fig. 2.7c that by moving the head up or down or side to side the observer can see a different part of the TFOV, although the IFOV is unchanged.

Fig. 2.7 Instantaneous and total FOVs

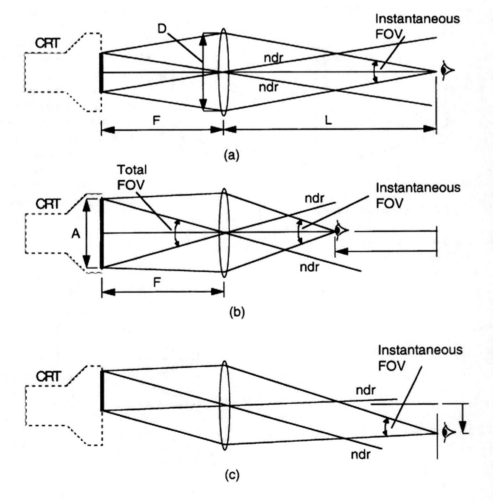

Fig. 2.8 HUD installation constraints and field of view

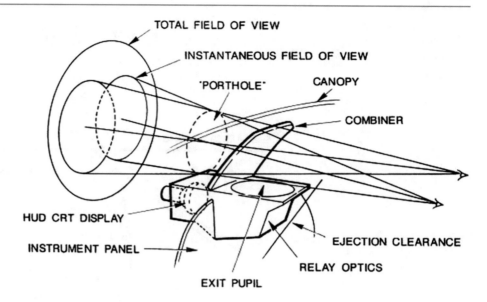

TOTAL FIELD OF VIEW

INSTANTANEOUS FIELD OF VIEW

'PORTHOLE'

CANOPY

COMBINER

HUD CRT DISPLAY

INSTRUMENT PANEL

EJECTION CLEARANCE

RELAY OPTICS

EXIT PUPIL

The effect is like looking through and around a porthole formed by the virtual image of the collimating lens as can be seen in Fig. 2.8. The diagram shows the IFOV seen by both eyes (cross-hatched), the IFOV seen by the left and right eyes respectively and the TFOV.

The analogy can be made of viewing a football match through a knot hole in the fence and this FOV characteristic of an HUD is often referred to as the 'knot hole effect'.

The constraints involved in the HUD design are briefly outlined below.

For a given TFOV, the major variables are the CRT display diameter and the effective focal length of the collimating lens system. For minimum physical size and weight, a small-diameter CRT and short focal length are desired. These parameters are usually balanced against the need for a large-diameter collimating lens to give the maximum IFOV and a large focal length, which allows maximum accuracy. The diameter of the collimating lens is generally limited to between 75 mm and 175 mm (3 inches and 7 inches approximately) by cockpit space constraints and practical considerations. Large lenses are very heavy and liable to break under thermal shock.

The HUD combiner occupies the prime cockpit location right in the centre of the pilot's forward line of view at the top of the glareshield. The size of the combiner is determined by the desired FOV and the cockpit geometry, especially the pilot's seating position.

The main body of the HUD containing the optics and electronics must be sunk down behind the instrument panel in order to give an unrestricted view down over the nose of the aircraft during high-attitude manoeuvres (refer to Fig. 2.8).

The pilot's design eye position for the HUD is determined by pilot comfort and the ability to view the cockpit instruments and head down displays and at the same time achieve the maximum outside world visibility. In the case of a combat aircraft, there is also the ejection line clearance to avoid the pilot being 'kneecapped' by the HUD on ejecting, which further constrains the design eye position.

Typical IFOVs range from about 13° to 18° with a corresponding TFOV of about 20–25°. The total vertical FOV of an HUD can be increased to around 18° by the use of a dual combiner configuration rather like a venetian blind. Effectively two overlapping portholes are provided, displaced vertically.

The effect of the cockpit space and geometry constraints is that the HUD design has to be 'tailor made' for each aircraft type and a 'standard HUD' which would be interchangeable across a range of different aircraft types is not a practical proposition.

The conventional combiner glass in a refractive HUD has multi-layer coatings which reflect a proportion of the collimated display imagery and transmit a large proportion of the outside world, so that the loss of visibility is fairly small. A pilot looking through the combiner of such an HUD sees the real world at 70% brightness upon which is superimposed the collimated display at 30% of the CRT brightness (taking typical transmission and reflection efficiencies). The situation is shown in Fig. 2.9 and is essentially a rather lossy system with 30% of the real-world brightness thrown away (equivalent to wearing sunglasses), as is 70% of the CRT display brightness.

In order to achieve an adequate contrast ratio so that the display can be seen against the sky at high altitude or against

Fig. 2.9 Conventional refractive HUD combiner operation

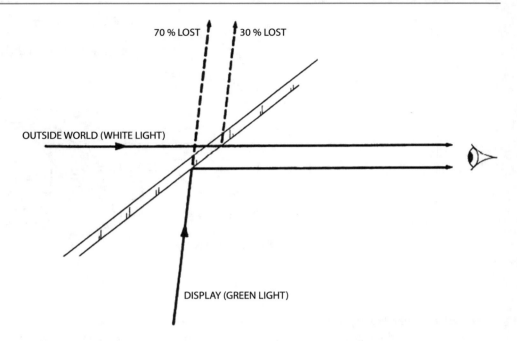

sunlit cloud, it is necessary to achieve a display brightness of 30,000 Cd/m^2 (10,000 ft. L) from the CRT.

In fact, it is the brightness requirement in particular which assures the use of the CRT as the display source for some considerable time to come, even with the much higher optical efficiencies which can be achieved by exploiting holographic optical elements.

The use of holographically generated optical elements can enable the FOV to be increased by a factor of two or more, with the instantaneous FOV equal to the total FOV. Very much brighter displays together with a negligible loss in outside world visibility can also be achieved, as will be explained in the next section. High optical transmission through the combiner is required so as not to degrade the acquisition of small targets at long distances.

It should be noted, however, that the development of 'Rugate' dielectric coatings applied to the combiners of conventional refractive HUDs has enabled very bright displays with high outside world transmission to be achieved, comparable, in fact, with holographic HUDs. A Rugate dielectric coating is a multi-layer coating having a sinusoidally varying refractive index with thickness which can synthesise a very sharply defined narrow wavelength band reflection coating, typically around 15 nm at the CRT phosphor peak emission. The coating exhibits similar high reflection and transmission values to holographic coatings but is not capable of generating optical power.

The IFOV of a refractive HUD using a combiner with a Rugate dielectric coating still suffers from the same limitations and cannot be increased like a holographic HUD. It can, nevertheless, provide a very competitive solution for applications where a maximum IFOV of up to 20° is acceptable.

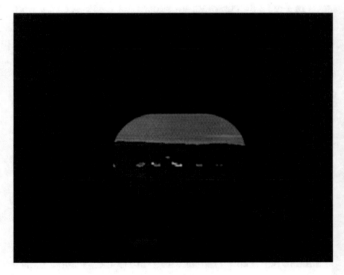

Fig. 2.10 Instantaneous FOV of conventional HUD

2.2.3 Holographic HUDs

The requirement for a large FOV is driven by the use of the HUD to display a collimated TV picture of the FLIR sensor output to enable the pilot to 'see' through the HUD FOV in conditions of poor visibility, particularly night operations. It should be noted that the FLIR sensor can also penetrate through haze and many cloud conditions and provide 'enhanced vision' as the FLIR display is accurately overlaid one to one with the real world. The need for a wide FOV when manoeuvring at night at low level can be seen in Figs. 2.10 and 2.11. The wider azimuth FOV is essential for the pilot to see into the turn (The analogy has been made of

Fig. 2.11 Instantaneous FOV of holographic HUD

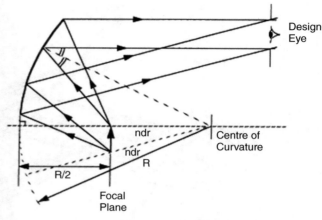

Fig. 2.13 Collimation by a spherical reflecting surface

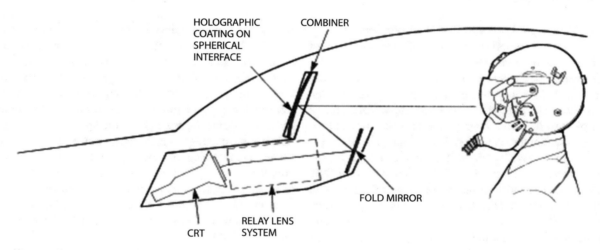

Fig. 2.12 Off-axis holographic combiner HUD configuration

trying to drive a car round Hyde Park Corner with a shattered opaque windscreen with your vision restricted to a hole punched through the window).

In a modern wide FOV holographic HUD, the display collimation is carried out by the combiner which is given optical power (curvature) such that it performs the display image collimation. Figure 2.12 shows the basic configuration of a modern single combiner holographic HUD. The CRT display is focused by the relay lens system to form an intermediate image at the focus of the powered combiner. The intermediate image is then reflected from the fold mirror to the combiner. This acts as a collimator as the tuned holographic coating on the spherical surface of the combiner reflects the green light from the CRT display and forms a collimated display image at the pilot's design eye position.

Figure 2.13 illustrates the collimating action of a spherical reflecting surface. The collimating action is, in fact, the optical reciprocal of Newton's reflecting telescope.

Because the collimating element is located in the combiner, the porthole is considerably nearer to the pilot than a comparable refractive HUD design. The collimating element can also be made much larger than the collimating lens of a refractive HUD, within the same cockpit constraints. The IFOV can thus be increased by a factor of two or more and the instantaneous and total FOVs are generally the same, as the pilot is effectively inside the viewing porthole.

This type of HUD is sometimes referred to as a 'Projected Porthole HUD' and the image is what is known as pupil forming. The display imagery can, in fact, only be seen within the 'head motion box'. If the eyes or head move outside a three-dimensional box set in space around the pilot's normal head position, then the display fades out. It literally becomes a case of 'now you see it – now you do not' at the head motion box limits. Modern holographic HUDs are designed to have a reasonably sized head motion box so that the pilot is not unduly constrained.

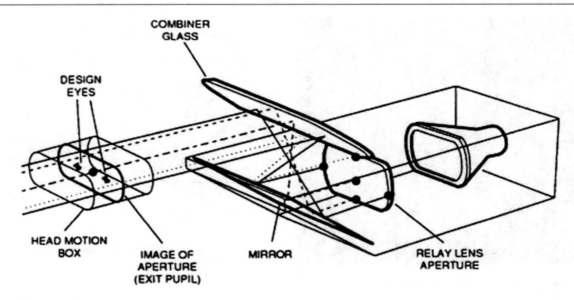

Fig. 2.14 The head motion box concept

Figure 2.14 illustrates the head motion box concept.

The combiner comprises a parallel-faced sandwich of plano-convex and plano-concave glass plates with a holographic coating on the spherical interface between them. The holographic coating is formed on the spherical surface of the plano-convex glass plate and the concave surface glass forms a cover plate so that the holographic coating can be hermetically sealed within the combiner. The holographic coating is sharply tuned so that it will reflect the green light of one particular wavelength from the CRT display with over 80% reflectivity but transmit light at all other wavelengths with around 90% efficiency (The CRT phosphors generally used are P43 or P53 phosphors emitting green light with a predominant wavelength of around 540 nm, and the hologram is tuned to this wavelength). This gives extremely good transmission of the outside world through the combiner (The outer faces of the combiner glass are parallel so that there is no optical distortion of the outside scene).

The outside world viewed through the combiner appears very slightly pink as the green content of the outside world with a wavelength of around 540 nm is not transmitted through the combiner. Holographic HUDs, in fact, are recognisable by the green tinge of the combiner.

The spherical reflecting surface of the combiner collimates the display imagery but there are large optical aberration errors introduced which must be corrected. These aberration errors are due to the large off-axis angle between the pilot's line of sight and the optical axis of the combiner, which results from the HUD configuration. Some corrections can be made for these aberrations by the relay lens system but there is a practical limit to the amount of correction which can be achieved with conventional optical elements without resorting to aspherical surfaces. This is where a unique

property of holographically generated coatings is used, namely the ability to introduce optical power within the coating so that it can correct the remaining aberration errors. The powered holographic coating produces an effect equivalent to local variations in the curvature of the spherical reflecting surface of the combiner to correct the aberration errors by diffracting the light at the appropriate points. The holographic coating is given optical power so that it behaves like a lens by using an auxiliary optical system to record a complex phase distribution on the combiner surface during the manufacturing process. This will be explained shortly.

A very brief overview of holographic optical elements is set out below to give an appreciation of the basic principles and the technology.

Holography was invented in 1947 by Denis Gabor, a Hungarian scientist working in the UK. Practical applications had to wait until the 1960s, when two American scientists, Emmet Leith and Joseph Upatnieks, used coherent light from the newly developed laser to record the first hologram.

Holographic HUDs use reflection holograms which depend for their operation on refractive index variations produced within a thin gelatin film sandwiched between two sheets of glass. This is really a diffraction grating and hence a more accurate name for such HUDs is diffractive HUDs. A holographic reflection coating is formed by exposing a thin film of photo-sensitive dichromated gelatin to two beams of coherent laser light. Due to the coherent nature of the incident beams a series of interference fringes are formed throughout the depth of the gelatin film. During the developing process these fringes are converted into planes of high and low refractive index parallel to the film surface. To a first approximation, the refractive index change between adjacent planes is sinusoidal as opposed to the step function associated

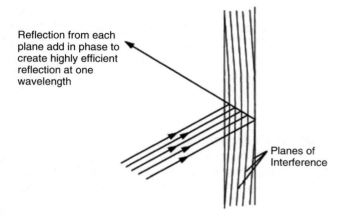

Reflection from each plane add in phase to create highly efficient reflection at one wavelength

Planes of Interference

Fig. 2.15 Holographic coating

with multi-layer coatings. During the developing process the gelatin swells, producing an increase in the tuned wavelength. Re-tuning the hologram is achieved by baking the film, which reduces the thickness and hence the spacing between planes of constant refractive index. The designer therefore specifies a construction wavelength at a given angle of incidence after baking. Figure 2.15 illustrates the planes or layers of varying refractive index formed in the holographic coating.

The bandwidth of the angular reflection range is determined by the magnitude of the change in refractive index. This variable can be controlled during the developing process and is specified as the hologram modulation.

At any point on the surface, the coating will only reflect a given wavelength over a small range of incidence angles. Outside this range of angles, the reflectivity drops off very rapidly and light of that wavelength will be transmitted through the coating.

The effect is illustrated in Figs. 2.16 and 2.17. Rays 1 and 3 are insufficiently close to the reflection angle, θ, for them to be reflected whereas Ray 2 is incident at the design angle and is reflected.

There is another more subtle feature of holograms, which gives particular advantages in an optical system. That is the ability to change the tuned wavelength uniformly across the reflector surface.

Figure 2.12 shows that the reflection coating must reflect the display wavelength at a different incident angle at the bottom of the combiner from that at the top. It is possible to achieve this effect with a hologram because it can be constructed from the design eye position.

The process for producing the powered holographic combiner is very briefly outlined below.

The process has three key stages:

- Design and fabricate the computer-generated hologram (CGH).

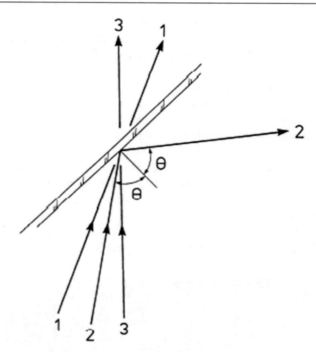

Fig. 2.16 Angularity selective reflection of monochromatic rays

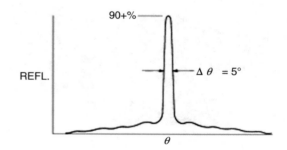

Fig. 2.17 Holographic coating performance

- Produce master hologram.
- Replicate copies for the holographic combiner elements.

The CGH and construction optics create the complex wavefront required to produce the master hologram. The CGH permits control of the power of the diffraction grating over the combiner thus allowing correction of some level of optical distortion resulting in a simplified relay lens system.

The CGH design starts with the specification performance and detailed design of the relay lens optics. This is used to define the combiner performance and then that of the construction optics, which finally leads to a set of wavefront polynomials defining the CGH. The whole process is iterative until a practical solution to all stages is achieved. The CGH is actually a diffraction grating etched into a chrome layer on a glass substrate by an electron beam.

The master hologram is created using this complex wavefront recorded onto a gelatin layer on a spherical glass substrate and is used as a template to produce the combiner elements.

The replication process uses a scanning exposure technique where a laser beam is raster scanned across the replication substrate which has been index-matched to the master substrate so as to produce a contact copy of the master hologram.

It should be pointed out that the design of holographic optical systems is a highly computer intensive operation and would not be possible without a very extensive suite of optical design software.

Development of the optical design software has required many man-years of effort and holographic optical design is very much a task for the experts and professionals. A very close liaison with the holographic optical manufacturing team and facility is essential.

Figure 2.18 is a photograph of a modern wide FOV off-axis single combiner holographic HUD with a powered holographic combiner.

2.2.4   HUD Electronics

The basic functional elements of a modern HUD electronic system are shown in Fig. 2.19. These functional elements may be packaged so that the complete HUD system is contained in a single unit, as in the Typhoon HUD above.

The system may also be configured as two units, namely the Display Unit and the Electronics Unit. The Display Unit contains the HUD optical assembly, CRT, display drive electronics and high- and low-voltage power supplies. The Electronics Unit carries out the display processing, symbol generation and interfacing to the aircraft systems.

The new generation of aircraft with integrated modular avionics use standardised cards/modules to carry out the HUD display processing and symbol-generation tasks. These are housed in a rack in one of the environmentally controlled cabinets.

The basic functional elements of the HUD electronics are described briefly below.

The data bus interface decodes the serial digital data from the aircraft data bus (typically a Military Standard, MIL STD 1553B data bus) to obtain the appropriate data from the aircraft sub-systems and inputs these data to the display processor.

Fig. 2.18 A modern wide FOV off-axis single combiner holographic HUD with a powered holographic combiner.

Fig. 2.19 HUD electronics

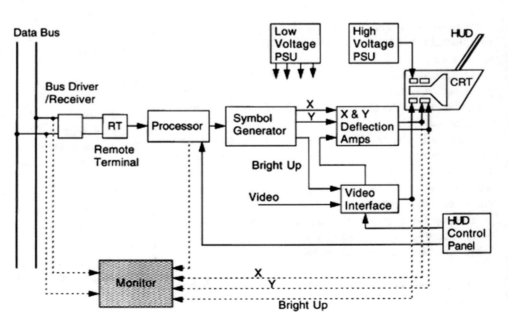

The input data include the primary flight data from the air data system and the INS, such as height, airspeed, vertical speed, pitch and bank angles, heading and flight path velocity vector.

Other data include microwave landing system (MLS) or instrument landing system (ILS) guidance signals, stores management and weapon aiming data in the case of a combat aircraft, discrete signals such as commands, warnings etc.

The display processor processes these input data to derive the appropriate display formats, carrying out tasks such as axis conversion, parameter conversion and format management. In addition the processor also controls the following functions:

- Self-test.
- Brightness control (especially important at low brightness levels).
- Mode sequencing.
- Calibration.
- Power supply control.

The symbol generator carries out the display waveform-generation task (digitally) to enable the appropriate display symbology (e.g. lines, circles, alpha-numerics) to be stroke written on the HUD CRT. The symbols are made up of straight line segments joining the appropriate points on the display surface in an analogous manner to a 'join the dots' child's picture book. Fixed symbols such as alpha-numerics, crosses, circles (sized as required) are stored in the symbol generator memory and called up as required. The necessary D to A conversions are carried out in the symbol generator, which outputs the appropriate analogue x and y deflection voltage waveforms and 'bright up' waveforms to control the display drive unit of the HUD CRT.

The CRT beam is thus made to trace the outline of the line drawn symbols, the process being known as stroke or cursive writing. Slow and fast cursive display writing systems are required for a day and night mode operation HUD. Daytime operation requires a slow cursive writing system to produce a sufficiently bright display which can be viewed against a 30,000 Cd/m^2 (10,000 ft. L) white cloud background.

Night time operation requires identical appearance symbology to the daytime symbology to be overlaid on the raster TV FLIR picture displayed on the HUD. The appearance of raster-generated symbology can differ significantly from stroke written symbology and pilots wish to see identical symbology night or day. A 'fast cursive' display system is thus used in night mode operations whereby the cursive symbology is drawn at high speed during the raster fly back interval. The display is written at about ten times the slow cursive speed and this makes significant demands on the bandwidth and power requirements of the deflection system. By taking a small number of lines from the top and bottom of the picture, typically around 20 lines total, the whole of the daytime symbology can be superimposed over the TV picture for night use.

Because of the high writing speeds, the brightness of the raster and fast cursive display was initially only suitable for night or low-ambient brightness use. However, improvements to CRTs, development of more efficient phosphors such as P53 and high-efficiency wavelength selective combiners have allowed raster and fast cursive brightness to be increased to around 3000 Cd/m^2 (1000 ft. L), which is useable in daytime.

The very wide range of ambient brightness levels which a day/night mode HUD must operate over is worthy of note. CRT luminance levels range from 30,000 Cd/m^2 to 0.3 Cd/m^2, a range of 10^5:1.

The display drive unit contains all the display drive electronics for the HUD CRT including the high-voltage and low-voltage power supply units. High-linearity and high-bandwidth x and y deflection amplifiers are required to meet the HUD accuracy and cursive writing requirements.

Video interface electronics are incorporated in the display drive electronics for the TV raster mode.

Modern HUD systems are used as the prime display of primary flight data such as height, airspeed, attitude and heading. The HUD thus becomes a safety critical system when the pilot is flying head up at low altitude and relying on the HUD. Modern HUD systems for both military and civil aircrafts are, therefore, being designed so that the probability of displaying hazardous or misleading information is of the order of less than 1×10^{-9}. The self-contained single unit HUD makes it an ideal system configuration for a safety critical HUD. Forward and backward path monitoring techniques can be employed to ensure that what is on the display is what was actually commanded by the aircraft system. The monitoring features incorporated in the HUD electronics are shown in Fig. 2.19.

2.2.5 Worked Example on HUD Design and Display Generation

A simple worked example based on typical parameters for a refractive HUD is set out below to show how:

1. The total and instantaneous FOVs are determined by the HUD CRT, collimating optics and cockpit geometry.
2. The attitude display symbology is generated to be conformal with the outside world.

The objectives are to give an engineering appreciation by deriving realistic numerical values and to reinforce the descriptive content of the preceding sections.

The following data are provided:

HUD CRT diameter (measured across display face)	50 mm (2 in. approx.)
Required total FOV	20°
Collimating lens diameter	125 mm (5 in. approx.)
Collimating lens centre from combiner glass (measured along axis of lens)	100 mm (4 in. approx.)
Pilot's design eye position from combiner glass (measured along pilot's LOS through centre of combiner)	400 mm (16 in. approx)

Assume x and y deflection drive waveform amplitudes of ±100 units produces full-scale deflection of the CRT spot in both positive and negative directions along the OX and OY axes (origin, O, at the centre of the display).

Questions

1. What is the required focal length of the collimating lens?
2. What is the instantaneous FOV?
3. How much up or down, or sideways head movement must the pilot make to see all the total FOVs of 20°?
4. How much nearer to the combiner must the pilot's eye position be in order to see the total FOVs?
5. Derive the equations for displaying the horizon line in terms of the pitch angle, θ, bank angle, Φ, and the angle θ_0 by which the horizon line is depressed from the centre of the display in straight and level flight.
6. What are the co-ordinates of the end points of the horizon line when the aircraft pitch angle is $-3°$ and bank angle $+10°$, given that the horizon line takes up 50% of the azimuth FOV and that the horizon line is depressed by 4° from the centre of the display in straight and level flight?
7. Sketch the display and show the aircraft velocity vector which is inclined at an angle of 4° below the horizontal.
8. Sketch output waveform of symbol generator to generate the horizon line for the above values.

Solution to Question 1 TFOV $= 2\tan^{-1} A/2F$, where A = display diameter and F = effective focal length of collimating lens (refer to Fig. 2.7b).

Hence, for a TFOV of 20° and a CRT diameter A of 50 mm, the required effective focal length F is given by

$$F = \frac{50}{2\tan 10°}$$

i.e. $F = 141.8$ mm

Solution to Question 2 IFOV $= 2\tan^{-1} D/2L$, where D = diameter of collimating lens and L = distance of virtual image of collimating lens from pilot's eyes (refer to Fig. 2.8).

From data given $L = (100 + 400) = 500$ mm and $D = 125$ mm.

Hence, IFOV $= 2\tan^{-1} 125/2 \times 500$ rads $= 14.25°$

Solution to Question 3 To see the TFOV of 20°, the required up, down or sideways head movement of the pilot is given by

$$500\tan\left(\frac{20° - 14.25°}{2}\right)$$

that is, 25 mm (1 inch approximately).

Solution to Question 4 To see the total FOV of 20°, the pilot's eye position must be brought nearer to the combiner by a distance, Z, which is given by

$$\frac{125}{2(500 - Z)} = \tan 10°$$

From which $Z = 145$ mm (5.7 inches approximately).

Solution to Question 5 Referring to Fig. 2.20, the co-ordinates (x_1, y_1) and (x_2, y_2) of the end points of a horizon line of length $2a$, which in straight and level flight is depressed by an amount θ_0 from the centre of the display, are given by

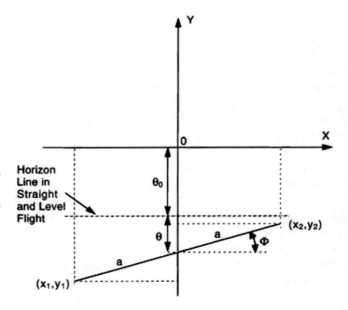

Fig. 2.20 Horizon line

Fig. 2.21 HUD display of
horizon line and velocity vector

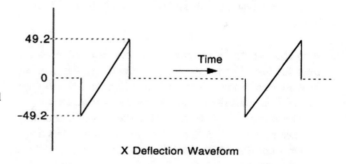

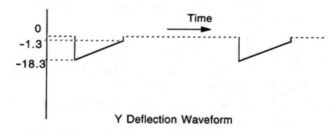

$$\begin{cases} x_1 = -a \cos \Phi \\ y_1 = -(\theta_0 + \theta + a \sin \Phi) \end{cases} \quad \begin{cases} x_2 = a \cos \Phi \\ y_2 = -(\theta_0 + \theta - a \sin \Phi) \end{cases}$$

Solution to Questions 6 & 7 Given $\theta_0 = 4°$, $\theta = -3°$ and
$\Phi = 10°$.

$2a = 0.5 \times$ azimuth FOV. Hence $a = 5°$.
From which

$$\begin{cases} x_1 = -5 \cos 10° = -4.92 \\ y_1 = -[4 + (-3) + 5 \sin 10°] = -1.87 \\ x_2 = 5 \cos 10° = 4.92 \\ y_2 = -[4 + (-3) - 5 \sin 10°] = -0.13 \end{cases}$$

Horizon line end point co-ordinates are thus $(-4.92, -1.87)$ and $(4.92, -0.13)$.
Display is shown in Fig. 2.21.

Solution to Question 8 The X and Y deflection waveforms
are shown in Fig. 2.22.

Full-scale deflection of CRT spot is produced when X and
Y deflection amplifiers are supplied by signals of ±100 units
from the symbol generator. Hence, to draw horizon line for
values given, the X axis deflection waveform is a ramp from
-49.2 units to $+49.2$ units, and the Y axis deflection waveform is a ramp from -18.7 units to -1.3 units.

2.2.6 Civil Aircraft HUDs

The application of HUDs in civil aircraft has been mentioned
earlier in the chapter. This section explains their potential and

Fig. 2.22 X and Y deflection waveforms

importance in greater detail and covers the progress that has
been made in producing viable systems for civil aircraft
operation.

The use of an HUD by civil aviation operators is still a
relatively novel practice and it was estimated in 2009 that
there were about 2000 HUD installations in revenue service
worldwide. This compares with over 15,000 military HUD
systems in service worldwide. The future for civil HUD
systems looks good, however, and the numbers are growing.
The potential market over the next decade is estimated to be
of the order of 5000 systems.

Fig. 2.23 Wind shear. This shows the simplest and most common fatal wind shear scenario. The aircraft enters a downdraught at A and the pilot experiences a rapidly increasing head wind. The indicated airspeed increases and the additional lift causes the aircraft to climb above the glide slope and so the pilot reduces power to recapture the approach. Suddenly, at B, the head wind disappears and, worse, the aircraft is in an intense downdraught. Worse still, at C, a rapidly increasing tail wind further reduces the probability of recovery

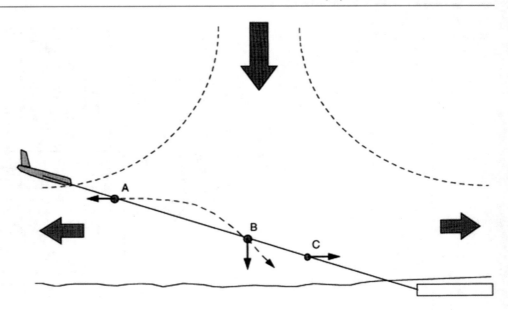

The main advantages of an HUD in a civil aircraft are as follows:

1. Increased safety by providing a better situational awareness for the pilot to control the aircraft by the head up presentation of the primary flight information symbology so that it is conformal with the outside world scene.

   - The problems of severe wind shear conditions have been mentioned earlier in the introduction to this chapter. Figure 2.23 shows how wind shear arises and the problems it creates. Over 700 passengers were killed in the USA alone in accidents in recent years caused by wind shear.
   - The HUD can also provide a flight path director display, which allows for the effect of wind shear from a knowledge of the aircraft's velocity vector, airspeed, height and aerodynamic behaviour.
   - The HUD can also increase safety during terrain, or traffic avoidance manoeuvres.
   - Ground Proximity Warning Systems (GPWS) are a very valuable aid in avoiding terrain, and enhanced GPWS will extend this protection still further.
   - In circumstances, however, during a terrain escape manoeuvre where the terrain is visible, the flight path vector (FPV) displayed on the HUD provides an unambiguous presentation on whether or not the terrain will be missed. If the FPV is overlaid on terrain ahead, the aircraft will hit it and in this situation the crew must decide on another course of action as opposed to 'holding on' to see what happens.
   - Traffic Collision Avoidance Systems (TCAS) provide traffic conflict and escape guidance and are of great benefit in preventing mid-air collisions. Guidance, however, is provided head down so the pilot is

manoeuvring on instruments whilst trying to acquire conflicting traffic visually. The HUD enables the escape manoeuvre and search for traffic to be accomplished head up.

   - Safety improvements in the course of each stage of a typical flight were predicted in a study published by Flight Safety Foundation (see Further Reading). The study concluded that an HUD would likely have had a positive impact on the outcome of 30% of the fatal jet transport incidents documented over the period 1958–1989.

   Increased revenue earning ability by extending operations in lower weather minima. The HUD is used to monitor the automatic landing system and to enable the pilot to take over on aircrafts which are not equipped to full Category III automatic landing standards (automatic landing systems are covered in Chap. 8).

2. Use of the HUD as part of an enhanced vision system to enable operations in lower weather minima at airfields not equipped with automatic landing aids (ILS/MLS). For example, the number of Type II and Type III ILS facilities in the USA is very limited – typically less than 70. The revenue earning potential of enhanced vision systems is thus very considerable. As explained earlier, the HUD displays a video image of the real world ahead of the aircraft derived from a millimetric wavelength radar sensor and an FLIR sensor installed in the aircraft together with the overlaid primary flight information and flight path parameters.

3. The use of the HUD for displaying ground taxiway guidance is being actively investigated, and is considered a likely extension to the HUDs roles. Ground taxiway guidance could be provided by differential global positioning system (GPS).

Very encouraging trials were carried out in the USA by the Federal Aviation Administration (FAA) in 1992. The enhanced vision system used a millimetric wavelength radar operating at 35 GHz and a 3–5 micron infrared imaging sensor (see Further Reading). The infrared sensor provided useful imaging information in haze and high-humidity conditions at a range which substantially exceeded the range of the pilots' eyes. However, descending through typical cloud and fog conditions, the range was no better than could be seen by the eye. The 35 GHz radar sensor, however, was able to operate successfully under all weather conditions and enabled a synthetic picture of the runway to be generated from the processed radar image at a range of 2–2.5 miles. The pilots were able to identify the runway on which they intended to land at a range of about 1.7 miles. They were routinely able to fly to Category IIIa minima (50 ft. decision height/700 ft. runway visual range) using the HUD display of the radar-generated runway image and the overlaid primary flight symbology as their sole means to conduct the published approach. The display of the aircraft's velocity vector on the HUD was found to be particularly helpful throughout the approach, when used in conjunction with the electronic image of the outside world, and enabled the control of the aircraft's flight path to be precisely managed.

The evaluation pilots reported Cooper-Harper ratings of 3 (satisfactory without improvement and with minimal compensation required by the pilot for desired performance).

A 95 GHz radar was also evaluated with similar successful results; the resolution was better, however, than the 35 GHz radar due to the shorter wavelength.

Active development of both radar and infrared enhanced vision sensor systems is continuing and their introduction into civil operational service is anticipated in the next decade. Figure 2.24 shows the effectiveness of an enhanced vision

HUD display with overlaid primary flight information symbology.

An overhead mounted HUD installation is practical in a civil aircraft and is the configuration generally adopted (Although glareshield HUDs are practical, there is rarely space to retrofit such a system). Figure 2.3 shows a civil HUD installation. The HUD elements are precisely located in a rigid mounting tray which provides the necessary mechanical/optical stability and the combiner can fold out of the way when not required. The field of view is 30 degrees by 25 degrees.

The optical configuration of the HUD is shown in Fig. 2.25. The thin spherical combiner incorporates a Rugate dielectric coating (referred to earlier). These highly efficient selective wavelength coatings are very hard and are not affected by the atmosphere and permit a relatively thin combiner to be realised, rather than the thick sandwich necessary with combiners exploiting biologically based holograms. The mass of the combiner is greatly reduced, which in turn reduces the mechanical structure required to support it with sufficient rigidity both in normal operation and in avoiding head injury during crash conditions. It should be noted that the thin spherical combiner does not introduce any optical distortion of the outside world. As mentioned earlier, the transmission and reflectivity values of a Rugate coating are comparable with holographic coatings.

The off-axis angle of the combiner is around 17° and is considerably smaller with the overhead installation than it is with a military cockpit (refer Fig. 2.12), which is typically around 30°. The off-axis errors can be corrected with a conventional relay lens train, and the optical power which can be provided by a holographic coating is not required.

The HUD incorporates electronics which monitor performance in both forward and reverse path (as described earlier)

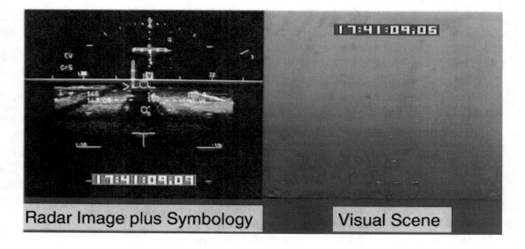

Fig. 2.24 Effectiveness of enhanced vision HUD display. (Courtesy of BAE Systems). Landing in Category III minima conditions at Point Mugu naval air station, California, USA. Photos show conditions at 150 ft. altitude and Enhanced Vision HUD display. (Note: Point Mugu has only Category I facilities)

Radar Image plus Symbology   Visual Scene

Fig. 2.25 Civil HUD optical configuration. (Courtesy of BAE Systems)

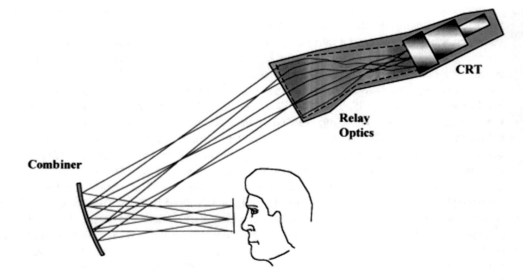

such that the integrity of the system, defined in terms of displaying hazardously misleading information to the pilot, equals that of conventional head down displays at a value of 10^{-9}.

Blind landing guidance can be derived from the ILS system and enhanced by mixing positional information derived from the Global Positioning System data using appropriate algorithms. This guidance information can be drawn cursively on the HUD together with primary flight information using suitable symbology. Such a system is known as 'synthetic vision' as distinct from enhanced vision. The latter system uses radar or infrared scanning systems, as described, and the imagery is raster drawn (with stroke written symbology overlaid).

Completely blind landings (Category IIIa and ultimately Category IIIb) can be performed with an HUD system displaying synthetic vision guidance icons (and primary flight information), given that the required integrity is provided. This can be achieved by the addition of a second means of monitoring, which can be either a second HUD channel or, more realistically, the addition of a fail-passive autopilot. A combination of the HUD and a fail-passive autopilot provides a cost-effective solution for extending the poor visibility operation of many Air Transport Category aircraft.

2.3 Helmet Mounted Displays

2.3.1 Introduction

The advantages of head up visual presentation of flight and mission data by means of a collimated display have been explained in the preceding section. The HUD, however, only presents the information in the pilot's forward field of view, which even with a wide FOV holographic HUD is limited to

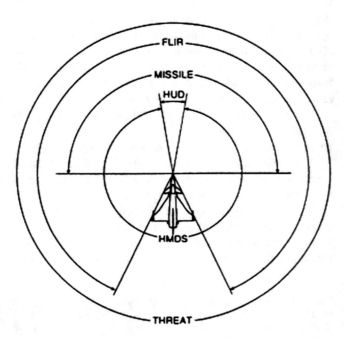

Fig. 2.26 Comparison of HUD and HMD fields of view

about 30° in azimuth and 20–25° in elevation. Significant increases in this FOV are not practicable because of the cockpit geometry constraints.

The pilot requires visual information head up when he is looking in any direction and this requirement can only be met by a helmet mounted display (HMD).

Figure 2.26 shows the comparison between HUD and HMD fields of view.

In its simplest form the HMD can comprise a simple helmet mounted sighting system which displays a collimated aiming cross or circle and some simple alpha-numeric information, with a typical FOV of around 5°. An addressable light-emitting diode (LED) matrix array is generally used as the display source, enabling a very light weight and compact HMD to be achieved. The helmet mounted sight (HMS) enables targets to be attacked at large off-boresight angles, as will be explained (It should be noted that the term 'boresight' is used to describe the direction in which the aircraft fore and aft axis is pointing and is based on the analogy of sighting through the bore of a gun-barrel mounted on the aircraft datum axis).

In its most sophisticated form the HMD can provide, in effect, an 'HUD on the helmet'. This can display all the information to the pilot, which is normally shown on an HUD but with the pilot able to look in any direction (attitude information is referenced to his line of sight [LOS]). The HMD can also have a wider FOV ranging from 35° to 40° for a fighter aircraft application to over 50° for a helicopter application. It should be appreciated that the FOV moves with the head, unlike the HUD, and a larger FOV reduces scanning head movement in an HMD. It is still very useful, however, to allow peripheral information to enter the eyes, so that the FOV is not just that of the optics.

The HMD also enables a very effective night/poor visibility viewing system to be achieved by displaying the TV picture from a gimballed infrared sensor unit which is slaved to follow the pilot's line of sight. The pilot's LOS with respect to the airframe is measured by a head position sensing system.

Such a helmet can also incorporate image intensification devices with a video output to provide a complementary viewing system.

2.3.2 Helmet Design Factors

It is important that the main functions of the conventional aircrew helmet are appreciated as it is essential that the integration of a helmet mounted display system with the helmet does not degrade these functions in any way. The basic functions are:

1. To protect the pilot's head and eyes from injury when ejecting at high airspeeds. For example, the visor must stay securely locked in the down position when subjected to the blast pressure experienced at indicated airspeeds of 650 knots. The helmet must also function as a crash helmet and protect the pilot's head as much as possible in a crash landing.
2. To interface with the oxygen mask attached to the helmet. Combat aircrafts use a special pressurised breathing system for high g manoeuvring.
3. To provide the pilot with an aural and speech interface with the communications radio equipment. The helmet incorporates a pair of headphones which are coupled to

the outputs of the appropriate communications channel selected by the pilot. The helmet and earpieces are also specifically designed to attenuate the cockpit background acoustic noise as much as possible. A speech interface is provided by a throat microphone incorporated in the oxygen mask.
4. In addition to the clear protective visor, the helmet must also incorporate a dark visor to attenuate the glare from bright sunlight.
5. The helmet must also be compatible with nuclear–biological–chemical (NBC) protective clothing and enable an NBC mask to be worn.

Any coatings on the visor, when it is used for projecting the display, must not reduce the 'Tally-ho' distance, that is, the distance at which the target can first be seen. Colouration effects from the coatings are also disliked by the pilots and neutral density coatings are preferred.

The optical system must not produce any visible witness lines, for example, should the two halves of a binocular system abut each other. This is to eliminate the possibility of such a witness line (even if vertical) being mistaken for the horizon during extreme manoeuvres.

Integrating an HMD system into the helmet must thus consider the communications and breathing equipment requirement, the protection, comfort and cooling aspects as well as the visual performance of the display.

First-generation HMDs have been based on modifying and effectively 'bolting on' the HMD optical module, display source and associated electronics to an existing aircrew helmet. This can result in a bit of a 'bodge' and the total helmet weight can be high.

More recent second-generation HMDs are totally integrated with the helmet design starting from a 'clean sheet of paper' and optimised to meet all the above requirements for the minimum overall weight. The combined mass of the integrated helmet and HMD system must be as low as possible to reduce the strain on the pilot's neck muscles during manoeuvres. Modern conventional aircrew helmets weigh about 1 kg (2.2 lb) so that the weight the pilot feels on his head during a 9 g turn is equal to 9 kg (20 lb. approximately), and this is just acceptable. The typical weight for an integrated helmet incorporating a binocular HMD (BHMD) system is around 1.7–2.5 kg (3.8–5.5 lb. approximately). This is limited by the current technology and lower weight is sought. The helmet CG should be in line with the pivoting point of the head on the spine so that there are minimal out of balance moments exerted on the neck muscles.

The moment of inertia of the integrated helmet system about the yaw and pitch axes of the head should be as low as possible. This is to minimise the inertia torques experienced by the pilot due to the angular acceleration (and deceleration) of the head and helmet in scanning for threats from behind or above. Peak head motion rates of 30°/s can occur in combat.

The helmet must fit securely on the pilot's head and not move about so that the display is visible under all conditions and the pilot's LOS can be inferred from measuring the attitude of the head with the head tracker system.

The HMD optical system must be rigidly fixed to the helmet with adequate mechanical adjustments to accommodate the physical variations within the normal pilot population (interocular distance and eye relief). There should be at least 25 mm (1 inch) clearance, or 'eye relief', between the nearest optical surface and the eye.

The display is only fully visible when the eyes are within a circle of a given diameter known as the exit pupil diameter, the centre of the exit pupil being located at the design eye position. The exit pupil diameter should be at least 12 mm (0.5 inches approximately). It should be noted that the helmet will be pressed down on the pilot's head during high g manoeuvres as the helmet lining is resilient, hence the need for a reasonable exit pupil so that the pilot can still see the display.

2.3.3 Helmet Mounted Sights

A helmet mounted sight (HMS) in conjunction with a head tracker system provides a very effective means for the pilot to designate a target. The pilot moves his head to look and sight on the target using the collimated aiming cross on the helmet sight.

The angular co-ordinates of the target sight line relative to the airframe are then inferred from the measurements made by the head tracker system of the attitude of the pilot's head.

In air to air combat, the angular co-ordinates of the target line of sight (LOS) can be supplied to the missiles carried by the aircraft. The missile seeker heads can then be slewed to the target LOS to enable the seeker heads to acquire and lock on to the target (A typical seeker head needs to be pointed to within about 2° of the target to achieve automatic lock on). Missile lock on is displayed to the pilot on the HMS and an audio signal is also given. The pilot can then launch the missiles.

This enables attacks to be carried out at large off-boresight angles, given agile missiles with high g manoeuvring capabilities. The pilot no longer needs to turn the aircraft until the target is within the FOV of the HUD before launching the missile. The maximum off-boresight angle for missile launch using the HUD is less than 15° even with a wide FOV HUD, compared with about 120° with an HMS (and highly agile missiles). This gives a major increase in air combat effectiveness – factors of three to four have been quoted in terms of the 'kill' probability, a very effective 'Force Multiplier' in military parlance. Figure 2.27 illustrates the concept together with a simple HMS.

Fig. 2.27 Helmet mounted sight and off-boresight missile launch. (Courtesy of BAE Systems)

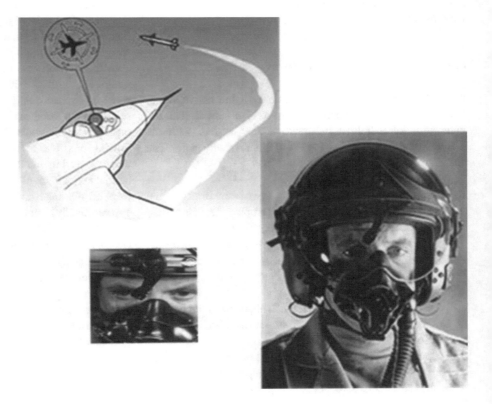

2.3.4 Helmet Mounted Displays

As mentioned in the introduction, the HMD can function as an 'HUD on the helmet' and provide the display for an integrated night/poor visibility viewing system with flight and weapon aiming symbology overlaying the projected image from the sensor.

Although monocular HMDs have been built, which are capable of displaying all the information normally displayed on an HUD, there can be problems with what is known as 'monocular rivalry'. This is because the brain is trying to process different images from each eye and rivalry can occur between the eye with a display and that without. The problems become more acute at night when the eye without the display sees very little, and the effects have been experienced when night-flying with a monocular system in a helicopter.

It has been shown that a binocular (or biocular) system whereby the same display is presented to both eyes is the only really satisfactory solution. Hence, current HMD designs are generally binocular systems.

Miniature 0.5 inch diameter CRTs are currently used as the display sources. The development of miniature, high-resolution liquid crystal displays (LCDs), however, has changed this situation. HMDs exploiting these devices are now being produced.

Night viewing applications require the widest possible FOV. This is particularly important for helicopters operating at night at very low-level (e.g. 30 ft) flying 'Nap of the Earth' missions taking every possible advantage of terrain screening. Binocular HMDs for helicopter applications have minimum FOVs of 40° and FOVs of over 50° have been produced.

Fighter/strike aircraft HMDs generally have FOVs of 35–40°, trading off FOV for a lower-weight optical system (Helmet weight must be kept to a minimum because of the loads exerted on the pilot's head during high g manoeuvres).

A lower-weight HMD system can be achieved by the use of a visor projection system in conjunction with a high-efficiency optical design. This allows a standard spherically curved aircrew visor to be used to carry out the combiner and collimation function by the addition of a neutral density reflection coating. The visor coating provides high display brightness whilst maintaining high real-world transmission (>70% can be achieved) with no colouration. Display accuracy is also insensitive to visor rotation because of the spherical shape and partial raising of the visor is possible, for example to carry out the 'Valsalva procedure' (pinching the nostrils and blowing to equalise the pressure in the ears).

The combiner and collimation function is thus achieved without adding additional weight to the optical system as the visor is essential anyway. FOVs of over 40° can be achieved with a binocular visor projection system; the visor, however, is a long way from the eyes and creating an extreme FOV is increasingly difficult over 60°. The alternative eyepiece systems can give a larger FOV but at very high weight penalties.

Although a binocular HMD is inherently a better solution for night/poor visibility operations, there is a large market for a low-cost, light-weight HMD for fighter aircrafts which are primarily involved in daytime operations to provide a helmet mounted cueing and sighting system. The lower weight and cost achievable with a visor-projected monocular HMD can make this a cost-effective solution in such applications. Although configured for daytime use, such a system is capable of displaying video from a sensor, providing the pilot with an enhanced cueing system after dark or in bad weather.

It is appropriate at this point to discuss the use of image intensification devices as these provide a night vision capability which essentially complements an infrared viewing system; each system being better in some conditions than the other. The wavelengths involved are shown below.

Sensor	Operating wavelength
Human eye	0.4–0.7 microns (blue-red-green)
Gen 2 image intensifier tube	0.6–0.9 microns (approximately)
FLIR	8–13 microns

The image from the Image Intensifier Tube (IIT) is a phosphor screen which emits green light in the centre of the visual band where the eye is most sensitive.

Night vision goggles (NVGs) incorporating image intensifiers have been in operational use in military helicopters since the late 1970s, and in fast jet aircraft from around the mid-1980s.

NVGs basically comprise two image intensifiers through which the observer views the scene at night by means of a suitable optical arrangement. Image intensifiers are highly sensitive detectors which operate in the red end of the visible spectrum amplifying light about 40,000 times from less than starlight up to visible level.

It should be noted that special cockpit lighting is necessary as conventional cockpit lighting will saturate the image intensifiers. Special green lighting and complementary filtering are required.

Most night-flying currently undertaken by combat aircrafts is carried out using NVGs, which are mounted on a bracket on the front of a standard aircrew helmet. The weight of the NVGs and forward mounting creates an appreciable out of balance moment on the pilot's head and precludes the pilot from undertaking manoeuvres involving significant g. The NVGs must also be removed before ejecting because of the g experienced during ejection (A quick release mechanism is incorporated in the NVG attachment).

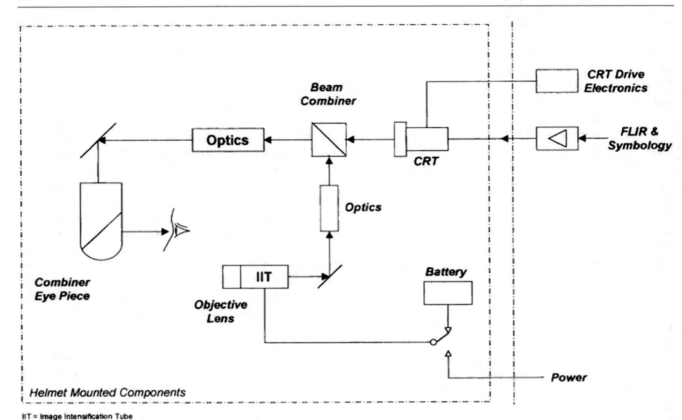

Fig. 2.28 Optical mixing of IIT and CRT imagery

Fig. 2.29 Helicopter HMD with integrated NVGs (Courtesy of BAE Systems). This HMD forms part of the mission systems avionics for the Tiger 2 attack helicopter for the German army

An integrated helmet design with the NVGs located near the natural pivot point of the head so that they do not create significant out of balance moments avoids this problem. NVGs can be incorporated into HMDs, and optically combined with the CRT displays so that the FLIR video display overlays the image intensifier display on an accurate one to one basis, enabling the two systems to be combined in a complementary manner.

Figure 2.28 illustrates the basic optical mixing of IIT and CRT imagery. The image intensifier devices can be battery powered from a small battery carried by the pilot. This enables the pilot to see in pitch darkness conditions on the ground and walk out to the aircraft/helicopter without any lights being required during covert operations.

Figure 2.29 illustrates a binocular HMD system with integrated IITs, which are optically combined with the CRT displays (as in Fig. 2.28).

In the case of fighter/strike aircraft, the weight and bulk penalty of optically combining the IITs with the display optics has led to the adoption of image intensification devices which image on to small Charge-Coupled Device (CCD) cameras. The output of the image intensifier CCD camera is a video signal which is fed back to the remote display drive electronics where it is electronically combined with the

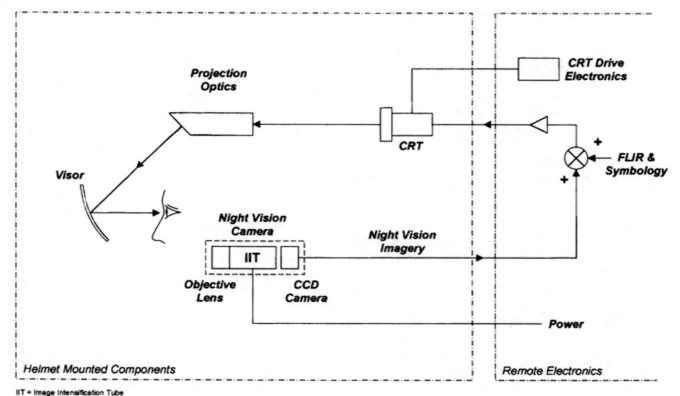

IIT = Image Intensification Tube

Fig. 2.30 Electronic combination of IIT and CRT

symbology and displayed on the helmet mounted CRTs. Figure 2.30 illustrates the concept of electronic combination of IIT and CRT.

The advantage of this approach is that it maximises the performance of the CRT display by removing the need to mix IIT and CRT imagery optically. The intensified imagery is now presented in a raster format, allowing the potential to enhance the night vision image, improving image contrast and reducing some of the unsatisfactory characteristics of directly viewed IITs, such as image blooming.

The ability to enhance the image contrast electronically allows the user to see undulations in terrain, such as featureless desert not visible using conventional NVGs. Similarly, a bright point source of light viewed by a conventional NVG results in a halo effect that blots out the surrounding scene. This effect can be eliminated, allowing the user to view clearly a scene containing bright sources of light, such as street lights.

2.3.4.1 The BAE Systems 'Striker® I' Binocular HMD

Figure 2.31 illustrates the BAE Systems 'Striker® I' Binocular HMD. Note the two night vision cameras mounted on each side of the helmet at the pilot's eye level. This is a

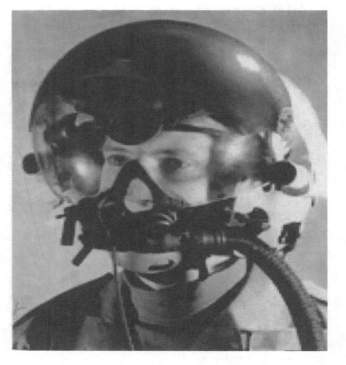

Fig. 2.31 'Striker® I' binocular HMD. (Courtesy of BAE Systems)

binocular visor-projected helmet mounted display that is used by the pilots flying the Typhoon. It has a 40-degree binocular FOV and has integral night vision cameras which are electronically combined with the CRTs, as shown in Fig. 2.30. FLIR or IIT imagery can be displayed with flight and weapon aiming symbology overlaid. An optical head tracker system employing helmet mounted LEDs and cockpit mounted cameras is used to determine the pilot's LOS. The LEDs are pulsed and the pattern and sequence is observed by the cameras. A computerised head tracker processor captures data from these cameras and rapidly calculates the angle and position of the pilot's head. This information is used to correctly position the display of vital symbology on the pilot's helmet mounted display.

The night vision imagery from the cameras is electronically combined with the display symbology and hence to the CRT in the Display Unit as shown in Fig. 2.30.

The Display Units are located in the brow of the helmet on the left and right sides respectively and are basically similar to those used in an HUD. The display source is a 0.5 inch diameter CRT with a relay lens train to correct the off-axis errors and collimation errors and a ray folding relay mirror.

Figure 2.32 illustrates the optical configuration. A ray folding relay mirror is centrally mounted on the brow of the helmet to convey the display image to the spherical visor, which has a reflective coating to collimate to display image and functions as a combiner, as explained earlier.

The 'Striker® I' is installed in the Eurofighter Typhoon and the Gripen. It has been in service for nearly two decades and had extensive operational use.

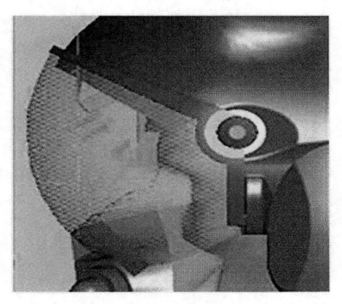

Fig. 2.32 Optical Configuration of 'Striker® I' Binocular HMD. (Courtesy of BAE Systems)

2.3.5 Head Tracking Systems

2.3.5.1 General Description

The need to measure the orientation of the pilot's head to determine the angular co-ordinates of the pilot's line of sight with respect to the airframe axes has already been explained. It should be noted that the problem of measuring the angular orientation of an object which can both translate and rotate can be a fundamental requirement in other applications such as robotics. In fact the solutions for head tracking systems can have general applications.

Space does not permit a detailed review of the various head tracking systems which have been developed. Most of the physical effects have been tried and exploited such as optical, magnetic and acoustic methods. The potential of using rate gyros as a stand-alone technique or in addition to one of the other methods is described.

Optical tracking systems work in a number of ways, for example:

(a) Pattern recognition using a CCD camera.
(b) Detection of pulsed LEDs mounted on the helmet. An example of this is the system described above used with the 'Striker® I' HMD on the Typhoon.
(c) Sophisticated measurement of laser-generated fringing patterns.

Magnetic tracking systems measure the field strength at the helmet from a magnetic field radiator located at a suitable position in the cockpit. There are two types of magnetic head tracker system:

(a) An Alternative Current (AC) system using an alternating magnetic field with a frequency of around 10 kHz.
(b) A Direct Current (DC) system using a DC magnetic field which is switched at low frequency.

Both systems are sensitive to metal in the vicinity of the helmet sensor. This causes errors in the helmet attitude measurement, which are dependent on the cockpit installation. These errors need to be measured and mapped for the particular cockpit and the corrections stored in the tracker computer. The computer then corrects the tracker outputs for the in situ errors. The AC system is more sensitive to these errors than the DC system, which is claimed to be ten times less sensitive to metal than AC systems.

The sensors and radiators for both AC and DC magnetic tracker systems are small and light weight. Typical sensor volume is about 8 cm^3 (0.5 in^3) and weighs about 20 gm (0.7 oz). The radiator volume is typically about 16 cm^3 (1 in^3) and weighs about 28 gm (1 oz) (excluding installation).

Angular coverage of the magnetic type of head tracker is typically around ±135° in azimuth and ± 85° in elevation.

Accuracy on or near boresight is generally within two milliradians (mRad) (0.1° approximately) and around 0.5° at large off-boresight angles.

A reasonable head motion box can be achieved, which is operationally acceptable for most applications.

All the systems – optical, magnetic and acoustic – have their individual limitations although the magnetic systems are currently the most widely used. Higher accuracy at larger off-boresight angles is attainable, however, with the latest optical systems.

Each system is notable for the complexity of the algorithms needed to sort out the signal from the noise and the man-years of software programming in them.

It should be noted that other errors can be encountered in measuring the pilot's direction of gaze apart from the inherent errors which are present in the head tracker system. This is because the pilot's direction of gaze is inferred from the helmet attitude measurement. The pilot can, in fact, experience great difficulty in keeping the helmet sight on the target when subjected to severe low-frequency vibration. This can result from the buffeting which can be encountered at high subsonic speeds in turbulent air conditions at low altitudes. The power spectrum of this buffeting vibration can include significant levels in the 1–10 Hz frequency range, which the pilot is unable to follow and keep the aiming cross on the target. Tracking errors of over 2° Root Mean Square (RMS) have been measured under replicated buffeting tests carried out on the ground using tape recorded flight data. Some appreciation of the vibration environment can be gained from the fact that a sinusoidal vibration of 0.5 g amplitude at a frequency of 2 Hz corresponds to an up and down motion of ±31 mm (±1.2 inches). It has been shown, however, that the pilot's gaze can still be maintained on the target whilst the pilot is experiencing low-frequency buffeting-type vibration. This is because the control of the eyeball (and hence the direction of gaze) by the eye muscles has a much faster response than the control of the head position by the neck muscles. The 'eyeball servo' has, in fact, a bandwidth of several Hertz in control engineering terms.

2.3.5.2 Helmet Tracking Using Rate Gyros

The gyroscope has been continually developed worldwide over the last 150 years to measure the angular rotation of the platform on which it is mounted to very high accuracy (It should be noted that the gyro measures the rotation with respect to axes which are fixed in space. This information in turn can be converted to typically 'Local Vertical Axes' – North, East, Downwards – by applying the appropriate corrections for the Earth's rotation and the vehicle motion over the spherical surface of the Earth).

Until fairly recently the cost, size and weight of the available gyros rendered them unsuitable. Technology advances, however, have led to the wide-scale usage and availability of miniature solid-state rate gyros. These have been developed exploiting semiconductor fabrication technology, such as the vibrating quartz tuning fork rate gyro. These have the attributes of small size, low weight, very high reliability and above all relatively low cost, which has made it possible to produce a relatively low-cost solution for their use in helmet tracking.

Vibrating quartz tuning fork rate gyros are being exploited not only in avionic applications such as flight control, standby attitude and heading reference systems, and missile mid-course guidance, but also in the automobile industry in car stability enhancement systems. Other applications include robotics, factory automation, instrumentation, and GPS augmentation for vehicle location systems, navigation systems, precision farming, antenna stabilisation, camera stabilisation and autonomous vehicle control. In short, there is a very large market for these rate gyros, apart from aerospace applications, and this drives the costs down.

These developments in gyro fabrication technology have come into large-scale exploitation over the two decades and have had a revolutionary impact. For the first time, miniature solid-state rate gyros with Mean Time Between Failures (MTBFs) in excess of 100,000 hours became available at relatively low costs.

The vibrating quartz tuning fork rate gyro is described later in the book in Chap. 5, Sect. 5.2.2. Micro Electro-Mechanical Systems (MEMS) Technology Rate Gyros.

These sensors utilise the very stable piezo-electric properties of single crystal quartz to produce a monolithic quartz sensor element.

The vibrating quartz tuning fork rate gyro is illustrated in Figs. 5.2 and 5.3 (courtesy of System Donner Inertial Division).

The pair of tuning forks and their support flexures and frames are batch fabricated from thin wafers of single crystal piezo-electric quartz and are micro-machined using photo-lithographic processes similar to those used to produce millions of digital quartz wristwatches each year.

Referring to Fig. 2.33, the outputs of three rate gyros mounted on the helmet, to sense the rates of rotation of the pilot's helmet about the helmet forward, lateral and azimuth axes, are processed in a computer to derive the helmet sight line orientation.

The computation processes are explained in Chap. 5 under Sects. 5.3.2 and 5.3.3.

The rate gyros can be used to greatly enhance the performance of existing tracker systems. The data derived from the rate gyros has negligible lags and low noise.

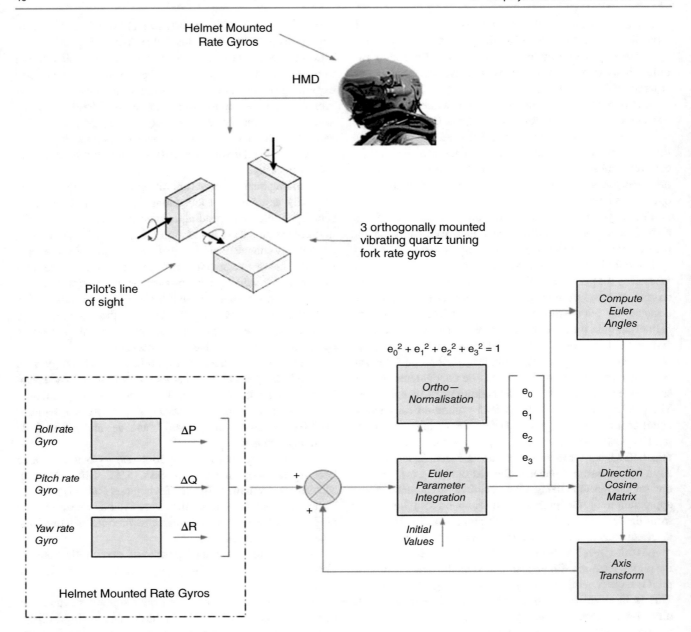

Fig. 2.33 Use of Helmet mounted rate gyros for Helmet tracking

Inertial sensors have no range limitations, no line-of-sight requirements and no risk of interference from any magnetic, acoustic, optical or Radio Frequency (RF) interference sources.

They can be sampled as fast as desired and provide relatively high bandwidth with motion measurement with negligible latency. Even tiny low-cost MEMS inertial sensors measure motion with very low noise resulting in jitter-free tracking, which looks very smooth to the eye.

Since they directly measure the motion derivatives, inertial sensors have the capability to perform high-quality prediction to compensate for graphics rendering and display rasterisation latencies without the noise problems of numerical differentiation.

The fundamental problem with any inertial system is that any sensor bias or noise is integrated with time and will cause the orientation and position to gradually drift with time.

It is thus essential to have high accuracy, alternative source of orientation and position to correct the drift errors in the inertial tracking system.

High-accuracy optical tracker systems have been developed using LEDs and cameras and exploiting pattern recognition or specific pattern areas around the cockpit or the use of correlation techniques using a preloaded frame.

Fig. 2.34 The BAE Systems 'Striker® II' binocular HMD. (Courtesy of BAE Systems)

The optical and the inertial data can be combined in a manner which exploits the best features of each system using a Kalman filtering process. Kalman Filters are explained in Chap. 6 under the subheading 6.3.

2.3.6 The BAE Systems 'Striker® II' Binocular HMD

Figure 2.34 shows the BAE Systems 'Striker® II' binocular HMD, which has demonstrated excellent performance characteristics.

The main advances over the 'Striker® I' are:

1. Replacement of the half inch diameter CRTs, with the accompanying requirement for a 13,000 Volt supply, with a high-resolution, solid-state digital display unit. This provides a lighter, compact, high-brightness colour display image.
   - The solid-state display unit provides a major improvement in reliability. It has an Mean Time Between Failures (MTBF) in the region of 100,000 hours compared with the CRT and its high voltage supply unit of around 2000-hour MTBF. (Basically, limited by the finite life of the heated cathode in the CRT.)
   - The high-resolution colour display enables important features to be enhanced by the use of colour to assist the pilot's situational awareness. (For example, hazards such as high-voltage electricity pylons and cables when flying at low level at night.)
   - Only limited information has been published by BAE Systems on 'Striker® II' design. It does however appear to use a similar optical configuration as the 'Striker® I' exploiting the spherical visor.
   - Figure 2.35 illustrates the basic optical configuration for a binocular HMD using conventional optical elements, bearing in mind the complex relay lens train required to correct the off-axis errors and residual collimation errors.
2. The major development is the elimination of the two night vision cameras and their replacement with a single new technology night vision sensor. This is after the Greek mythological one-eyed monster.
   - The 'Striker® II' HMD brings its high-performance digital night vision camera inside the helmet, which helps reduce G-force effects on the pilot's head and neck to improve comfort, and eliminates the need to manually configure and adjust night vision goggles (NVG) hardware for day-to-night transitions. With its high-resolution binocular visor-projected full-colour display, the new system integrates a centre-mounted ISE11™ (see note below) sensor based on Intevac Photonic patented advanced imaging sensor technology, known as electron bombardment active pixel sensor (EBAPSTM). This advanced sensor increases the display's 24-hour capability, bringing the system's night vision acuity to a level equivalent to or better than NVG performance but without the compromises that NVGs impose. Note: ISE11™: Independent Steaming Event, 11th trial of weapons and sensors on the Aircraft Carrier Gerald R Ford in August 2020. The programme tested E-O Lasers, Gyrocompasses Radar and Sonar for range and accuracy. Type approval was given to the sensor used by Intevac and incorporated into the Striker 11 HMD and the programme was Trade marked TM.
   - Figure 2.36 illustrates the high quality of the 'Striker® II' night vision provided by the new system. The quality of the collimated flight symbology data overlaying the visual scene, horizon line, velocity (or flight path) vector symbol and the aircraft forward (or boresight) axis, bank angle, pitch angle, height and airspeed.
3. Development of a new helmet tracking system of outstanding performance.
   - The information published by BAE Systems mainly concentrates on its performance and gives very limited information on the system details.

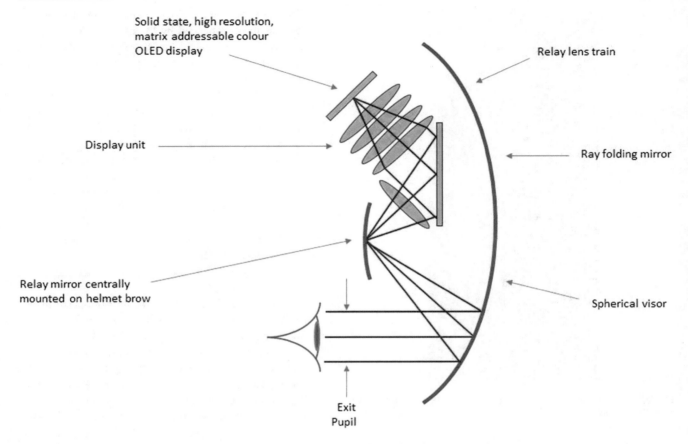

Solid state, high resolution, matrix addressable colour OLED display

Display unit

Relay mirror centrally mounted on helmet brow

Relay lens train

Ray folding mirror

Spherical visor

Exit Pupil

Fig. 2.35 Basic optical configuration of a binocular HMD using conventional optical elements

Fig. 2.36 'Striker® II' HMD night vision display. (Courtesy of BAE Systems)

The 'Striker® II' HMD builds on the already well-established 'Striker® I' HMD, which has decades of combat-proven experience on Typhoon and Gripen C/D aircraft. A full-colour solution, 'Striker® II' provides fixed- and rotary-wing pilots with remarkable situational awareness, next-generation night vision, 3D audio and target tracking technology; all within a fully integrated visor-projected HMD system.

The highly advanced 'Striker® II' supports the display of high-resolution sensor systems such as a distributed aperture system, which allows pilots to see through the body of the aircraft – giving them a vital advantage when it comes to split-second decision making. Using optical sensors embedded in the aircraft, 'Striker® II' immediately calculates the pilot's exact head position and angle. This means no matter where the pilot is looking, 'Striker® II' displays accurate targeting information and symbology, with near-zero latency.

2.3.7 HUD/HMD Optical Waveguide System Technology

The previous sections have covered the basic principles, functions and design of HUDs and HMDs which are currently in service and that will remain in service for many years to come.

Recent developments in holographic waveguide technology, however, have had a profound impact on new HUD and HMD design. Exploitation of this new technology offers a major improvement in terms of mass, cost, volume, simplicity and optical performance.

The basic principles and configuration of the HUD collimation system using conventional optical elements are explained earlier in the chapter in Sect. 2.2.2. Refer to Figs. 2.5 and 2.6 and to Fig. 2.7 illustrating instantaneous and total FOVs. Refer, also, to the cockpit geometry constraints illustrated in Fig. 2.8 and the 'see-through-mirror' combiner, Fig. 2.9.

It is also necessary to accommodate a reasonable range of pilot's head movement, or 'head motion box', otherwise it is a case of 'now you see it, now you do not'.

It can be seen that the FOV is determined by the diameter of the collimating lens in the projection aperture.

The problem with this basic system in an HUD is that an optical system is needed with a large exit pupil to accommodate two eyes and head motion. Large lenses or mirrors must be corrected for geometric distortions. A complex and expensive relay lens system is also required to correct the errors in the collimating lens.

Space and weight constraints severely limit the diameter of the relay lens aperture, which in turn determines the size of the exit pupil. Particularly large lenses will be very expensive if they are not to suffer from aberrations. They are also size limited by the severe thermal and vibration stresses experienced in military aircraft.

2.3.7.1 Basic Concepts

Consider, now, the concept of producing a smaller FOV collimated display image and generating an array of optical copies of the collimated display image and conveying the array to the HUD combiner, as shown in Fig. 2.37.

A smaller FOV collimated display image can be achieved with a simple lens system. The optical elements are small and of relatively low cost, the lenses are on axis and a simple relay lens system is used to correct the collimating lens errors.

Because the display image within an exit pupil is collimated, that is, focused at infinity, it does not matter if the pilot can see a particular symbol in two overlapping exit pupils. That symbol will still overlay the same point on the distant outside world scene. Figure 2.38 illustrates this fundamental principle.

It will be shown how optical waveguides and diffraction gratings can be combined to produce an array of identical collimated display images on the combiner.

Light introduced into a slab waveguide of optically more dense transparent material compared to its surroundings (e.g. glass, $n1$, in air, $n2$) will be totally internally reflected, provided the angle is greater than the critical angle $\ominus C$ (given by $\sin\ominus C = n2/n1$, Snell's law). Refer to Fig. 2.39.

If a collimated image comprising a range of angles greater than the critical angle is introduced into the waveguide, they will 'bounce' by total internal reflection (TIR) on their way along the waveguide suffering little attenuation. Refer to Fig. 2.40.

At each reflection on the internal surface of the waveguide the light will be wholly contained within the waveguide unless it can be disturbed to leave the waveguide. This is achieved with a diffraction grating. Refer to Fig. 2.41. The gratings can be constructed holographically in a light-sensitive medium with the use of interference fringes generated at the intersection of two laser beams. This allows the grating to be conveniently produced within the waveguide itself.

A diffraction grating is an optical component with a periodic structure that splits and diffracts light into several beams travelling in different directions, $\ominus$. The simplest type of grating is one with many evenly spaced parallel slits. The typical result is illustrated in Fig. 2.42.

The diffractions are called 'orders', by convention upper +ve and lower −ve (analogous to upper and lower sidebands of modulated radio frequency [RF] signals). The zeroth central order is un-diffracted light. Orders less than the critical angle will emerge towards the observer and those greater than critical angle including the critical angle will continue along the waveguide where they will encounter the diffraction grating again. This is shown in Fig. 2.43.

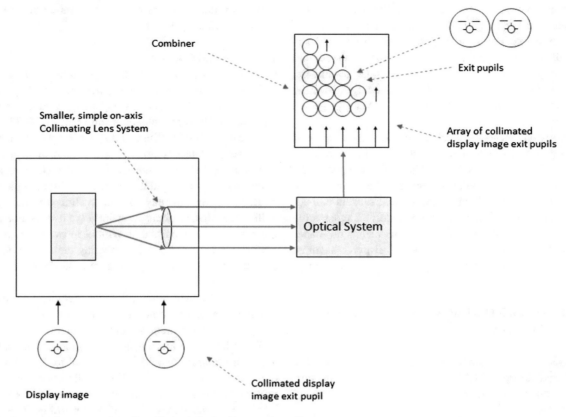

Fig. 2.37   Concept of an array of small, collimated display image exit pupils

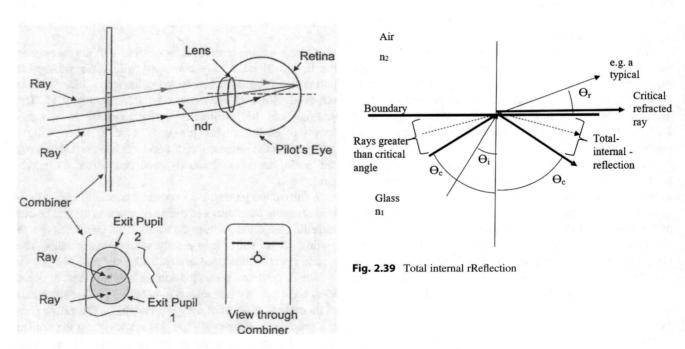

Fig. 2.38   Basic principle of array of multiple exit pupils. *Rays A and B from the Velocity Vector symbol in Exit Pupils 1 and 2 are parallel and are focused onto the same point on the retina. Pilot sees only one symbol.*

Fig. 2.39   Total internal rReflection

Fig. 2.40 Waveguide transmission

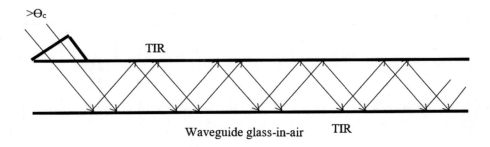

Fig. 2.41 Waveguide with diffraction grating

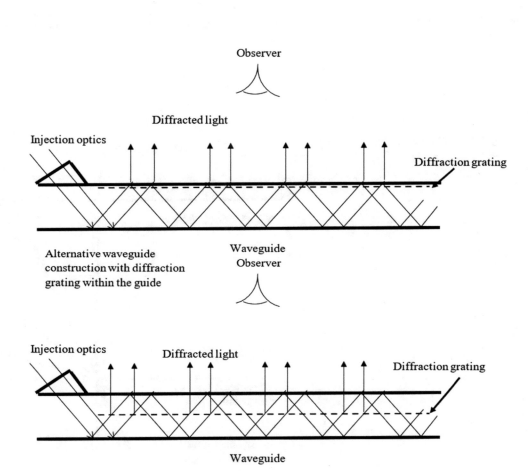

The design of the ideal diffraction grating for a display depends upon the optical system used and would be of commercial confidence and is not speculated upon here.

HMD and HUD systems exploiting holographic waveguide technology are being actively developed by BAE Systems, Rochester.

The systems exploit some of the original work in optical waveguide technology carried out by Professor Bill Crossland and Dr. Adrian Travis at the Cambridge University Engineering Department.

2.3.7.2 The Q Sight HMD

The first system exploiting holographic waveguide technology to be developed by BAE Systems, Rochester, was the 'Q Sight' HMD; the first prototype was test flown in a helicopter around 2008.

Figure 2.44 is a schematic illustration of the Q Sight HMD showing the basic elements.

The display image is first optically collimated with a simple lens system and then launched into the waveguide by an optical prism arrangement. The optical elements are small and of relatively low cost; the lenses are on axis and can be of optical quality plastic.

Fig. 2.42 Diffraction orders

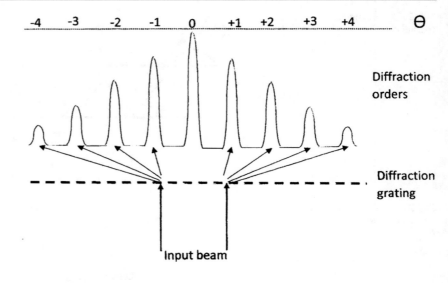

Fig. 2.43 Diffraction grating orders

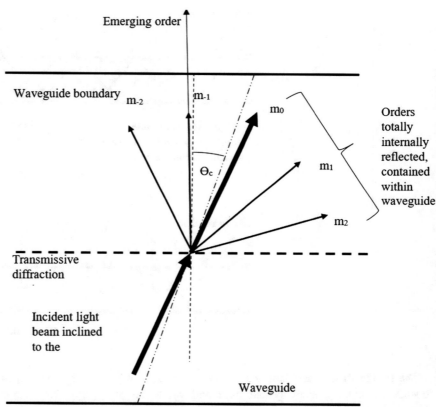

The optical waveguide comprises a sandwich of two thin rectangular glass plates with a holographically generated diffraction grating in the middle of the sandwich. The light rays from the collimated display image are then internally reflected along the waveguide as shown in Fig. 2.45.

When a reflected light wave encounters the diffraction grating, a very small fraction of the light is diffracted out, as shown. The diffracted light from the grating forms a continuous row of images of the display exit pupil, which is duplicated along the horizontal waveguide.

A second waveguide is optically coupled at right angles with the first waveguide.

The row of exit pupils from the first waveguide is then coupled into the second waveguide so as to form an array of rows and columns of exit pupils along the waveguide combiner, as shown in Fig. 2.46.

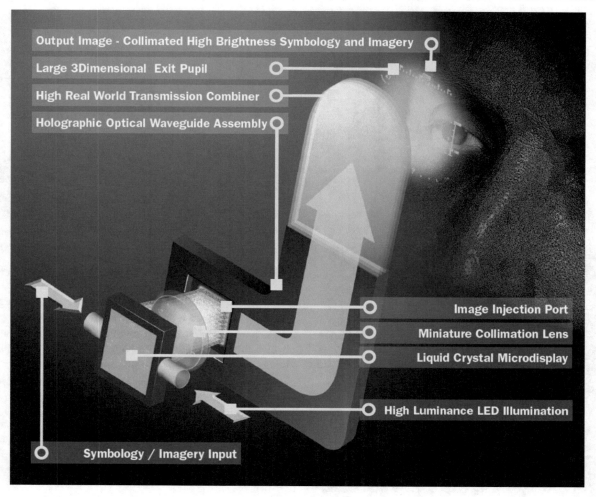

Fig. 2.44 Schematic illustration of the Q Sight HMD. (Courtesy of BAE Systems)

Fig. 2.45 Optical waveguide with central diffraction grating

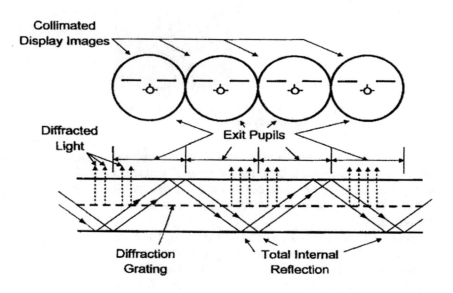

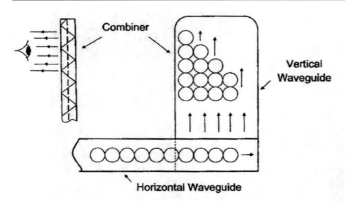

Fig. 2.46 Array of exit pupils on combiner

Fig. 2.47 The Q Sight HMD. Inset shows display symbology overlaying world. (Courtesy of BAE Systems)

The array of rows and columns of exit pupils are then conveyed along the waveguide to the combiner. The function of this second waveguide is to couple the exit pupils to the eye. The image is ejected normal to the glass and overlaid on the outside world.

Figure 2.47 is a photograph of a Q Sight HMD. The basic simplicity of the HMD can be readily seen. It is intrinsically inexpensive to manufacture and a modular approach enables it to be incorporated into existing pilot helmets as a 'bolt-on' attachment.

Fig. 2.48 LiteHUD® pilot display unit. (Courtesy of BAE Systems)

The simplicity of the Sight, low mass and the large eye position box make it an attractive solution. Symbology and, or, video can be displayed providing full head up operation. It is also fully compatible with Night Vision Goggles (NVGs), without modification or reconfiguration.

The Sight is mounted on its own mount and positioned approximately 25 mm from the eye. The holographic waveguide combiner is roughly credit card size and about 2.5 mm thick.

2.3.8 The LiteHUD®

LiteHUD®, shown in Fig. 2.48, is a glareshield mounted head up display that has been designed to be compatible with the future needs of trainer and fast jet cockpits. The waveguide is mounted horizontally within the HUD with the output being reflected off a combiner to the pilot's eyes.

LiteHUD® is a small and compact head up display (HUD), offering space and weight advantages by incorporating revolutionary waveguide optics with the latest digital display technology and highly reliable electronics. LiteHUD® is 60% smaller by volume and up to 50% lighter than a typical HUD with an extremely large eye motion box. In addition a fully integrated colour HUD camera capability is provided.

This fully qualified product has already been selected for use across a number of diverse platforms as its ultra-compact design can be integrated into almost any cockpit: from turbo-prop trainers, to the fighters of tomorrow featuring large area displays. The use of digital modular electronics provides a far more reliable design yielding lower life cycle costs.

Glareshield mounted waveguide-based HUDs are compatible with the trend towards the installation of large area

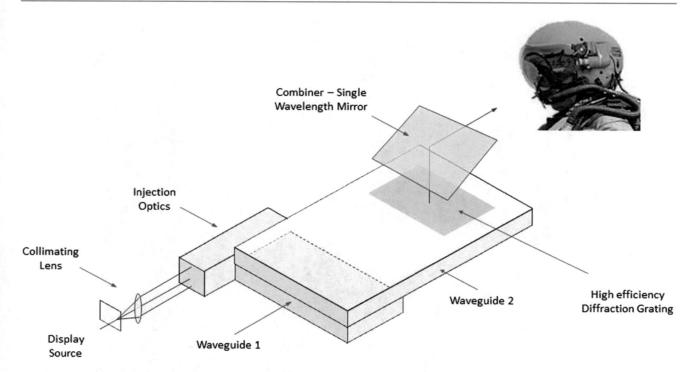

Fig. 2.49 Schematic optical configuration of a waveguide HUD

displays in place of the traditional multiple head down displays in military fast jet cockpits.

The need for a further holographic waveguide combiner to convey the collimated array of exit pupils to the pilot's eyes has been avoided in the LiteHUD® by using a high-efficiency diffraction grating to transmit the display image to a simple 45° conventional combiner.

Figure 2.49 is a schematic illustration to show the operation of the conventional combiner. Details of the LiteHUD® design have not been released and the schematic assumes the design exploits the earlier waveguide HUD configuration with possibly more advanced holographically generated diffraction gratings.

The combiner is basically similar to the combiners used in the first generation of head up displays. It consists of a thin parallel-faced glass plate with a multi-layer reflective coating deposited on one face. The combiner has a high transmissibility of the outside scene and the reflective coating is tuned to achieve maximum efficiency in reflecting the wavelength of the collimated display image exit pupils into the pilot's eyes and providing a high-brightness display which can be seen under bright sunlight conditions.

The requirement for a large collimated lens in the simple HUD, however, has been eliminated by the holographic waveguide and generation of the array of collimated display images in multiple exit pupils.

A 60% reduction in the volume occupied by the LiteHUD® compared with a conventional HUD has been achieved.

Figure 2.50 shows the LiteHUD® installed in the BAE Advanced Hawk concept demonstrator and Fig. 2.51 the HUD display symbology.

2.3.9 The LiteWave®

LiteWave® is a waveguide HUD designed to be mounted above the pilot. The combiner is the waveguide. The use of waveguide technology makes LiteWave® the smallest and most compact HUD available and provides higher reliability than other HUD technologies. This is shown in Fig. 2.52.

Figure 2.53 shows the LiteWave® HUD in comparison to the current installation in the A320 airliner, which uses traditional HUD technologies. LiteWave® is up to 70% smaller and lighter than conventional HUDs.

2.3.10 HMDs and the Virtual Cockpit

The concept of a 'virtual cockpit' where information is presented visually to the pilot by means of computer-generated 3D imagery is being very actively researched in a number of establishments both in the USA and the UK.

Only a brief introduction to the topic can be given in this chapter because of space constraints. There is no doubt, however, of its importance in the future and the revolutionary impact it is likely to have on the next generation of aircraft cockpits. The increasing use of remote piloted vehicles

Fig. 2.50 Installed LiteHUD®. (Courtesy of BAE Systems)

Fig. 2.51 HUD symbology. (Courtesy of BAE Systems)

Fig. 2.52 LiteWave® overhead HUD. (Courtesy of BAE Systems)

Fig. 2.53 LiteWave® is considerably smaller than existing conventional HUDs. (Courtesy of BAE Systems)

(RPVs) and their control from a 'parent' aircraft, or ground station, is another future application for HMDs and virtual cockpit technology. It should be noted that RPVs can include land vehicles or underwater vehicles as well as airborne vehicles.

A correctly designed binocular HMD (BHMD) is a key component in such systems because it is able to present both a display of information at infinity and also stereo images to each eye so that the pilot sees a 3D image. It should be emphasised that the BHMD optical system needs to be accurately designed and manufactured in order to achieve a good stereoscopic capability.

The ability to generate 3D displays opens up entirely new ways of presenting information to the pilot (and crew), the objectives being to present the information so that it can be visually assimilated more easily and in context with the mission. As an example, information related to the outside world can be viewed head up and information related to the mission and the aircraft 'health' can be viewed head down using only one display system, namely the BHMD.

Figure 2.54 illustrates the use of a BHMD to implement such a system. When head up, the pilot views the outside world directly, or indirectly by means of a TV display on the HMD from a head-steered gimballed electro-optical sensor unit (e.g. infrared imaging sensor, low-light TV camera or CCD camera).

When looking down at the instrument panel, the virtual cockpit computer system recognises the pilot's head down sight line from the head tracker output and supplies this information to the display-generation system. The display-generation system then generates a stereo pair of images of the appropriate instrument display on the panel, which corresponds to the pilot's sight line. Thus, looking down into the cockpit at the position normally occupied by a particular instrument display will result in the pilot seeing a 3D image of that instrument display appearing in the position it normally occupies – i.e. a virtual instrument panel display. Carried to its logical conclusion, the normal head down displays and instruments would no longer be required. Simple standby instruments of the 'get you home' type would be sufficient to cover the case of a failure in the HMD system. Such a bold step, however, is still several years away from implementation in a production aircraft.

Novel ways of presenting information to the pilot by means of the BHMD include displaying a 3D 'pathway in the sky' as a flight director display, which can be overlaid on the normal outside scene (directly or indirectly viewed) or on a computer-generated outside world scene created from a terrain database. A 'God's eye view' can also be presented; for example, what the pilot would see if he was outside the aircraft and looking down from a height of say 2000 ft. above

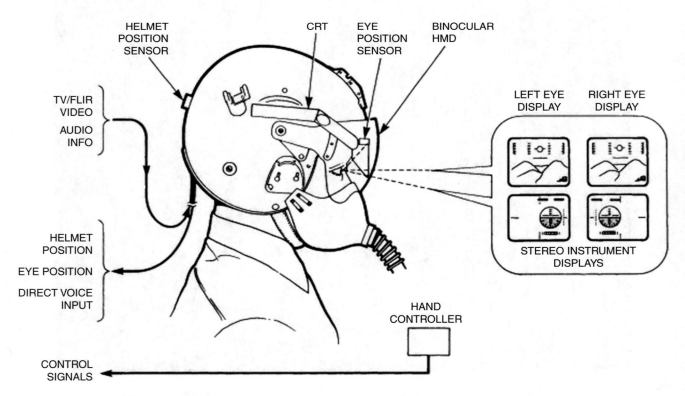

Fig. 2.54 Binocular HMD and virtual cockpit

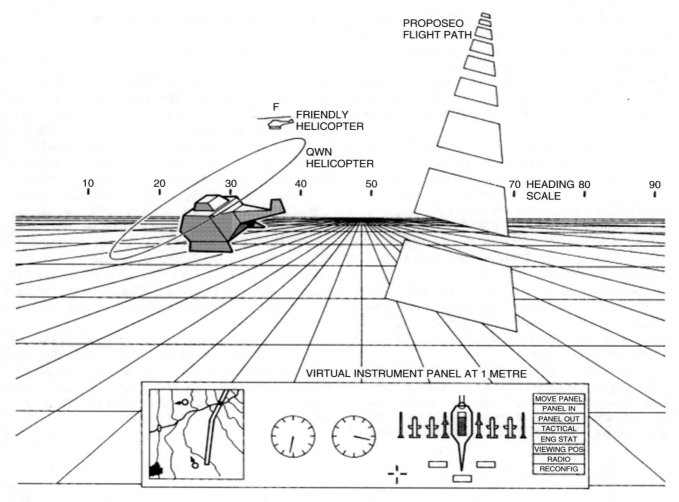

Fig. 2.55 God's eye view display concept. (Courtesy of BAE Systems)

and slightly behind his aircraft. ('Beam me up there, Scotty!'.) The threat situation can be presented on such a 'God's eye view' display – e.g. enemy aircraft, missile sites, anti-aircraft artillery ('Triple A') sites or enemy radar sites. Threat zones defining the missile (or gun) range, radar field of view etc. can be presented as 3D surfaces so that the pilot can assimilate the threat situation on a single display.

Figure 2.55 shows a 3D 'God's eye view' display, which provides the pilot with a stereoscopic view of his helicopter as well as its relationship to its proposed flight path. Further information can be provided by enabling the pilot to change his apparent position outside the aircraft.

With advances in display technology it is reasonable to anticipate light-weight high-resolution colour displays which would be capable of providing a helmet mounted colour display of comparable quality to the current colour head down displays. This, for example, would enable a colour map to be displayed on the HMD.

It is worth sounding a slightly cautionary note on virtual cockpit technology and the need for all the human factor issues to be addressed and satisfactory solutions achieved for any problems that may be encountered.

Some disquiet has, in fact, been expressed recently on the physiological and psychological effects of prolonged viewing of commercial 'virtual reality' systems and the ability of the user to switch back to the real world without delay. Poor-quality stereo optics requires the visual processing part of the brain to make major adjustments to fuse the resulting stereo images into a 3D picture. To adjust back to normal real-world viewing can then take some time.

The cockpit environment, however, is very different and requires frequent viewing of the outside world whether directly, or indirectly, by means of a head-steered gimballed camera system. The virtual reality problems referred to above may, therefore, not arise. The stereo quality of the optical system of the BHMD is also far higher than the current commercial virtual reality systems.

The hardware, in terms of binocular HMDs, graphics processing, symbology generation, together with the software to implement many of the virtual cockpit concepts, exists today, although system integrity still needs to be established. The relatively slow adoption of the concepts is due more to defence budget constraints both in the UK and the USA rather than technology limitations and technical risks.

Mention should be made of the potential of what are referred to as virtual retinal displays (VRDs) for future virtual cockpits. In such systems, images are not presented on a display surface but instead are projected directly on to the retina of the pilot's eye by a raster scanned laser light beam. The technology offers a number of advantages and is being actively developed by, for example, a company called Microvision Inc., Seattle, USA. References to VRD technology are given at the end of the chapter.

The current technology of implementing a virtual retinal display involves lasers, optical fibres and miniature vibrating mirrors for raster scanning the retina of the eye, together with coupling optics. A binocular VRD thus imposes a significant weight delta to the helmet.

At the present time, the miniature reflective active matrix colour liquid crystal display (AMLCD) is able to provide a lower weight, less complex and lower cost solution for a fast jet HMD where minimal weight is essential. The AMLCD display is not as 'crisp' as a VRD, but is fully acceptable.

2.4 Computer Aided Optical Design

2.4.1 Introduction

The use of computer aided optical design methods has been referred to earlier and is briefly discussed in this section to show the reader its importance. None of the modern HUDs and HMDs, in fact, would have been feasible without its use.

Modern optical design software enables very sophisticated (sounds better than complex) optical designs to be optimised and the performance determined. For example, ray tracing for an optical train including de-centred and tilted optical elements, aspheric elements and holographic elements can be carried out. An automatic optimisation program can be carried out iteratively whereby the curvature of all the individual elements, their spacing, tilt, de-centring, refractive index can be varied to seek an optimal solution using 'hill climbing' methods. This enables the optical designer to evaluate a very wide range of configurations to achieve the performance goals.

These programs take several hours to run even on the most modern mainframe computers. Figure 2.56 shows a computer-generated ray trace for a wide FOV holographic HUD as an example of the optical complexity involved in such designs. It also allows quick feasibility studies of a range of optical configurations to be run.

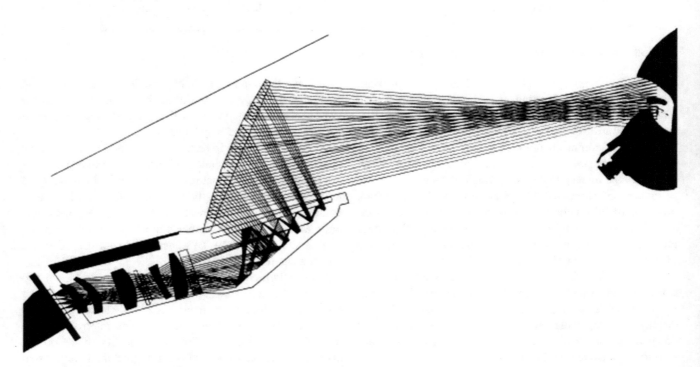

Fig. 2.56 Computer-generated optical ray tracing. (Courtesy of BAE Systems)

There are, of course, many more optical analysis functions which can be carried out using modern optical design software, but space constraints limit further discussion.

2.5 Discussion of HUDs Versus HMDs

2.5.1 Introduction

The current and future roles of the HUD and HMD in military aircraft are discussed in the next subsection.

The question of an HMD for civil aircraft applications does not arise as civil airline pilots have no need to wear helmets. The pilot's communication requirements are met by a simple push-on headset with an attached microphone. The acoustic noise levels in the civil aircraft are much lower than in military aircraft and the head protection requirements do not arise. An oxygen mask is only required in the event of the loss of cabin pressurisation, whereas the helmet forms part of a special oxygen breathing system for high g manoeuvring in a combat aircraft.

2.5.2 Military Aircraft HUDs and HMDs

The ability of the HMD to display all the information that is displayed on the HUD over a wider FOV, which is not confined to the forward sector, raises the question of whether the HUD will still be necessary in the future.

The answer is that the HUD is likely to remain a key system for some considerable time as the prime display of primary flight information to the pilot, and to enable high-accuracy weapon aiming with guns, unguided bombs and rockets to be achieved.

Dealing first with the question of integrity, the display of primary flight information must meet an integrity figure of 10^{-9} for the probability of occurrences of displaying potentially misleading or hazardous information.

A single unit HUD system (combined processor, symbol generator and display) can meet this integrity figure using the monitoring techniques described earlier.

The HMD system, however, is not a 'one box' system and comprises helmet, cockpit interface unit, head tracker and a processor/symbol generator, which are all dispersed around the cockpit and requires a relatively large number of connectors. Critically, the need to minimise the head-supported weight results in the CRT drive circuits being located remotely from the helmet at least at the end of an umbilical cable. Such a widely dispersed architecture cannot achieve as high an integrity as a single unit HUD system.

The integrity issue, however, is not an insuperable one with the advances being made in display technology and the incorporation of adequate redundancy and monitoring into the system. Clearly, the integrity of the display of primary flight data such as attitude information and heading on the HMD is dependent on the correct functioning of the head tracker system. It appears feasible to devise self-monitoring techniques to check the integrity of the head tracker system. Alternatively, a dual tracker system could be installed as a monitor, or used in conjunction with a self-monitoring system. An alternative 'get you home' source of attitude could be obtained from a triad of miniature solid-state rate gyros and accelerometers mounted directly on the helmet, and computing the helmet attitude from their outputs.

Dealing with the question of accuracy, the HMD is entirely adequate for launching precision guided weapons such as missiles and laser guided bombs. These weapons are highly effective, so why is very high accuracy still required for weapon aiming? The answer is that guided weapons are highly expensive and in today's scenario of cost limitations, 'iron bombs', rockets and guns come cheaply. Despite the efficiency of beyond visual range (BVR) missiles, the difficulty of identifying friend from foe still exists and the near-in use of guns is still considered necessary.

It is not yet feasible to target unguided weapons with an HMD to the same accuracy as with an HUD.

HUD accuracy is derived from a chain of elements comprising:

- Input data.
- Processing and symbol generation.
- Analogue deflection and video.
- Cathode ray tube.
- Installation.

All these factors have now become highly sophisticated to the extent that the final element in the chain, the pilot, gives equal or greater errors. Overall, from data input to HUD optical output, an accuracy of better than 0.5–1 mRad can be obtained for placing a symbol over the target at the centre of the field of view. Even at the edge of a typical refractive HUD, at 10 degrees off axis, a figure of 1.5–2 mRad is achievable.

The HUD is rigidly attached to the airframe and aligned to a fraction of mRad by mechanical and/or electronic means. The alignment of optics and CRT also benefits from the rigidity of the HUD chassis such that the boresight errors can be as low as 0.15 mRad in pitch and yaw, and 0.6 mRad in roll. The HUD is equally in a fixed relationship to the windshield and views the outside world only through a small and well-defined segment. Corrections can be readily applied in respect of windshield optical effects and symbol position distortion.

Conversely, the HMD cannot be as rigidly located with respect to the pilot's eye because of limitations of helmet design such as eye relief, weight and comfort. The pilot's line of sight with respect to the airframe axes has to be inferred by the head tracker and although accuracy on boresight and in the principal region of the head motion box is better than 2 mRad, this degrades to 8–10 mRad at large off-boresight angles. It is also affected by the ability of the pilot to hold his head steady under the buffeting experienced at high subsonic speeds in turbulent conditions. The advent of a practical eye tracker may eventually enhance the operational accuracy of the HMD. A further difficulty with the HMD is that it is not easy to compensate for windshield errors to a very high degree of accuracy over the whole head motion box.

The end to end accuracy chain for the HMD will, in addition to the HUD factors, have these additional factors:

- Helmet alignment.
- Tracker accuracy.
- Windshield.

Figure 2.57 shows the accuracy relationship to off-boresight angle for the HUD and HMD. The current limitations of the HMD in terms of unguided weapon delivery accuracy can be seen.

Development is being carried out to improve head tracking accuracy and it should be possible to achieve acceptable accuracy, albeit not as good as the HUD. The degradation in HMD aiming accuracy encountered if the aircraft experiences severe buffeting in high-speed low-level attacks has been mentioned, and is a concern.

The cost, weight and cockpit 'real estate' savings, which can be made by eliminating the HUD in a fighter/strike aircraft, have provided a strong incentive to improve the integrity and accuracy of the HMD system. The weight and bulk of the HMD must also be made as low as possible.

It is interesting to note that the new Lockheed-Martin Lightning 2, Joint Strike Fighter does not have an HUD and relies entirely on the HMD.

The new technology HMDs and HUDs described in Sect. 2.9, which exploit holographic waveguide technology, however, offer lower cost, lower weight and better performance than current technology systems. It may thus be possible to justify having both an HUD and an HMD with this new technology. This is because of the inherently higher accuracy the HUD provides for delivering unguided weapons and the increased integrity of displaying primary flight information which two independent systems can provide.

The existence of laser devices which can cause eye damage has been in the public domain since the early 1980s. These can be operated at relatively low power levels to dazzle the pilot of an approaching aircraft so that the pilot becomes disoriented and has to break off the attack. The laser can also be operated at higher power levels which are sufficient to cause permanent damage to the retina of the eye resulting in impaired vision and blindness if sufficiently high. In future conflicts it may, therefore, be necessary to fly with a suitably opaque visor installed on the helmet to protect the pilot's eyes from optical weapons when operating in areas where such a threat exists.

The pilot would then fly the aircraft by means of a head-steered indirect viewing system with overlaid flight and weapon aiming symbology.

The potential of the HMD and the exploitation of virtual cockpit concepts have already been briefly discussed and it is likely that such concepts will be exploited in the next-generation military aircraft.

Finally, it should be noted that there are civil applications for the HMD as part of an indirect viewing system for night/all weather operation of helicopters. These include air sea rescue operations, servicing offshore oil rigs, police helicopter surveillance and border patrols using helicopters.

2.6   Head Down Displays

2.6.1   Introduction

Electronic technology has exhibited an exponential growth in performance over four decades and is still advancing. By the early 1980s, it became viable to effect a revolution in civil flight-decks and military cockpits by replacing the majority

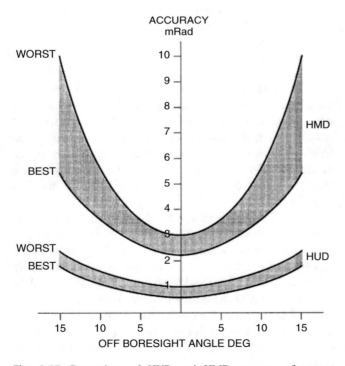

Fig. 2.57 Comparison of HUD and HMD accuracy of current in-service systems

of the traditional dial-type instruments with multi-function colour CRT displays.

'Wall to wall' colour displays have transformed the civil flight-deck – from large 'jumbo' jets to small commuter aircraft.

The only traditional dial instruments that have been retained are the electro-mechanical standby instruments such as the altimeter, airspeed indicator, artificial horizon and heading indicator. These electro-mechanical standby instruments, however, are now being replaced by all solid-state equivalents with a colour LCD display presentation, in the new generation of aircraft entering service, or in avionic update programmes.

The colour CRT is still a major display technology in terms of the number of aircraft equipped with colour multi-function CRT displays. The situation, however, has changed rapidly over the last decade with the development of high-resolution, active matrix colour liquid crystal displays (AMLCDs).

Flat-panel, high-resolution, colour displays exploiting commercial 'off the shelf' colour AMLCDs have displaced colour CRT displays in head down display applications in new build civil and military aircrafts and avionics system updates.

It should be noted that the current supremacy of the flat-panel AMLCD display may well be challenged in the near future by the organic light-emitting diode (OLED) display. OLED displays were still under development when the Second Edition of this book was published in 2003. Development problems have been overcome and high-resolution, super-thin, flat-panel, colour OLED displays are common-place in mobile phones and are appearing in top of the range TV sets. OLED displays have a number of significant advantages over AMLCDs. They are an emissive display with an inherently wide viewing angle and crisp display characteristics. They have a fast response time of 10–20 ns and have a wide operating temperature range, when properly packaged. They have active matrix addressing (like AMLCDs) and can have high brightness. They are thin and do not require an illuminating light source like the AMLCD, thus enabling a slim flat-panel display to be achieved.

Typically, an HDD comprises the display together with the drive electronics and power regulation. As with the HUD, there is frequently a simple processor and housekeeping software to control the brightness, self-test and mode changes. Some units now contain a full processor and symbol generator and data bus interface, but it is more usual for the HDDs to be driven by a central display processor(s).

The size of an HDD is typically defined by an Aeronautical Radio Incorporated (ARINC) standard of square format with one inch increments – for example, 5 × 5, 6 × 6 inch. This is not ideal for the normal 4:3 aspect ratio video and indeed rectangular format displays are found in both military and civil cockpits.

2.6.2 Civil Cockpit Head Down Displays

Figure 1.6 in Chap. 1 shows a modern civil flight-deck.

The displays are duplicated for the captain and second pilot and being multi-function it is possible to reconfigure the displayed information in the event of the failure of a particular display surface.

The electronic primary flight display (PFD) replaces six electro-mechanical instruments: altimeter, vertical speed indicator, artificial horizon/attitude director indicator, heading/compass indicator and Mach meter. PFD formats follow the classic 'T' layout of the conventional primary flight instruments, as mentioned in Chap. 7.

All the primary flight information is shown on the PFD thereby reducing the pilot scan, the use of colour enabling the information to be easily separated and emphasised where appropriate.

Figure 2.58 shows a representative primary flight display.

Airspeed is shown on a scale on the left with pressure altitude and vertical speed on the right-hand scales. Aircraft heading information is shown on a 'tape' scale-type format below the attitude display. The artificial horizon/attitude display has a blue background above the horizon line representing the sky and a brown background below the horizon line representing the ground. This enables 'which

Fig. 2.58 Primary flight display. (Courtesy of Airbus)

way is up' and the aircraft orientation to be rapidly assimilated by the pilot in recovering from an unusual attitude, for example, as the result of a severe jet upset.

It should be noted that the attitude direction indicator is a key display on any aircraft for recovering from unusual attitudes particularly when the normal outside world visual cues are not present (e.g. flying in cloud or at night) as the pilot can become disorientated. The normal display of attitude on the HUD is not suitable for this purpose, being too fine a scale that consequently moves too rapidly.

Figure 2.59 shows a typical navigation (or horizontal situation) display. Pilot selectable modes include the traditional compass rose heading display, expanded ILS or Very High Frequency Omnidirectional Range (VOR) formats, a map mode showing the aircraft's position relative to specific waypoints and a North-up mode showing the flight plan. Weather radar displays may be superimposed over the map. The vertical flight profile can also be displayed.

Figure 2.60 shows an engine/warning display.

Figure 2.61 shows a systems display for monitoring the aircraft systems, for example, the fuel system, hydraulic systems, electrical systems, air conditioning and other systems, showing current status.

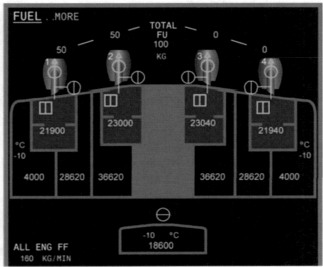

Fig. 2.60 Engine/Warning display. (Courtesy of Airbus)

Fig. 2.59 Navigation (or horizontal situation) displays. (Courtesy of Airbus)

Fig. 2.61 Systems display. (Courtesy of Airbus)

2.6.3 Military Head Down Displays

Video head down displays now include FLIR, Low Light Television (LLTV) and maps. All the HUD functions may be repeated overlaid on the video pictures. Fuel and engine data, navigation waypoints and a host of 'housekeeping' functions (e.g. hydraulics, pressurisation) may be displayed. A stores management display is also required showing the weapons carried and their status.

It is usual to have a bezel around the display with keys. Sometimes key functions are dedicated or they may be 'soft keys' where the function is written beside the key on the display. So-called tactile surfaces are being introduced using such techniques as infrared beams across the surface of the display or even surface acoustic waves where the pressure applied can also be measured to give *X*, *Y* and *Z* co-ordinates.

Typical advanced military cockpits are configured with four head down displays currently using AMLCD technology. There are two large colour displays; a horizontal situation display (HSD) providing a 6 × 8 inch map display in 'portrait' format with symbol overlay of routing and threat data, and a vertical situation display (VSD) providing an 8 × 6 inch IR video display showing targeting video at various magnifications.

The other two displays are smaller 5 × 5 inch monochrome displays comprising a systems status display (SSD) displaying systems status data, and a systems control display (SCD) displaying systems control data. Both displays have a tactile data entry overlay.

The advanced cockpits for the new generation of fighter/strike aircraft have just two large colour displays as the primary head down displays.

Figure 2.62 shows the instrument panel of the Lockheed Martin 'Lightning 2' Joint Strike Fighter, which features two large flat-panel colour display surfaces. The pilot can divide each screen into several windows, as can be seen in the illustration, enabling a very wide variety of information to be displayed at the same time. These flat-panel displays are the same type as used in commercial laptop computers and are hardened to withstand the fighter environment by mounting them in rugged bezels.

2.6.4 Display Symbology Generation

Symbology such as lines, circles, curves, dials, scales, alphanumeric characters, tabular information, map display features is drawn as a set of straight-line segment approximation, or vectors (like a 'join the dots' children's pictures). Figure 2.63 shows the symbology generation and display process.

2.6.5 Digitally Generated Moving Colour Map Displays

A map is by far the best way of visually assimilating the aircraft's horizontal situation. That is, the position of the aircraft relative to the chosen waypoints and destination, alternative routes, targets and location of specific terrain features such as mountains, hills, lakes, rivers, coast line, towns, cities, railways and roads. Colour maps can now be

Fig. 2.62 Lockheed Martin 'Lightning 2' Joint Strike Fighter cockpit. (Courtesy of Lockheed Martin)

Fig. 2.63 Symbology generation and display

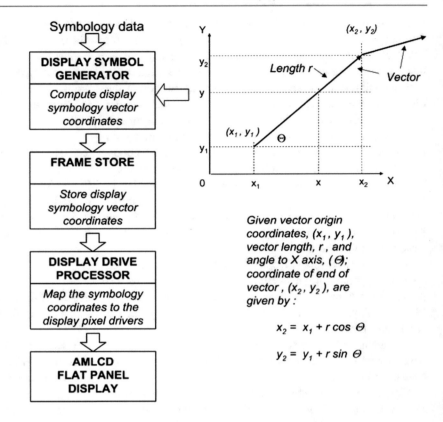

generated digitally from a map database stored in the computer memory and moved in real time with the aircraft so that the aircraft is at the centre of the map display. The map can be oriented 'track-up' or 'North-up' by selecting the appropriate mode.

The navigation system provides the basic information to move and rotate the map to maintain the selected orientation (usually track-up).

The system has major advantages in flexibility and the ability to present and overlay information, which is not possible with the earlier projected film map displays. Some of the advantages are:

- *Scale flexibility.* The map can be displayed at any selected scale. The time to re-draw the map at the new selected scale is typically 1–2 s.
- *Look ahead and zoom facility.* The pilot can look at the map some distance ahead of the aircraft's present position and 'zoom in' at a large scale to see the terrain details, for example in the target area or destination airfield etc.
- *Terrain clearance display.* The terrain above the pilot's present altitude can be distinctly coloured so that the pilot is aware of the location of potentially dangerous terrain such as hills and mountains relative to the aircraft. Appropriate avoiding action can then be taken in good time.

- *Terrain screening display.* The areas below the radar LOS of known (or likely) enemy radar sites can be displayed on the map so that advantage can be taken of terrain screening to minimise the probability of detection.
- *Decluttering.* Features and information (e.g. place names) that are not required can be deleted from the map display to declutter the map for a particular phase of the mission. Specific features can also be emphasised. The map display can be restored to show the full information whenever required.
- *Threat locations.* These may be enemy missile sites, 'triple A' sites, fighter airfields or radars. Threat envelopes showing, say, the missile range from a particular site, or sites, can be shown in a distinct colour.
- *Known low-altitude obstructions.* Electricity pylons, power cables, towers etc. can be emphasised on the map, for example by brightening them up, when flying at low altitude.
- *3D map displays.* A 3D map can be generated of, say, the target area to assist the pilot in visually assimilating the local terrain and hence assist in target acquisition.

Figure 2.64a, b illustrates digitally generated video map displays.

The information to construct the map is stored in a digital database as sets of specific terrain features such as: coast

Fig. 2.64 (**a**) Digitised map with information overlay (50,000:1 scale). (**b**) Colour moving map displays (by courtesy of BAE Systems). (Note: in the 'North-Up' mode, the map automatically aligns to the present position which is the centre of the display screen. When using 'Track/heading up' modes, the present position is offset to give maximum look ahead)

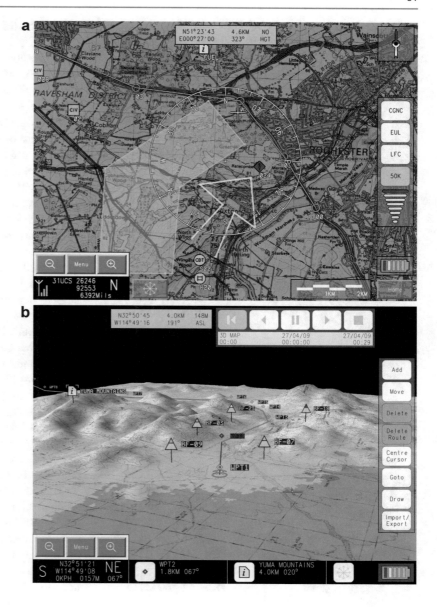

lines, rivers, lakes, contour lines, town outlines, railways, roads, woods, airfields and electricity pylons.

The feature outline is specified as a set of vector co-ordinates that form straight-line segment approximations to the feature outline. The colour infill for the feature is also specified in the data.

Map databases that are stored in this way are known as vector databases. The method gives an extremely powerful means of data compression to enable all the information depicted on a map to be stored in a practical size of computer memory for airborne applications. The map can easily be decluttered by deleting a particular feature set(s) from the display.

The task of compiling a world wide mapping database is a continuing activity by the various mapping agencies around the world, but is clearly a major task. There are vast areas where accurate ground survey data are not available and the map data have to be derived from surveillance satellites.

The task of drawing a map and translating and rotating it in real time with the aircraft's motion is a major processing task but can be readily accomplished with modern electronic hardware now available: microprocessors, graphics chip sets, semiconductor memory devices etc. Typically, the map data are loaded into a frame store, which holds the data for an area whose sides are 1.4 times the sides of the map area being displayed. Scrolling the map display is illustrated in Fig. 2.65. Map rotation is accomplished by rotating the addressing of the frame store as shown in Fig. 2.66 (The frame store size factor of 1.4:1 covers the worst case heading changes).

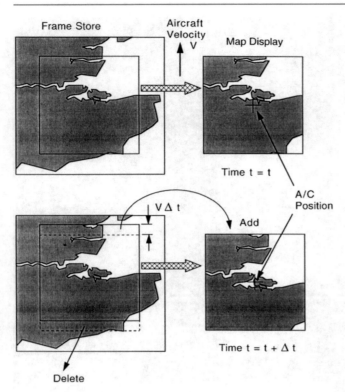

Fig. 2.65 'Scrolling' the map display

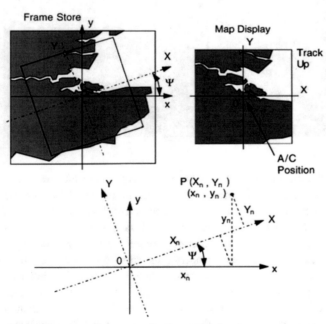

Coordinates of point P with respect to Map Display axes, OX & OY, are (X_n, Y_n), and with respect to Frame Store axes, Ox & Oy, are (x_n, y_n).

Data for point P (x_n, y_n) on Map Display is held in Frame Store coordinates
$$x_n = X_n \cos \Psi - Y_n \sin \Psi$$
$$y_n = Y_n \cos \Psi + X_n \sin \Psi$$

Fig. 2.66 Rotating the map display by rotating the addressing of the frame store

2.6.6   Solid-State Standby Display Instruments

Until recently, the standby instruments which provide primary flight information to the pilot(s) in the event of loss of the main displays were all electro-mechanical instruments. These are very well proven and reliable instruments having been progressively developed and improved over many years.

The cost of ownership of electro-mechanical instruments, however, is relatively high and is increasing as there is a high skilled labour content in their repair and maintenance, together with the need for high-standard clean room facilities. This is because these instruments depend on very-low-friction precision mechanisms and in the case of gyroscopic instruments there is also inevitable wear and deterioration in the spin axis bearing system. The skilled instrument technicians required to service and repair these instruments are also a declining group and are not being replaced by and large as new technology no longer requires their skills.

The technology now available has made it possible to produce an all 'solid-state' equivalent instrument with a colour AMLCD display, which is a 'form, fit and function' replacement for the electro-mechanical instrument, or instruments in some cases.

The solid-state equivalent instrument is packaged in a standard instrument case/housing and uses the same electrical connector interfaces. The initial cost of these instruments is competitive with the electro-mechanical version, but more importantly they offer a ten fold improvement both in reliability and cost of ownership.

This is because of the inherent reliability of solid-state sensors with no wear or degradation mechanism in their operation. They incorporate very comprehensive built-in test and monitoring, and repair is effected by replacing modules and the use of automated testing facilities. Highly skilled instrument technicians and clean room facilities are not required. The instruments are also of higher accuracy because of the use of solid-state sensors and the ability to apply complex corrections using the embedded microprocessor – electro-mechanical instruments are limited in the complexity of the corrections that can be made.

A recent development is to replace the standby attitude indicator, standby altimeter and standby airspeed indicator with a single solid-state integrated standby instrument system packaged in a 3 Unit of measure for Avionics instrument size (ATI) case (see note). Figure 2.67 illustrates a typical solid-state integrated standby instrument system displaying attitude, altitude and airspeed. These integrated standby instrument systems comprise a triad of orthogonally mounted solid-state rate gyros and accelerometers, and solid-state pressure transducers together with an embedded microprocessor and colour AMLCD display and drive electronics. Note: ATI is a unit of measurement for Avionics instrument size. A 3 ATI case is an instrument size that is 3.2 inches.

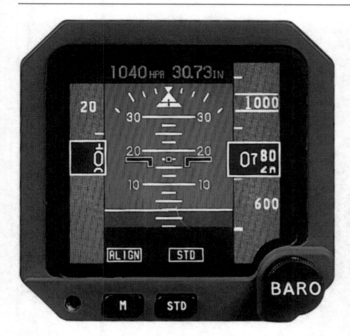

Fig. 2.67 Solid-state integrated standby instrument. (Courtesy of Smiths Industries Aerospace)

The microprocessor computes the aircraft attitude from the gyro and accelerometer outputs and altitude and airspeed from the pressure transducer outputs (refer to Chap. 6, Sect. 6.4.2; and Chap. 7, Sect. 7.4.3). The system operates from the 28 Volt supply, which is supported by the emergency batteries. The integral power conditioning electronics are designed to cope with the variations in the battery supply voltage.

2.7 Data Fusion

Data fusion is the name given to the process of combining the data from a number of different sources to provide information that is not present in the individual sources. For example, a synthetic 3D picture can be derived of the terrain in front of the aircraft from an accurate terrain database and accurate information on the aircraft's position and attitude. The aircraft's position and attitude information are provided by a GPS and an INS, or a Terrain Referenced Navigation (TRN) system and an INS. The synthetic picture of the terrain can be overlaid one to one with the outside scene derived from a FLIR sensor and displayed on an HUD or a head down display. Ground features that may be hard to detect on the FLIR such as ridge lines can be accentuated together with electricity pylons. This enables the pilot to continue visual flight in conditions of marginal visibility where normally it

would be necessary to pull up and fly at a higher altitude. Figure 2.68 illustrates a display enhanced by data fusion.

It is appropriate at this point to discuss alternative ways of displaying primary flight information and guidance which are being evaluated in both the USA and Europe. Such displays are sometimes referred to as '3D or 4D' displays or 'pathways in the sky'. These displays provide a pictorial presentation of the aircraft's spatial situation and flight path using synthetically generated ground imagery, and exploit data fusion.

The presentation of primary flight information on colour head down displays by and large follows the format established by the blind flying panel of electro-mechanical instruments in the classical 'T' layout.

All pilots learn to fly using this primary flight information format and its retention in colour head down displays has enabled their acceptance to be achieved in a smooth and straightforward manner with very little learning curve required.

The display of information, however, in any application involves a process of abstraction, and a 'mental translation' is required to build up a situational awareness and mental picture of what is happening. The objective of these new pictorial displays is to provide the pilot with a more intuitive and natural appreciation of the situation in terms of the aircraft's state, its actual flight path and the desired flight path with respect to the outside world.

Figure 2.69 illustrates a 3D pathway in the sky display as an example of a military strike mission display.

There is a very large number of viable display presentations for civil cockpits and reaching general agreement on the most suitable will take time. It is clear that a more easily assimilated pictorial-type display presentation can be produced compared with the present symbolic presentations of flight and navigational information. It would seem reasonable to predict that these new pictorial presentations will be adopted some time in the future.

To generate these pictorial displays is not a problem with the technology now available. There is, however, a big difference between demonstrating an experimental system and the development of a production system approved by the airworthiness authorities. The cost involved in achieving certification by the civil airworthiness authorities is a significant sum (as with any airborne system); the software is safety critical and the flight trials to prove the system do not come cheaply.

We could, nevertheless, have such displays 'tomorrow' if the need was agreed.

References to articles on new types of display presentations are given at the end of this chapter

Fig. 2.68 The illustration shows the HUD display whilst carrying out an erroneous descent in a mountainous scenario. The upper picture shows an enhanced vision display; the lower picture shows a synthetic vision display. (Courtesy of BAE Systems)

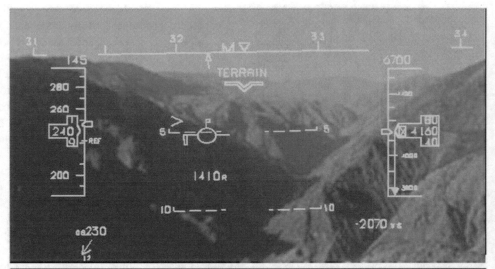

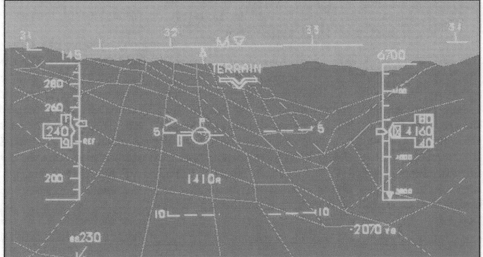

2.8   Intelligent Displays Management

The exploitation of intelligent knowledge-based systems (IKBS) technology, frequently referred to as 'expert systems', to assist the pilot in carrying out the mission is the subject of a number of very active research programmes, particularly in the USA. One of the US programmes is the 'pilot's associate program', which aims to aid the pilot of a single seat fighter/attack aircraft in a similar way to the way in which the second crew member of a two crew aircraft assists the pilot. The prime aim is to reduce the pilot workload in high workload situations.

Space constraints do not permit more than a very brief introduction to this topic, which is of major importance to the next generation of military aircrafts as these will use a single pilot to carry out the tasks which up to now have required a pilot and a navigator/weapons systems officer.

The exploitation of IKBS technology on the civil flight-deck will follow as the technology becomes established in military applications.

A subset of all the proposed expert systems on an aircraft is an intelligent displays management system to manage the information which is visually presented to the pilot in high workload situations.

It is the unexpected or uncontrollable that leads to an excessive workload, examples being:

- The 'bounce': interception by a counter attacking aircraft with very little warning.
- Evasion of ground threat: surface–air missile (SAM).
- Bird strike when flying at low altitude.
- Engine failure.
- Weather abort or weather diversion emergency.

Fig. 2.69 'Pathway in the sky' pictorial display. (Courtesy of BAE Systems). Red indicates threat zones

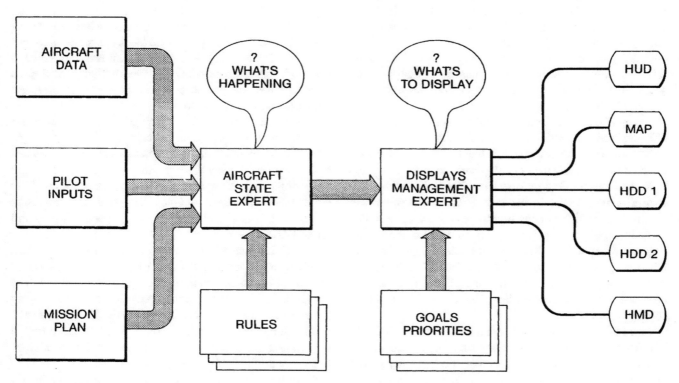

Fig. 2.70   Intelligent displays management. (Courtesy of BAE Systems)

A block diagram illustrating the concepts of an intelligent displays management system is shown in Fig. 2.70. The system comprises an 'aircraft state expert' which deduces 'what is happening' from the aircraft data, pilot inputs, threat warning systems and the mission plan by the application of an appropriate set of rules. The aircraft state expert in turn controls the 'displays management expert', which determines the information displayed on the various display surfaces: HUD, map, head down displays or HMD according to an appropriate set of goals and priorities.

2.9   Displays Technology

This section deals with two recent developments in HUD and HMD displays technology, which will have a major influence on future HUD and HMD performance, size, weight, reliability, and both initial cost and cost of ownership.

2.9.1   Replacing the CRT: 'New Lamps for Old'

It should be noted that for nearly 60 years, the cathode ray tube (CRT) was the only display source available which could meet the very exacting requirements for a head up display.

Major advances, however, have been made in solid-state technology over the last decade (2010–2020), and are continually being made, and from around 2015 new designs of HUDs and HMDs have started to become available. These will gradually supplant the CRT-based HUDs in new build and updates of existing aircraft.

There are, however, a very large number of CRT-based HUDs in service and likely to remain in service for many years. The description of their operation has thus been retained in this Edition.

There is clearly a large existing market for a solid-state CRT replacement, which enables the existing optical system to be retained as this is a major element in the overall cost of the HUD system, and the optical system never fails. In effect a 'New Lamps for Old' replacement of the CRT with a solid-state display source together with updated electronics and eliminating the high-voltage supply – a cost-effective update.

It is appropriate at this point to review the salient characteristics of the HUD CRT in order to appreciate why it has been such a 'hard act to follow'. The CRT is basically a very simple and technically elegant device which has truly been one of the greatest enabling inventions of the twentieth century.

Its main limitations are its reliability (around 2000-hour MTBF), which impacts adversely on life cycle costs in civil airline operation. CRT failure modes are generally benign and catastrophic failures are rare, so a planned replacement is possible. Ageing is a slow process associated with the cathode degradation and phosphor burning. In practice, the highly stressed high-voltage supply has been the usual cause of failure.

The resolution of IR and FLIR sensors has also now overtaken that possible with a CRT, at the luminance required.

The CRT, however, has major advantages which have been hard to match such as sharp, crisp cursive display symbology and the very wide viewing angle inherent with an emissive display, and extremely high display integrity; the probability of the CRT itself displaying misleading information is less than one part in a billion per hour.

The CRT provides image drawing and the light source in one package. It can operate in a cursive, raster or mixed mode giving crisp images in all of these modes free from dynamic artefacts such as aliasing and speckle, which can occur with other display sources. It can also provide a stable and controllable luminance over the range > 0.6 to $<12{,}000$ ft-lamberts. The CRT phosphor colour can be easily optimised to the peak eye response normalised to 555 nm, where the PI, P43 and P53 phosphors are highly efficient. (It is interesting to note that an HUD CRT can operate well in excess of 10,000 ft-lamberts [$>34{,}000$ cd/m^2], whereas a typical domestic TV which uses a flat-panel transmissive LCD is around 500 cd/m^2).

The CRT gun is basically handmade and only a very-high-volume programme can justify any automation, with the result that gun variations have to be corrected. There are few engineers around now, however, who understand analogue deflection and video circuits.

The ideal solution would be to produce a laser equivalent to the CRT and scan the laser beam like the electron beam in a CRT. In a further facsimile of the CRT, it is also possible to scan a phosphor screen with a laser beam to give the desired colour. It has not proved possible, however, to meet the required scanning speeds to date.

The current solution to replacing the CRT is to use a projected-type display with a very high luminance light source to illuminate a light modulator, such as a flat-panel high-resolution display device. Although losses in the light modulation process constrain the display efficiency, the HUD requirements can be met and the reliability greatly increased.

Figure 2.71 is a block diagram of a projected display unit for an HUD.

A high-luminance light source is used to illuminate a flat-panel display device comprising a matrix addressable x–y array of electrically switchable, light reflecting mirror elements. The display image so created is then projected on to the display screen and relayed by the HUD optics to the HUD combiner.

The luminance of the light source is controllable over a very wide dynamic range to meet all ambient light conditions from very bright sunlight to night operation.

High-luminance LEDs which meet the HUD luminance requirements have also become available fairly recently. LED-based HUDs are being actively developed by a number of HUD manufacturers. High-luminance LEDs cost less than an equivalent high-luminance laser, and they can be switched at high speed enabling a very wide range of luminance levels to be readily achieved by controlling the mark-space ratio of the switching frequency.

Fig. 2.71 Block diagram of an HUD projected display unit

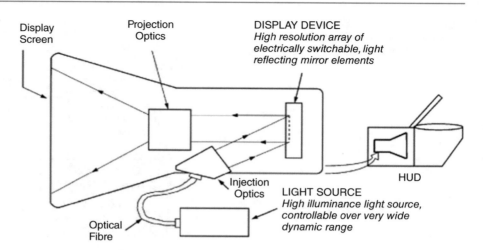

The Texas Instruments Digital Micromirror

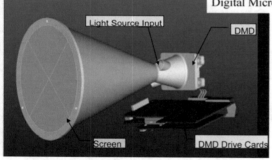

Fig. 2.72 'Digital Light Engine' projector display unit (by courtesy of BAE Systems)

The flat-panel display device can be a high-resolution reflective-type LCD. These are widely used in commercial (and military) display applications and are of modest cost because of the very large sales volume.

The reflection efficiency of reflective-type LCDs, however, is only about 15% and requires a very high luminance light source to meet the required display brightness levels. (Transmissive-type LCDs are unsuitable as their transmission efficiency is too low.)

An alternative display device to the reflective LCD in a projected display system is the Texas Instruments 'Digital Micro-Mirror Device', or 'DMD™'.

These devices are available in high-resolution x–y arrays; for example, the Texas Instruments SXGA 1440 × 1050 DMD™ device has over 1.5 million micro-mirrors in a 4 × 3 aspect ratio package. They have high reflectivity (over 80%) and have demonstrated very high reliability in a

wide variety of civil and military display applications. They are also widely used in very large screen projection systems to display high-definition video.

Essentially, the Texas Instruments DMD™ is an array of over a million hinged micro-mirrors which can be individually deflected mechanically through (12°). Light modulation in a dark field projection system is achieved by tilting the micro-mirror to reflect light from the illuminating source on to, or away from the display screen. The micro-mirrors are matrix addressed and deflected by the electrostatic forces created by applying a voltage to the appropriate mirror electrodes; each micro-mirror produces a display pixel. The micro-mirrors are very small indeed, around 10 microns square (1 micron is a thousandth of 1 mm). The small size can be appreciated by the fact that the tip of the mirror moves through just 2 microns in rotating 20 degrees. The micro-mirrors can be switched at several kHz.

Space constraints limit further discussion of these devices and references are given in 'Further Reading'.

Figure 2.72 illustrates a projected display unit for an HUD, which BAE Systems refer to as a 'Digital Light Engine' (DLE). The inset illustration shows the Texas Instruments DMD™ used in the DLE.

The BAE Systems Rochester 'Digital Light Engine', DLE, has excellent resolution, over twice that of a CRT, and meets all the ambient light requirements. A high-radiance LED provides the illuminating source for a Texas Instruments Digital Micro-Mirror Device. BAE Systems Rochester have already been awarded contracts to update the HUDs in a number of aircraft, such as the Typhoon, and the DLE is of course used in the new LiteHUD® and LiteWave® designs.

Figure 2.73 illustrates the HUD with the DLE and associated electronics installed. It is fully compliant with Form, Fit, Function requirements and provides a major improvement in MTBF over the CRT-based version as well as increased capability. Display surface resolution: 1280 × 1024 pixels; Field of view: 25 × 22 degrees.

Fig. 2.73 Digital Light Engine replacement HUD for CRT version. (Courtesy of BAE Systems)

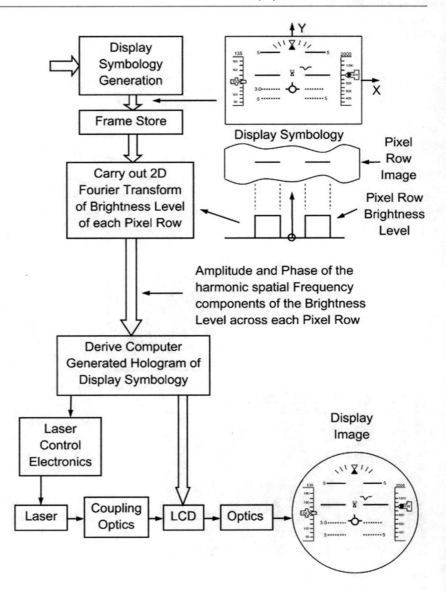

It should be noted that the probability of displaying misleading information is less than one part in a billion per hour where the HUD is designated a Primary Flight Instrument.

Pixel-based displays, however, have an inherent problem in meeting this extremely high display integrity requirement. A typical concern, for example, is that a figure '8' may be degraded to read a figure '3' by the loss of a vertical row of pixels.

Although this has a low probability of occurring and not being noticed, it is still too high to meet the required integrity.

Different approaches to address this problem include detailed examination and monitoring of the drive circuits and a camera to monitor the display.

The camera system is one way of monitoring what is actually displayed in terms of both symbol form and positional placement. A symbol from the camera is compared to the original data generated.

2.10 Control and Data Entry

2.10.1 Introduction

As mentioned in the introduction to this chapter, the pilot (and crew) must be able to control the information being displayed, for example, to switch modes and information sources at the various phases of the mission, or in the event of malfunctions, failures, emergencies, threats etc.

It is also essential for the pilot to be able to enter data into the various avionic systems (e.g. navigation way points). Control and data entry are thus complementary to the various

Fig. 2.74 Touch panel concept

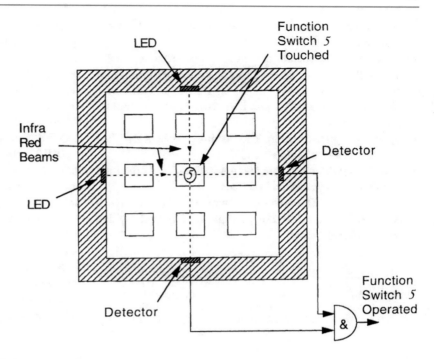

displays and enable the pilot (and crew) to interact with the various avionic systems in the aircraft. The means for effecting control and data entry must be as easy and as natural as possible, particularly under high workload conditions. This section gives a brief overview of tactile control panels, direct voice input and eye trackers as a means of control and data entry.

2.10.2 Tactile Control Panels

These have already been briefly described in Sect. 2.6.3. A typical tactile control panel uses a matrix array of infrared beams across the surface of the display which displays the various function keys. Touching a specific function key on the display surface interrupts the x and y infrared beams, which intersect over the displayed key function, and hence signals the operation of that particular key function.

Figure 2.74 illustrates the basic principles.

2.10.3 Direct Voice Input

Direct voice input (DVI) control is a system which enables the pilot to enter data and control the operation of the aircraft's avionic systems by means of speech. The spoken commands and data are recognised by a speech recognition system which compares the spoken utterances with the stored speech templates of the system vocabulary. The recognised commands, or data, are then transmitted to the aircraft sub-systems by means of the interconnecting data bus (e.g. MIL STD 1553B data bus).

As examples:

(a) To change a communication channel frequency, the pilot says – 'radio' (followed by) 'select frequency three four five decimal six'.

(b) To enter navigation data, the pilot says – 'navigation' (followed by) 'enter waypoint latitude 51 degrees 31 minutes 11 seconds North; longitude zero degree 45 minutes 17 seconds West'.

Feedback that the DVI system has recognised the pilot's command correctly is provided visually on the HUD and HMD (if installed), and aurally by means of a speech synthesiser system. The pilot then confirms the correctly recognised command by saying 'enter' and the action is initiated.

The pilot can thus stay head up and does not have to divert attention from the outside world in order to operate touch panels, switches, push buttons, keyboards etc. DVI can thus reduce the pilot's workload in high workload situations. It should be noted that the vocabulary required is not extensive and pilots (and crew) make good DVI subjects. This is because they are trained to speak clearly and concisely in a strongly structured way when giving commands and information over the communication channels to fellow crew members, other aircrafts and ground control.

The main characteristics and requirements for an airborne DVI system are briefly summarised below:

• Fully connected speech. The speech recognition system must be able to recognise normal fully connected speech with no pauses required between words. (Systems which require a pause between each word are known as 'isolated word recognisers'.)

- Must be able to operate in the cockpit noise environment. The background noise level can be very high in a fast jet combat aircraft.
- Vocabulary size. The required vocabulary is around 200–300 words.
- Speech template duration. The maximum speech template duration is around five seconds.
- Vocabulary duration. The maximum duration of the total vocabulary is around 160 seconds.
- Syntax nodes. The maximum number of syntax nodes required is about 300. An example of a typical 'syntax tree' is shown below:

- Duration of utterance. There must be no restrictions on the maximum duration of an input utterance.
- Recognition response time. This must be in real time.

Only a very brief outline of speech recognition systems can be given because of space constraints. The basic principles are to extract the key speech features of the spoken utterance and then to match these features with the stored vocabulary templates. Sophisticated algorithms are used to select the best match and to output the recognised words, if the confidence level is sufficiently high. Figure 2.75 illustrates the speech features of an individual word which can be extracted by spectral analysis of the spoken utterances. The distinctive features can be seen in the 3D 'spectrogram'.

Very extensive research and development has been carried out and is a continuing activity worldwide to produce speech recognition systems which are 'speaker independent', that is, they will recognise words spoken clearly by any speaker.

Various recognition algorithms are used including Markov pattern matching and neural-net techniques.

Commercial systems with an impressive performance are now widely available; for example, automated telephone enquiry and data entry systems. Although these systems only require a limited vocabulary, the accurate recognition of strings of numbers is a good test for any speech recognition system.

The airborne environment, however, poses particular requirements such as the ability to operate to a very high confidence level in a high background noise situation. Recognition accuracies of at least 96% is required in the cockpit environment to minimise having to repeat a command as this

would defeat the objective of easing pilot workload. The speech recognition system must also recognise commands spoken during the physical stress of manoeuvring. For these reasons, the stored vocabulary templates are currently derived directly from the pilot to characterise the system to his particular speech patterns. As already mentioned, the stored vocabulary requirements are not extensive (about 200–300 words).

IKBS technology can also be used to improve the recognition accuracy by deducing the context in which the words are spoken and ruling out words which are out of context or unlikely.

In the case of numerical data, numbers which are outside the likely range for that quantity, for example radio frequencies or latitude/longitude co-ordinates, can be rejected.

Knowledge of the context is essential when we carry out intelligent conversations in a noisy environment where not every word can be heard clearly, for instance at a party where there are many conversations taking place at the same time.

2.10.4 Speech Output Systems

Audio warning systems using speech synthesisers to provide voiced warning messages to the pilot/crew of system malfunctions and dangers/threats are now well established. They are also complementary to a DVI system to provide the essential feedback that a spoken command/data input has been correctly recognised.

Development of digitally generated speech synthesisers has been taking place over many years and high-performance systems are now available and in wide-scale use in many everyday applications, apart from aircraft.

A very interesting recent development for combat aircraft application is the use of stereo sound warning signals to indicate the direction of the threat to the pilot – like behind and to the left or right, above or below etc. Trials in advanced cockpit simulators with fast jet pilots from front-line squadrons have produced favourable and positive responses and it is considered that stereo warning systems are a likely future development.

2.10.5 Display Integration with Audio/Tactile Inputs

The integration and management of all the display surfaces by audio/tactile inputs enables a very significant reduction in the pilot's workload to be achieved in the new generation of single seat fighter/strike aircraft.

Figure 2.76 illustrates the basic concepts of an audio/tactile management system. Such a system is installed in the Eurofighter Typhoon and the effectiveness of the system in

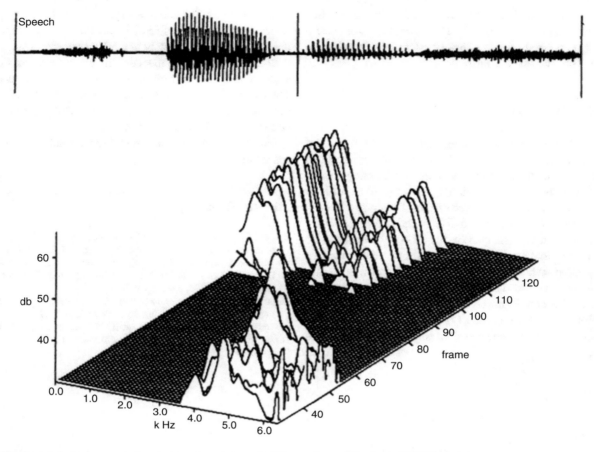

Fig. 2.75   Speech features – spectral analysis – a 'spectroscape' of the word spar. (Courtesy of BAE Systems)

Fig. 2.76 Audio/tactile management systems. (Note: Tactile control panels can be integrated with HDDs)

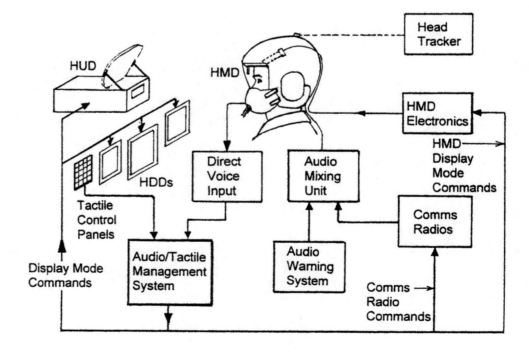

reducing pilot workload when combined with the carefree manoeuvring resulting from the FBW flight control system and the automated engine control system is referred to as 'Voice, Throttle, Stick' control.

2.10.6 Eye Trackers

Eye tracking systems are being fairly widely used and evaluated in ground simulators for such future applications as improved target designation accuracy by enabling a more accurate measurement of the pilot's gaze angle to be made in conjunction with a head tracker system. Data entry in conjunction with a helmet mounted display can also be achieved. A keyboard can be displayed on the HMD and data can be entered by looking at the appropriate data symbol (e.g. function switch and the digits 0–9) and then operating a simple push button. The pilot's gaze angle is measured by the eye tracker and displayed by a simple cross on the HMD. The pilot can thus see his LOS is on the chosen symbol. A fast and accurate data entry system can be achieved by this means.

Prototype helmet mounted eye trackers have been built which exploit the principle of corneal reflection. These have demonstrated 0.5° accuracy at a 50 Hz iteration rate.

It appears, however, that current helmet mounted sighting systems and also current data entry systems are adequate. There does not appear to be a requirement for an eye tracking system, at the moment, which would justify the increased helmet mounted mass and cost.

Further discussion has thus been curtailed because of space constraints.

Further Reading

Bartlett, C.T.: A practical application of computer generated holography to head up display design. Displays. **15**(2), 124–130 (1994)

Burgess, M.A., Hayes, R.D.: Synthetic vision – a view in the fog. In: Proceedings of IEEE/AIAA 11th Digital Avionics System Conference, Seattle (1992, 5–8), pp. 603–610

Flight Safety Foundation study of 543 total loss and 536 partial loss jet transport accidents, including 30 years prior to 1989

Rowntree, T.: The intelligent aircraft. IEEE Rev. **39**, 23–27 (1993)

Tomorrow's Cockpit Displays: Aerospace (1991, November), pp. 12–14

Aerodynamics and Aircraft Control

3.1 Introduction

The objective of this chapter is to provide the reader with an introduction to aerodynamics, aircraft stability and control and the dynamic response and behaviour of an aircraft so as to be able to understand the role and tasks of automatic flight control systems. The aim is to establish the essential background to fly-by-wire (FBW) flight control (Chap. 4) and autopilots (Chap. 8).

3.2 Basic Aerodynamics

Most readers will be aware of the basic principles of aircraft flight and how the wings generate lift to support the weight of the aircraft. The aerodynamic aspects of aircraft flight are thus covered briefly, principally to recap and define the terms and aerodynamic parameters which are used.

3.2.1 Lift and Drag

An aerofoil inclined at an angle to a moving air stream will experience a resultant force due to aerodynamic effects. This resultant aerodynamic force is generated by the following:

(a) A reduction in the pressure over the upper surface of the wing as the airflow speeds up to follow the wing curvature, or camber.
(b) An increase in pressure on the undersurface of the wing due to the impact pressure of the component of the air stream at right angles to the undersurface.

About two-thirds of this resultant aerodynamic force is due to the reduction in pressure over the upper surface of the wing and about one-third due to the increase in pressure on the lower surface. Fig. 3.1 illustrates the pressure distribution over a typical aerofoil.

The aerodynamic force exerted on an aerofoil can also be explained as being the reaction force resulting from the rate of change of momentum of the moving air stream by the action of the aerofoil in deflecting the airstream from its original direction (see Fig. 3.2).

The ability of an aerofoil to deflect a stream of air depends on the angle between the aerofoil and the airstream and on the curvature, or camber, of the aerofoil. The aerodynamic force is increased by increasing the camber or by increasing the angle between the aerofoil and the airstream. The thickness of the aerofoil also determines how efficiently the aerofoil produces a force.

This resultant aerodynamic force can be resolved into two components. The component at right angles to the velocity of the air relative to the wing (generally referred to in aerodynamic textbooks as *relative wind*) is called the *lift force*.

The component parallel to the air velocity is known as the *drag force* and, as its name implies, acts on the wing in the opposite sense to the velocity of the wing relative to the air mass (or alternatively in the direction of the relative wind). It should be noted that the drag force component of the aerodynamic force acting on the wing is known as the *wing drag* and constitutes a large part of the total drag acting on the aircraft particularly at high angles of incidence.

3.2.2 Angle of Incidence/Angle of Attack

Referring to Fig. 3.3, the angle between the direction of the air velocity relative to the wing (relative wind) and the wing chord line (a datum line through the wing section) is known both as the *angle of incidence* and as the *angle of attack*. Angle of incidence is the term generally used in the UK and angle of attack in the USA. Angle of incidence is used in this book.

R. P. G. Collinson, *Introduction to Avionics Systems*, https://doi.org/10.1007/978-3-031-29215-6_3

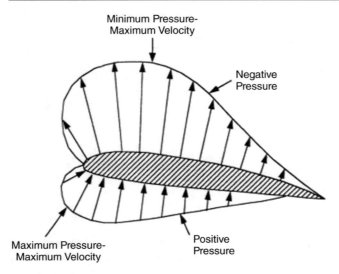

Fig. 3.1 Pressure distribution over an aerofoil

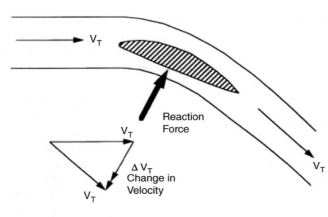

Fig. 3.2 Change in airflow momentum by aerofoil

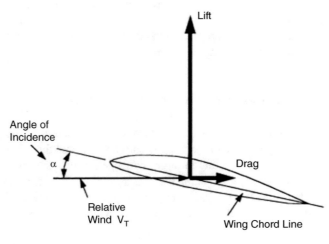

Fig. 3.3 Angle of incidence

3.2.3 Lift Coefficient and Drag Coefficient

Aerodynamic forces are dependent on the impact pressure, which, as explained in Chap. 2, is the pressure created by the change in kinetic energy of the airstream when it impacts on the surface. Aerodynamicists use the term dynamic pressure, Q, to denote the impact pressure which would result if air were incompressible.

$$\text{Dynamic pressure } Q = \frac{1}{2}\rho V_T{}^2 \qquad (3.1)$$

where ρ is air density and V_T is air speed.

The aerodynamic lift force, L_W, acting on the wing is a function of the dynamic pressure, Q, the surface area of the wing, S, and a parameter which is dependent on the shape of the aerofoil section and the angle of incidence, α. This parameter is a non-dimensional coefficient, C_L, which is a function of the angle of incidence and is used to express the effectiveness of the aerofoil section in generating lift.

$$C_L = \frac{L_W}{\frac{1}{2}\rho V_T^2 S} \qquad (3.2)$$

Thus

$$L_W = \frac{1}{2}\rho V_T{}^2 S C_L \qquad (3.3)$$

The C_L versus α relationship is reasonably linear up to a certain value of angle of incidence when the airflow starts to break away from the upper surface and the lift falls off very rapidly – this is known as stalling. The slope of the C_L versus α relationship and the maximum value, $C_{L\,max}$, are dependent on the aerofoil section. $C_{L\,max}$ typically ranges from about 1.2 to 1.6. Maximum values of the angle of incidence before the onset of stalling range from 15 to 20° for conventional wing plans; for delta wing aircraft, the figure can be up to 30–35°. The C_L versus α relationship also changes as the aircraft speed approaches the speed of sound, and changes again at supersonic speeds.

Similarly, a non-dimensional characteristic called the *drag coefficient*, C_D, which is a function of the angle of incidence, α, is used to enable the drag characteristics of an aerofoil to be specified:

$$C_D = \frac{D_W}{\frac{1}{2}\rho V_T^2 S} \qquad (3.4)$$

That is, drag force

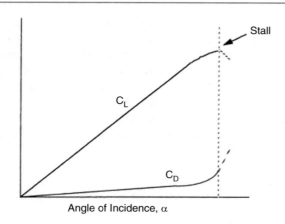

Fig. 3.4 C_L and C_D versus α

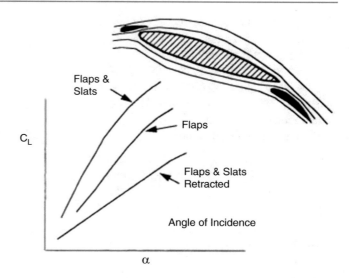

Fig. 3.5 Leading edge slats and trailing edge flaps

$$D_W = \frac{1}{2}\rho V_T^2 S C_D \qquad (3.5)$$

The lift and drag characteristics of an aerofoil are related, the relationship being approximately

$$C_D = C_{DO} + k C_L{}^2 \qquad (3.6)$$

where C_{DO} and k are constants for a particular aerofoil section.

Thus C_D increases rapidly at high values of C_L (and α). It also increases very rapidly indeed as the aircraft's speed approaches the speed of sound due to compressibility effects. The C_D versus α characteristics of a wing designed for supersonic operation are again very different at supersonic speeds compared with subsonic speeds.

Figure 3.4 shows the general shape for a typical subsonic aerofoil, of the relationship of C_L and C_D with α. These characteristics have been determined for a very wide range of standardised aerofoil sections. The selection of a particular aerofoil section in the aircraft design process is determined by the optimum combination of characteristics for the aircraft mission and range of operating height and speed conditions.

Matching the wing lift requirements over the flight envelope, for example, high lift at take-off and low drag at high speed, or increased lift for manoeuvring in combat, is achieved with the use of retractable leading edge flaps or slats and trailing edge flaps (see Fig. 3.5). These enable relatively large increases in C_L to be obtained with maximum values of C_{Lmax} between 3 and 4.

3.2.4 Illustrative Example on Basic Aerodynamics

A simple example of the application of these basic aerodynamic relationships is set out below. The object is to show how lift, incidence and manoeuvring capability can be estimated for various height and speed combinations so as to give an 'engineering feel' for the subject. The basic parameters for a hypothetical aircraft are as follows:

Aircraft mass, m 30,000 kg
Wing area, S 75 m^2
Maximum lift coeff, C_L max   1.2
Maximum angle of incidence, α_{max} 15°
C_L versus α relationship is assumed to be linear.
Air data: Air density, ρ_0, at sea level and standard temperature (15 °C) and pressure (1013.25 mbar) is 1.225 kg/m^3.
   Air density, ρ, at 30,000 ft. (9144 m) is 0.4583 kg/m^3.
Question (i) What is the wing incidence, α, when flying straight and level at a speed of 80 m/s (160 knots approx.) at a height of 200 ft. in the approach to the airfield?
Question (ii) What is the corresponding wing incidence when flying straight and level at a speed of 200 m/s (400 knots approx.) at a height of 1000 ft.?
   Note: Neglect change in density 0–1000 ft. for simplicity.
Question (iii) What is the maximum normal acceleration that can be achieved at a height of 30,000 ft. when flying at a speed of 225 m/s (450 knots)? Neglect compressibility effects and assume adequate thrust is available to counteract the increase in drag and maintain speed of 450 knots at α_{max}.

Case 1 Low Level/Low Speed

Dynamic pressure $Q = \frac{1}{2}\rho V_{\mathrm{T}}^2 =$
$0.5 \times 1.225 \times 80^2 = 3920 \text{ N/m}^2$

Wing lift $= QSC_{\mathrm{L}} = 3920 \times 75 \times C_L$ N

Aircraft weight $= 30{,}000 \times 9.81$ N ($g = 9.81 \text{ m/s}^2$)

Hence, $3920 \times 75 \times C_L = 30{,}000 \times 9.81$

and required $C_{\mathrm{L}} = 1.0$.

C_{L} versus α is linear, hence

$$\frac{dC_{\mathrm{L}}}{d\alpha} = \frac{C_{\mathrm{L\,max}}}{\alpha_{\mathrm{max}}} = \frac{1.2}{15} \text{ degrees}^{-1}$$

$$C_{\mathrm{L}} = \frac{dC_{\mathrm{L}}}{d\alpha} \cdot \alpha$$

Hence,

$$\alpha = 1.0 \times \frac{15}{1.2} = 12.5°$$

Case 2 Low Level/High Speed

Lift is proportional to V_{T}^2, hence required incidence for straight and level flight at 200 m/s:

$$= \left(\frac{80}{200}\right)^2 \times 12.5 = 2°$$

Case 3 High Altitude/High Speed

Maximum achievable lift $= \frac{1}{2} \times 0.4583 \times 225^2 \times 75 \times 1.2$
$= 1044 \times 10^3$ N

Aircraft weight $= 294 \times 10^3$ N

Hence, lift available for manoeuvring $= 750 \times 10^3$ N

Achievable normal acceleration
$$= \frac{\text{Normal force}}{\text{Aircraft mass}} = \frac{750 \times 10^3}{30{,}000}$$
$$= 25 \text{m/s}^2 = 2.5g \text{ (approx.)}$$

3.2.5 Pitching Moment and Aerodynamic Centre

The *centre of pressure* is the point where the resultant lift and drag forces act and is the point where the moment of all the forces summed over the complete wing surface is zero. There will thus be a *pitching moment* exerted at any other point not at the centre of pressure. The centre of pressure varies with angle of incidence and for these reasons *aerodynamic centre* is now used as the reference point for defining the pitching moment acting on the wing. The aerodynamic centre of the wing is defined as the point about which the pitching moment does not change with angle of incidence (providing the

velocity is constant). It should be noted that all aerofoils (except symmetrical ones) even at zero lift tend to pitch and experience a pitching moment or couple. The aerodynamic centre is generally around the quarter chord point of the wing (measured from the leading edge). At supersonic speeds, it tends to move aft to the half chord point. This pitching moment or couple experienced at zero lift, M_0, is again expressed in terms of a non-dimensional coefficient, pitching moment coefficient, C_{M_0}

$$C_{M_0} = \frac{M_0}{\frac{1}{2}\rho V_T^2 Sc} \tag{3.7}$$

where c is the *mean aerodynamic chord*, equal to wing area/wing span. Thus, pitching moment

$$M_0 = \frac{1}{2}\rho V_{\mathrm{T}}^2 Sc\, C_{M_0} \tag{3.8}$$

3.2.6 Tailplane Contribution

The total pitching moment about the aircraft's centre of gravity (CG) and its relationship to the angle of incidence is of prime importance in determining the aircraft's stability. This will be discussed in the next section. However, it is appropriate at this stage to show how the resultant pitching moment about the CG is derived, particularly with a conventional aircraft configuration with a horizontal tailplane (refer to Fig. 3.6). The tailplane makes a major contribution to longitudinal stability and provides the necessary downward lift force to balance or trim the aircraft for straight and level

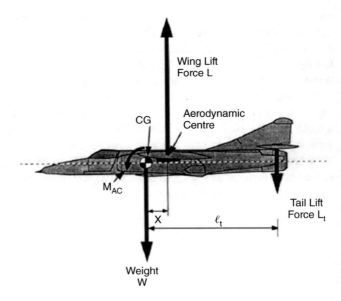

Fig. 3.6 Tailplane contribution

flight. The moment about the CG due to the tailplane lift balances the nose-down pitching moment due to the wing lift and the inherent wing pitching moment or couple, M_0. It should be noted that the trim lift exerted by the tailplane is in the opposite sense to the wing lift thereby reducing the total lift acting on the aircraft.

Resultant moment about CG, M, is given by

$$M = -L_W x - M_0 + L_t l_t \qquad (3.9)$$

(nose-up momesnts are defined as positive).

Wing lift:

$$L_W = \frac{1}{2}\rho V_T{}^2 S C_L$$

Wing pitching moment:

$$M_0 = \frac{1}{2}\rho V_T{}^2 S c\, C_{M_0} \qquad (3.10)$$

Tailplane lift:

$$L_t = k_t \frac{1}{2}\rho V_T{}^2 S_t C_{Lt} \qquad (3.11)$$

where S_t is the tailplane area and C_{Lt} is the tailplane lift coefficient.

k_t is the ratio of the dynamic pressure at the tailplane to the freestream dynamic pressure and is known at the *tailplane efficiency factor*. This factor takes into account the effects of downwash from the airflow over the wings on the tailplane. The tailplane efficiency factor varies from 0.65 to 0.95 and depends on several factors such as the location of the tailplane with respect to the wing wake, etc.

Moment about CG due to tailplane lift is $k_t \frac{1}{2}\rho V_T{}^2 C_{Lt}(S_t l_t)$.

The term $(S_t\, l_t)$ is often called the *tailplane volume*.

Dividing both sides of Eq. (3.9) by $\frac{1}{2}\rho V_T{}^2 S c$ yields

$$C_M = -\frac{x}{c}C_L - C_{M_0} + k_t \frac{(S_t l_t)}{S c}C_{Lt} \qquad (3.12)$$

The equation shows the respective contributions to the total pitching moment coefficient. However, the most important characteristic from the aerodynamic stability aspect is the variation of the overall pitching moment coefficient with incidence. (This will be clarified in the next section.) Figure 3.7 shows the variation with incidence of the wing and tailplane pitching moments and the combined pitching moment. It can be seen that the effect of the tailplane is to increase the negative slope of the overall pitching moment with incidence characteristics.

That is, $dC_M/d\alpha$ is more negative and this increases the aerodynamic stability.

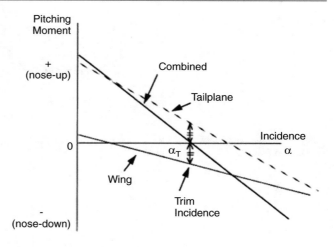

Fig. 3.7 Pitching moment versus incidence

3.3 Aircraft Stability

A stable system is one which returns to its original state if disturbed; a neutrally stable system remains in its disturbed state and an unstable system will diverge from its state on being subjected to the slightest disturbance. This is illustrated in Fig. 3.8, which shows a ball bearing placed on concave, flat and convex surfaces, respectively, as an analogous example of these three stability conditions. An aircraft is said to be stable if it tends to return to its original position after being subjected to a disturbance without any control action by the pilot. Figure 3.9a, b illustrates various degrees of stability, both statically and dynamically.

3.3.1 Longitudinal Stability

As already mentioned, to balance or trim the aircraft to achieve straight and level flight requires that the total pitching moment about the CG is zero and the total lift force equals the aircraft weight. However, to achieve static stability it is necessary that the total pitching moment coefficient, C_M, about the CG changes with angle of incidence, α, as shown in Fig. 3.10. The value of the angle of incidence at which the pitching moment coefficient about the CG is zero is known as the *trim angle of incidence*, α_T.

For values of angle of incidence less than α_T, the pitching moment coefficient is positive so the resulting pitching moment is in the nose-up sense and tends to restore the aircraft to the trim incidence angle, α_T. Conversely, for angles of incidence greater than α_T, the pitching moment coefficient is negative so the resultant nose-down moment tends to restore the aircraft to the trim incidence angle, α_T.

Thus $dC_M/d\alpha$ must be negative for longitudinal static stability and $C_M = 0$ at $\alpha = \alpha_T$.

Fig. 3.8 Simple example of stability conditions

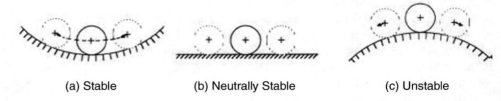

(a) Stable   (b) Neutrally Stable   (c) Unstable

Fig. 3.9 (**a**) Static stability. (**b**) Dynamic stability

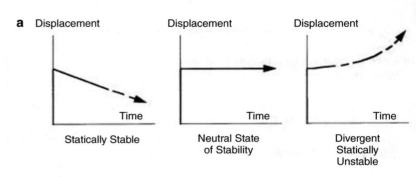

a Displacement   Displacement   Displacement

Time   Time   Time

Statically Stable   Neutral State   Divergent
                       of Stability   Statically
                                         Unstable

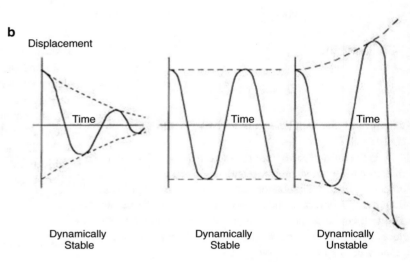

b

Displacement

Time   Time   Time

Dynamically   Dynamically   Dynamically
Stable   Stable   Unstable

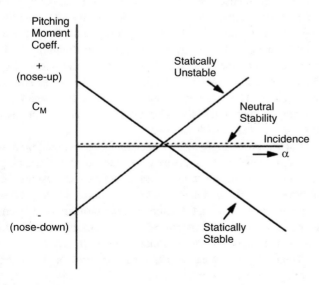

Fig. 3.10 C_M versus α

It should also be noted that the aircraft centre of gravity must be forward of the aerodynamic centre of the complete aircraft for static stability.

Static Margin and Neutral Point Movement of the CG rearwards towards the aerodynamic centre decreases the static stability.

The position of the CG where $dC_M/d\alpha = 0$ is known as the *neutral point* and corresponds to the position of the aerodynamic centre of the complete aircraft. The static margin of the aircraft is defined as the distance of the CG from the neutral point divided by the mean aerodynamic chord, c. This is positive when the CG is forward of the neutral point. Thus, the static margin is always positive for a stable aircraft. As an example, an aircraft with a static margin of 10% and a mean aerodynamic chord of, say, 4 m (13.3 ft) means that a rearward shift of the CG of more than 40 cm (16 in) would result in instability.

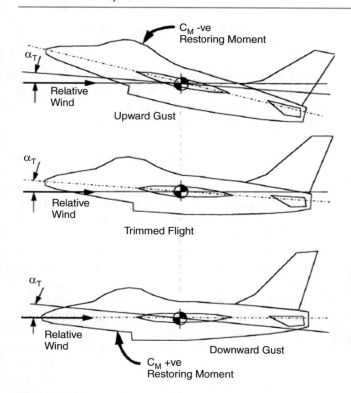

Fig. 3.11 Natural longitudinal stability

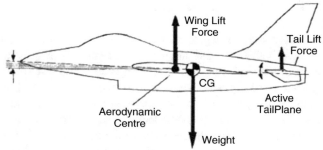

Fig. 3.12 Aerodynamically unstable aircraft

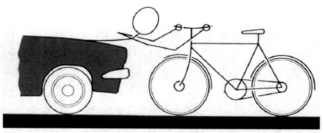

Fig. 3.13 Analogy of controlling aerodynamically unstable aircraft

Figure 3.11 illustrates the restoring moments when a stable aircraft is disturbed, for instance, by being subjected to an upward or downward gust.

3.3.2 Aerodynamically Unstable Aircraft

Figure 3.12 shows an aerodynamically unstable aircraft with the CG aft of the aerodynamic centre. The tailplane trim lift is acting in the same sense as the wing lift and so is more aerodynamically efficient. The angle of incidence required for a given lift is lower thereby reducing the induced drag – the induced drag being proportional to the square of the angle of incidence ($C_D = C_{Do} + kC_L^2$, and C_L is approximately linear with α).

The tailplane volume ($S_t l_t$) can also be reduced further lowering the weight and drag and hence improving the performance, subject to other constraints such as rotation moment at take-off.

The pilot's speed of response to correct the tendency to divergence of an unstable aircraft is much too slow and the tailplane must be controlled automatically. The time for the divergence to double its amplitude following a disturbance can be of the order of 0.25 seconds or less on a modern high-agility fighter, which is aerodynamically unstable. The problem of flying an unstable aircraft has been compared to trying to steer a bicycle backwards (see Fig. 3.13).

Higher performance and increased agility can be obtained with an aerodynamically unstable aircraft and relying totally on an automatic stability system. The technology to implement such an automatic stability system with a manoeuvre command flight control system is now sufficiently mature to meet the very exacting safety and integrity requirements and system availability.

Such a system has become known as a 'fly-by-wire' flight control system because of its total dependence on electrical signal transmission and electronic computing, and will be covered in the next chapter. The difference between a negative tailplane lift and positive tailplane lift to trim the aircraft is strikingly illustrated in Fig. 3.14, which shows an aerodynamically stable aircraft (Tornado) flying in formation with an aerodynamically unstable aircraft (fly-by-wire Jaguar experimental aircraft).

3.3.3 Body Lift Contributions

It should be noted that lift forces are also generated by the fuselage. These fuselage (or body) lift forces are significant and they can also seriously affect both lateral and longitudinal aircraft stability. The engine casings in aircraft configurations with 'pod' type engine installations can also generate significant lift forces, which can again affect the aircraft stability.

3.4 Aircraft Dynamics

An aircraft has six degrees of freedom because its motion can involve both linear and angular movement with respect to three orthogonal axes. Deriving the dynamic response of the aircraft to disturbances or control surface movements first involves the generation of the differential equations which describe its motion mathematically. The solution of these equations then gives the response of the aircraft to a disturbance or control surface input. This enables the control that has to be exerted by the automatic flight control system to be examined on the basis of the aircraft's dynamic behaviour.

Fig. 3.14 Negative tailplane lift on Tornado. Positive tailplane lift on FBW Jaguar. (Photograph by Arthur Gibson provided by British Aerospace Defence Ltd.)

3.4.1 Aircraft Axes – Velocity and Acceleration Components

The aircraft motion is normally defined with respect to a set of orthogonal axes, known as body axes, which are fixed in the aircraft and move with it (see Fig. 3.15).

These axes are chosen for the following reasons:

(a) The equations of motion with respect to body axes are simpler.

(b) The aircraft motion can be readily measured with respect to body axes by robust motion sensors such as rate gyros and accelerometers, which are body mounted (i.e. 'strapped down').

(c) Body axes are natural ones for the pilot and are the axes along which the inertia forces during manoeuvres (e.g. the normal acceleration during turns) are sensed.

(d) The transformation of motion data with respect to body axes to fixed space axes is not difficult with modern processors. Broadly speaking, body axes are used in deriving the control dynamics and short-term behaviour of the aircraft. Space axes are more suitable for the longer-period guidance aspects in steering the aircraft to follow a particular flight path with respect to the Earth.

Referring to Fig. 3.15, the origin of the axes, O, is located at the centre of gravity (CG) of the aircraft with OX and OZ in the plane of symmetry of the aircraft (longitudinal plane), with OZ positive downwards and OY positive to starboard (right). The figure shows the longitudinal and lateral planes of the aircraft. A fixed frame of axes is assumed to be

Fig. 3.15 Aircraft axes and velocity components

AXIS	Linear Velocity Component	Angular Velocity Component
Forward or Roll Axis, 0X	Forward Velocity U	Roll Rate P
Sideslip or Pitch Axis, 0Y	Sideslip Velocity V	Pitch Rate q
Vertical or Yaw Axis, 0Z	Vertical Velocity W	Yaw Rate r

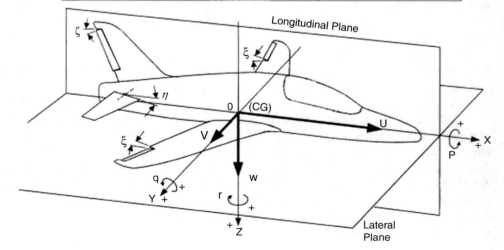

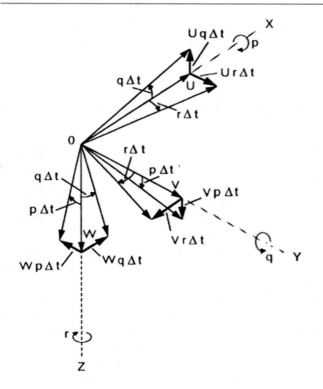

Fig. 3.16 Vector change in velocity components due to angular rotation

$\Delta U/\Delta t$, $\Delta V/\Delta t$, $\Delta W/\Delta t$, which in the limit becomes dU/dt, dV/dt, dW/dt.

The linear acceleration components are thus:

$$\text{Acceleration along OX} = \dot{U} - Vr + Wq \qquad (3.13)$$

$$\text{Acceleration along OY} = \dot{V} + Ur - Wp \qquad (3.14)$$

$$\text{Acceleration along OZ} = \dot{W} - Uq + Vp \qquad (3.15)$$

The angular acceleration components about OX, OY and OZ are $\dot{p}$, $\dot{q}$ and $\dot{r}$, respectively. The dot over the symbols denotes d/dt and is Newton's notation for a derivative. Thus, $\dot{U} = dU/dt$ and $\dot{p} = dp/dt$, etc.

The motion of the aircraft can then be derived by solving the differential equations of motion obtained by applying Newton's second law of motion in considering the forces and moments acting along and about the OX, OY and OZ axes, respectively. Namely, the rate of change of momentum is equal to the resultant force acting on the body, that is,

$$\text{Force} = \text{mass} \times \text{acceleration}$$

In the case of angular motion, this becomes:

$$\text{Moment (or torque)} = (\text{moment of inertia}) \times (\text{angular acceleration})$$

instantaneously coincident with the moving frame, and the velocity components of the aircraft CG along the OX, OY, OZ axes with respect to this fixed frame are the forward velocity, U, the sideslip velocity, V, and the vertical velocity, W, respectively. The corresponding angular rates of rotation of the body axes frame about OX, OY, OZ are the roll rate, p, the pitch rate, q, and the yaw rate, r, with respect to this fixed frame of axes. The derivation of the acceleration components when the aircraft's velocity vector is changing both in magnitude and direction is set out below as it is fundamental to deriving the equations of motion. Basically, the components are made up of centrifugal (strictly speaking centripetal) acceleration terms due to the changing direction of the velocity vector as well as the components due to its changing magnitude.

In Fig. 3.16, at time $t = 0$, the axes OX, OY, OZ are as shown in dotted lines. At time $t = \Delta t$, the angular rotations of the vector components are $p\Delta t$, $q\Delta t$, $r\Delta t$, respectively. The corresponding changes in the vectors are thus $-Uq\Delta t$, $Ur\Delta t$, $Vp\Delta t$, $-Vr\Delta t$, $-Wp\Delta t$, $Wq\Delta t$. The rates of change of these vectors, that is, the centripetal acceleration components, are thus: $-Uq$, Ur, Vp, $-Vr$, $-Wp$, Wq.

The changes in the velocity components due to the change in magnitude of the velocity vector are ΔU, ΔV, ΔW, respectively. The rates of change of these velocity components are

3.4.2 Euler Angles – Definition of Angles of Pitch, Bank and Yaw

The orientation of an aircraft with respect to a fixed inertial reference frame of axes is defined by the three Euler angles. Referring to Fig. 3.17, the aircraft is imagined as being oriented parallel to the fixed reference frame of axes. A series of rotations bring it to its present orientation:

(i) Clockwise rotation in the horizontal plane, through the yaw (or heading) angle ψ, followed by
(ii) A clockwise rotation about the pitch axis, through the pitch angle θ, followed by
(iii) A clockwise rotation about the roll axis, through the bank angle Φ

The order of these rotations is very important – a different orientation would result if the rotations were made in a different order.

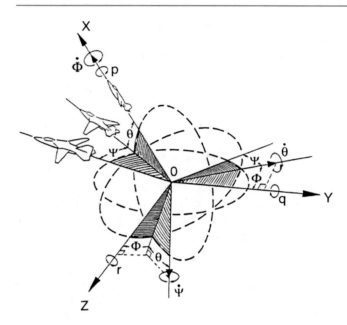

Fig. 3.17 Euler angles

The relationship between the angular rates of roll, pitch and yaw, p, q, r (which are measured by body mounted rate gyros) and the Euler angles, ψ, θ, Φ and the Euler angle rates $\dot{\psi}, \dot{\theta}, \dot{\Phi}$ are derived as follows:

Consider Euler bank angle rate, $\dot{\Phi}$

$$\text{Component of } \dot{\Phi} \text{ along} \begin{cases} OX = \dot{\Phi} \\ OY = 0 \\ OZ = 0 \end{cases}$$

Consider Euler pitch angle rate, $\dot{\theta}$

$$\text{Component of } \dot{\theta} \text{ along} \begin{cases} OX = 0 \\ OY = \dot{\theta}\cos\Phi \\ OZ = -\dot{\theta}\sin\Phi \end{cases}$$

Consider Euler yaw angle rate, $\dot{\psi}$

$$\text{Component of } \dot{\psi} \text{ along} \begin{cases} OX = -\dot{\Psi}\sin\theta \\ OY = \dot{\Psi}\cos\theta\sin\Phi \\ OZ = \dot{\Psi}\cos\theta\cos\Phi \end{cases}$$

Hence

$$p = \dot{\Phi} - \dot{\Psi}\sin\theta \qquad (3.16)$$

$$q = \dot{\theta}\cos\Phi + \dot{\Psi}\cos\theta\sin\Phi \qquad (3.17)$$

$$r = \dot{\Psi}\cos\theta\cos\Phi - \dot{\theta}\sin\Phi \qquad (3.18)$$

It should be noted that these relationships will be referred to in Chap. 5, in the derivation of attitude from strap-down rate gyros.

3.4.3 Equations of Motion for Small Disturbances

The six equations of motion which describe an aircraft's linear and angular motion are non-linear differential equations. However, these can be linearised provided the disturbances from trimmed, straight and level flight are small. Thus in the steady state the aircraft is assumed to be moving in straight and level flight with uniform velocity and no angular rotation. That is, with no bank, yaw or sideslip and the axes OX and OZ lying in the vertical plane. The incremental changes in the velocity components of the CG along OX, OY, OZ in the disturbed motion are defined as u, v, w, respectively, and the incremental angular velocity components are p, q, r about OX, OY, OZ. The velocity components along OX, OY, OZ are thus $(U_0 + u)$, v, $(W_0 + w)$ in the disturbed motion where U_0 and W_0 are the velocity components along OX, OZ in steady flight ($V_0 = 0$, no sideslip). u, v, w, p, q, r are all small quantities compared with U_0 and W_0 and terms involving the products of these quantities can be neglected.

The acceleration components thus simplify to:

$$\dot{u} + W_0 q \qquad (3.19)$$

$$\dot{v} - W_0 p + U_0 r \qquad (3.20)$$

$$\dot{w} - U_0 q \qquad (3.21)$$

The external forces acting on the aircraft are shown in Fig. 3.18 and comprise lift, drag, thrust and weight. These forces have components along OX, OY, OZ and can be separated into aerodynamic and propulsive force components and gravitational force components.

In the steady, straight and level trimmed flight condition, the resultant forces acting along OX, OY, OZ due to the lift, drag, thrust and weight are zero. The resultant moments about OX, OY, OZ are also zero. For small disturbances, it is therefore only necessary to consider the incremental changes in the forces and moments arising from the disturbance, in formulating the equations of motion (see Fig. 3.19). The incremental changes in the aerodynamic forces and moments are functions of the linear and angular velocity changes u, v, w and p, q, r, respectively. The resultant incremental changes in the aerodynamic forces during the disturbance along OX, OY, OZ are denoted by X_a, Y_a, Z_a, respectively.

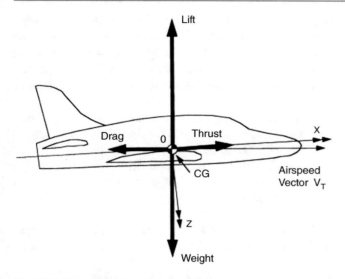

Fig. 3.18 External forces acting on aircraft

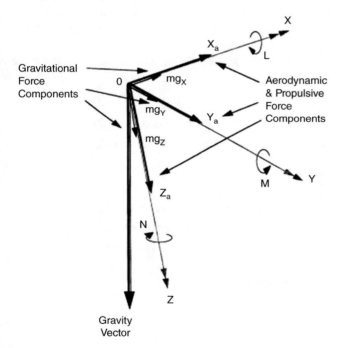

Fig. 3.19 Force components and moments

In the disturbed motion, the incremental changes in the components of the weight along OX, OY, OZ are approximately equal to $-mg\theta$, $mg\Phi$ and zero, respectively, assuming θ_0 is only a few degrees in magnitude.

From which the resultant incremental forces acting along OX, OY, OZ are

$$\text{along OX} = X_a - mg\theta$$
$$\text{OY} = Y_a + mg\Phi$$
$$\text{OZ} = Z_a$$

The equations of linear motion are thence derived from the relationship

$$\text{Force} = \text{mass} \times \text{acceleration}$$

$$X_a - mg\theta = m(\dot{u} + W_0q) \tag{3.22}$$

$$Y_a + mg\Phi = m(\dot{v} - W_0p + U_0r) \tag{3.23}$$

$$Z_a = m(\dot{w} - U_0q) \tag{3.24}$$

The resultant external moments acting on the aircraft due to the aerodynamic forces about OX, OY, OZ resulting from the disturbance are denoted by L, M, N, respectively.

Assuming principal axes are chosen for OX, OY, OZ so that the inertia product terms are zero.

The equations of angular motion are as follows:

$$L = I_x\dot{p} \tag{3.25}$$

$$M = I_y\dot{q} \tag{3.26}$$

$$N = I_z\dot{r} \tag{3.27}$$

It should be noted that the equations deriving the aircraft body rates in terms of Euler angles and Euler angle rates (refer to Sect. 3.4.2) can be simplified for small disturbances from straight and level flight. For small disturbances ψ, θ, Φ will all be small quantities of the first order and p, q, r (and $\dot{\psi}$, $\dot{\theta}$, $\dot{\Phi}$) can be taken to be small quantities of the first order. Thus neglecting second-order quantities, for small disturbances $p = \dot{\Phi}$, $q = \dot{\theta}$ and $r = \dot{\psi}$.

3.4.4 Aerodynamic Force and Moment Derivatives

The use of aerodynamic derivatives enables the incremental aerodynamic forces (X_a, Y_a, Z_a) and moments (L, M, N) to be expressed in terms of the force, or moment, derivatives multiplied by the appropriate incremental velocity change from the steady flight conditions. This linearises the equations of motion as the derivatives are assumed to be constant over small perturbations in the aircraft's motion. The equations of motion then become differential equations with constant coefficients and are much easier to analyse and solve. This assumption of constant derivatives is reasonable for small disturbances from a given flight condition. However, it is necessary to establish the variation in these derivatives over the range of height and speed and Mach number conditions over the entire flight envelope and mission.

These variations can be large and must be allowed for in estimating the responses.

The concept of the force and moment derivatives is based on estimating the change resulting from a change in an individual variable with all the other variables held constant. The resultant change is then the sum of all the changes resulting from the individual variable changes.

Thus if a quantity, F, is a function of several independent variables $x_1, x_2, x_3,..., x_n$, that is, $F = f(x_1, x_2, x_3,..., x_n)$, then the resultant change, ΔF, is given by

$$\Delta F = \frac{\partial F}{\partial x_1} \cdot \Delta x_1 + \frac{\partial F}{\partial x_2} \cdot \Delta x_2 + \frac{\partial F}{\partial x_3} \cdot \Delta x_3 \ldots + \frac{\partial F}{\partial x_n} \cdot \Delta x_n$$

$\partial F / \partial x_1$, $\partial F / \partial x_2$, $\partial F / \partial x_3$,..., $\partial F / \partial x_n$ are the partial derivatives of F with respect to $x_1, x_2, x_3,...,x_n$.

The notation adopted for derivatives is to indicate the partial derivative variable by means of a suffix. Thus

$$\frac{\partial F}{\partial x_1} = F_{x_1}, \quad \frac{\partial F}{\partial x_2} = F_{x_2}, \quad \frac{\partial F}{\partial x_3} = F_{x_3} \ldots \frac{\partial F}{\partial x_n} = F_{x_n}$$

hence

$$\Delta F = F_{x_1} \Delta x_1 + F_{x_2} \Delta x_2 + F_{x_3} \Delta x_3 \ldots + F_{x_n} \Delta x_n$$

The aerodynamic forces and moments acting on the aircraft are functions of the linear and angular velocity components u, v, w, p, q, r of the aircraft (all small quantities of the first order).

The forces acting along OX, OY, OZ are X, Y, Z, respectively, and the corresponding moments about OX, OY and OZ are L, M and N. The derivatives of these forces and moments due to a change in a particular variable are thus denoted by the appropriate suffix, for example X_u, X_w, Y_v, Z_w, L_p, M_q, N_r, N_v, corresponding forces and moments being $X_u u$, $X_w w$, $Y_v v$, $Z_w w$, $L_p p$, $M_q q$, $N_r r$, $N_v v$, etc. It should be noted that some text books and reference books use a 'dressing' (°) over the top of the symbol to denote a derivative, the (°) indicating the value is expressed in ordinary SI (*Le Systèm Internationale d'Unités*) units, for example, $\circ X_u$, $\circ Y_v$, $\circ M_q$.

There are also forces and moments exerted due to the movement of the control surfaces from their trimmed position. Thus in the case of controlling the aircraft's motion in the longitudinal plane by movement of the tailplane (or elevator) from its trimmed position by an amount, η, the corresponding force exerted along the OZ axis would be $Z_\eta \eta$ and the corresponding moment exerted about the OY axis would be $M_\eta \eta$.

Because the disturbances are small, there is no cross-coupling between longitudinal motion and lateral motion. For example, a small change in forward speed or angle of pitch does not produce any side force or rolling or yawing moment. Similarly, a disturbance such as sideslip, rate of roll and rate of yaw produces only second-order forces or moments in the longitudinal plane. Thus the six equations of motion can be split into two groups of three equations.

(a) Longitudinal equations of motion involving linear motion along the OX and OZ axes and angular motion about the OY axis.
(b) Lateral equations of motion involving linear motion along the OY axis and angular motion about the OX and OZ axes.

Thus each group of three equations can be solved separately without having to deal with the full set of six equations.

3.4.4.1 Longitudinal Motion Derivatives

These comprise the forward and vertical force derivatives and pitching moment derivatives arising from the changes in forward velocity, u, vertical velocity, w, rate of pitch, q, and rate of change of vertical velocity, $\dot{w}$.

The main longitudinal derivatives due to u, v, q and w are shown below.

Forward velocity, u

$$\begin{cases} \text{Forward force derivative } X_u \\ \text{Vertical force derivative } Z_u \end{cases}$$

Vertical velocity, w

$$\begin{cases} \text{Forward force derivative } X_w \\ \text{Vertical force derivative } Z_w \\ \text{Pitching moment derivative } M_w \end{cases}$$

Rate of pitch, q

$$\begin{cases} \text{Vertical force derivative } Z_q \\ \text{Pitching moment derivative } M_q \end{cases}$$

Rate of change of vertical velocity, $\dot{w}$

$$\begin{cases} \text{Vertical force derivative } Z_{\dot{w}} \\ \text{Pitching moment derivative } M_{\dot{w}} \end{cases}$$

Force Derivatives Due to Forward Velocity $\mathbf{X_u}$, $\mathbf{Z_u}$ ***and Vertical Velocity*** $\mathbf{X_w}$, $\mathbf{Z_w}$ The changes in the forward velocity, u, and vertical velocity, w, occurring during the disturbance from steady flight result in changes in the incidence angle, α, and air speed, V_T. These changes in incidence and air speed result in changes in the lift and drag forces. The derivatives depend upon the aircraft's lift and drag coefficients and the rate of change of these coefficients with incidence and speed.

The forward force derivative X_u due to the change in forward velocity, u, and the forward force derivative, X_w, due to the change in vertical velocity, arise from the change in drag resulting from the changes in airspeed and incidence, respectively. The resulting forward force components are $X_u u$ and $X_w w$. The *vertical force derivative* Z_u, due to the forward velocity change, u, and the *vertical force derivative* Z_w, due to the vertical velocity change, w, arise from the change in lift due to the changes in airspeed and incidence, respectively. The resulting vertical force components are equal to $Z_u u$ and $Z_w w$.

Pitching Moment Derivative Due to Vertical Velocity, M_w The change in incidence resulting from the change in the vertical velocity, w, causes the pitching moment about the CG to change accordingly. The resulting pitching moment is equal to $M_w w$ where M_w is the pitching moment derivative due to the change in vertical velocity, w. This derivative is very important from the aspect of the aircraft's longitudinal stability.

It should be noted that there is also a pitching moment derivative M_u due to the change in forward velocity, u, but this derivative is generally small compared with M_w.

Force and Moment Derivatives Due to Rate of Pitch Z_q, M_q These derivatives arise from the effective change in the tailplane angle of incidence due to the normal velocity component of the tailplane. This results from the aircraft's angular rate of pitch, q, about the CG and the distance of the tailplane from the CG, l_t (see Fig. 3.20). This normal velocity component is equal to $l_t q$ and the effective change in the tailplane angle of incidence is equal to $l_t q/V_T$. This incidence change results in a lift force acting on the tailplane, which is multiplied by the tail moment arm, l_t, to give a significant damping moment to oppose the rate of rotation in pitch.

The vertical force (acting on the tailplane) is equal to $Z_q q$ where Z_q is the vertical force derivative due to rate of pitch, q. The resultant pitching moment about the CG is equal to $M_q q$ where M_q is the pitching moment derivative due to the rate of pitch, q. The M_q derivative has a direct effect on the damping of the aircraft's response to a disturbance or control input and is a very important derivative from this aspect. The use of auto-stabilisation systems to artificially augment this derivative will be covered later in this chapter.

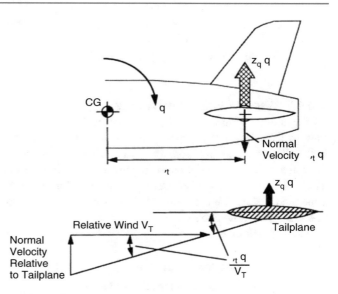

Fig. 3.20 Damping moment due to pitch rate

Force and Moment Derivatives Due to the Rate of Change of Vertical Velocity $Z_{\dot{w}}$, $M_{\dot{w}}$ These force and moment derivatives result from the lag in the downwash from the wing acting on the tailplane and are proportional to the rate of change of downwash angle with wing incidence. The resulting vertical force component is equal to $Z_{\dot{w}} \dot{w}$ and the resulting pitching moment is equal to $M_{\dot{w}} \dot{w}$ where $Z_{\dot{w}}$ is the vertical force derivative due to rate of change of vertical velocity, $\dot{w}$, and $M_{\dot{w}}$ is the pitching moment derivative due to the rate of change of vertical velocity, $\dot{w}$.

Control Derivatives Z_η, M_η Control of the aircraft in the longitudinal plane is achieved by the angular movement of the tailplane (or elevator) by the pilot. The lift force acting on the tailplane as a result of the change in tailplane incidence creates a pitching moment about the aircraft's CG because of the tail moment arm. The pitching moment is proportional to the tailplane (or elevator) angular movement. The vertical force resulting from the tailplane (or elevator) movement is equal to $Z_\eta \eta$ where Z_η is the vertical force derivative due to the angular movement of the tailplane, η, from its steady, trimmed flight position. The pitching moment is equal to $M_\eta \eta$ when M_η is the pitching moment derivative due to the change in tailplane angle, η.

Longitudinal Forces and Moments
The main forces and moments acting along and about the OX, OY, OZ axes affecting longitudinal motion are set out below:

$$X_a = X_u u + X_w w \qquad (3.28)$$

$$Z_a = Z_u u + Z_w w + Z_{\dot{w}}\dot{w} + Z_q q + Z_\eta \eta \qquad (3.29)$$

$$M = M_w w + M_q q + M_{\dot{w}}\dot{w} + M_\eta \eta \qquad (3.30)$$

3.4.4.2 Lateral Motion Derivatives

Changes in the sideslip velocity, v, rate of roll, p, and rate of yaw, r, following a disturbance from steady flight produces both rolling and yawing moments. This causes motion about the roll axis to cross-couple into the yaw axis and vice versa. The sideslip velocity also results in a side force being generated. The main lateral motion derivatives due to v, p and r are shown below:

Sideslip velocity v

$$\begin{cases} \text{Side force derivative } Y_v \\ \text{Yawing moment derivative } N_v \\ \text{Rolling moment derivative } L_v \end{cases}$$

Rate of roll, p

$$\begin{cases} \text{Rolling moment derivative } L_p \\ \text{Yawing moment derivative } N_p \end{cases}$$

Rate of yaw, r

$$\begin{cases} \text{Yawing moment derivative } N_r \\ \text{Rolling moment derivative } L_r \end{cases}$$

Side Force Derivative Due to Sideslip Velocity $\mathbf{Y}_v$ The change in sideslip velocity, v, during a disturbance changes the incidence angle, β, of the aircraft's velocity vector, V_T, (or relative wind) to the vertical surfaces of the aircraft comprising the fin and fuselage sides (see Fig. 3.21). The change in incidence angle v/V_T results in a sideways lifting force being generated by these surfaces. The net side force from the fuselage and fin combined is equal to $Y_v v$ where Y_v is the side force derivative due to the sideslip velocity.

Yawing Moment Derivative Due to Sideslip Velocity $\mathbf{N}_v$ The side force on the fin due to the incidence, β, resulting from the sideslip velocity, v, creates a yawing moment about the CG, which tends to align the aircraft with the relative wind in a similar manner to a weathercock (refer to Fig. 3.21).

The main function of the fin is to provide this directional stability (often referred to as weathercock stability). This yawing moment is proportional to the sideslip velocity and is dependent on the dynamic pressure, fin area, fin lift coefficient and the fin moment arm, the latter being the distance between the aerodynamic centre of the fin and the yaw axis

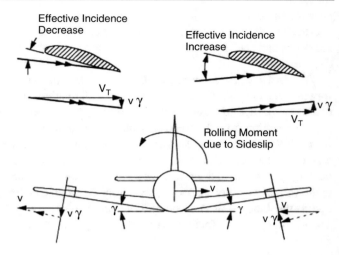

Fig. 3.21 Lateral forces

through the CG. However, the aerodynamic lateral forces acting on the fuselage during sideslipping also produce a yawing moment which opposes the yawing moment due to the fin and so is destabilising. The net yawing moment due to sideslip is thus dependent on the combined contribution of the fin and fuselage. The fin area and moment arm, known as the *fin volume*, is thus sized to provide good directional stability under all conditions and subject to other constraints such as engine failure in the case of a multi-engine aircraft.

The yawing moment due to sideslip velocity is equal to $N_v v$ where N_v is the yawing moment derivative due to sideslip velocity.

Rolling Moment Derivative Due to Sideslip Velocity $\mathbf{L}_v$ When the aircraft experiences a sideslip velocity, the effect of wing dihedral is for this sideslip velocity to increase the incidence on one wing and reduce it on the other (see Fig. 3.22). Thus if one wing tends to drop whilst sideslipping, there is a rolling moment created which tends to level the wings.

The effect of wing sweepback is also to give a differential incidence change on the wings and a rolling moment is generated, even if the dihedral is zero. There is also a contribution to the rolling moment due to sideslip from the fin. This is because of the resulting lift force acting on the fin and the height of the aerodynamic centre of the fin above the roll axis. There is also a contribution from the fuselage due to flow effects round the fuselage, which affect the local wing incidence. These are beneficial in the case of a high wing and detrimental in the case of low wing. It is in fact necessary to incorporate considerably greater dihedral for a low wing than that required for a high wing location. The rolling moment derivative due to sideslip is denoted by L_v and the rolling moment due to sideslip is equal to $L_v v$.

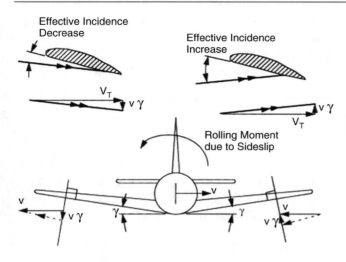

Fig. 3.22 Effect of dihedral and sideslip

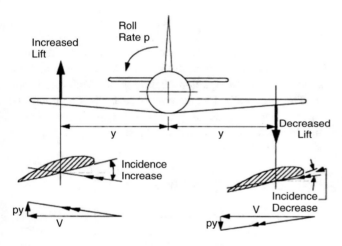

Fig. 3.23 Rolling moment due to rate of roll

Rolling Moment Derivative Due to Rate of Roll $\mathbf{L}_p$ When the aircraft is rolling, the angular velocity causes each section of wing across the span to experience a tangential velocity component which is directly proportional to its distance from the centre. Referring to Fig. 3.23, it can be seen that one section of wing experiences an increase in incidence whilst the corresponding section on the other wing experiences a decrease. The lift force exerted on one wing is thus increased whilst that on the other wing is decreased and a rolling moment is thus generated. The rolling moment due to the rate of roll, p, acts in the opposite sense to the direction of rolling and is equal to $L_p p$ where L_p is the rolling moment derivative due to rate of roll.

Yawing Moment Derivative Due to Rate of Roll $\mathbf{N}_p$ The rate of roll which increases the lift on the outer part of one wing and reduces it on the other also creates a differential drag effect. The increase in lift is accompanied by an increase in

drag in the forward direction and the decrease in lift on the other wing by a corresponding reduction in drag. A yawing moment is thus produced by the rate of roll, p, which is equal to $N_p p$ where N_p is the yawing moment derivative due to rate of roll.

Yawing Moment Derivative Due to Rate of Yaw $\mathbf{N}_r$ The rate of yaw, r, produces a tangential velocity component equal to $l_f\, r$ where l_f is the distance between the aerodynamic centre of the fin and the yaw axis through the CG. The resulting change in the effective fin incidence angle, $l_f\, r/V_T$, produces a lift force which exerts a damping moment about the CG opposing the rate of yaw. The yawing moment due to the rate of yaw is equal to $N_r r$ where N_r is the yawing moment derivative due to rate of yaw.

Rolling Moment Derivative Due to Rate of Yaw $\mathbf{L}_r$ When the aircraft yaws, the angular velocity causes one wing to experience an increase in velocity relative to the airstream and the other wing a decrease. The lift on the leading wing is thus increased and the trailing wing decreased thereby producing a rolling moment. The rolling moment derivative due to rate of yaw is denoted by L_r and the rolling moment due to rate of yaw is equal to $L_r r$.

Lateral Control Derivatives Due to Ailerons and Rudder   The ailerons and rudder are illustrated in Fig. 3.15. The angle through which the ailerons are deflected differentially from their position in steady, trimmed flight is denoted by ξ and the angle the rudder is deflected from the position in steady, trimmed flight is denoted by ζ.

Differential movement of the ailerons provides the prime means of lateral control by exerting a controlled rolling moment to bank the aircraft to turn; this will be covered later in this chapter.

Rolling Moment Derivative Due to Aileron Deflection L_ξ The effect of the differential deflection of the ailerons from their steady, trimmed flight position, ξ, is to increase the lift on one wing and reduce it on the other thereby creating a rolling moment. The rolling moment due to the aileron deflection is equal to $L_\xi \xi$ where L_ξ is the rolling moment derivative due to aileron deflection.

Yawing Moment Derivative Due to Aileron Deflection $\mathbf{N}_\xi$ The differential lift referred to above is also accompanied by a differential drag on the wings which results in a yawing moment being exerted. The yawing moment due to the aileron deflection is equal to $N_\xi \xi$ where N_ξ is the yawing moment derivative due to aileron deflection.

Deflection of the rudder creates a lateral force and yawing moment to counteract sideslipping motion and the yawing moment due to the movement of the ailerons. (The control function of the rudder will be covered in more detail later in this chapter).

Side Force Derivative Due to Rudder Deflection Y_ζ The sideways lift force acting on the fin due to the deflection of the rudder, ζ, from its steady, trimmed flight position is equal to $Y_\zeta \zeta$ where Y_ζ is the side force derivative due to rudder deflection.

Yawing Moment Derivative Due to Rudder Deflection $\underline{N_\zeta}$ The yawing moment exerted by the rudder is equal to $N_\zeta \zeta$ where N_ζ is the yawing moment derivative due to the deflection of the rudder from its steady, trimmed flight position.

Lateral Forces and Moments
The main sideways force, Y_a, the rolling moment, L, and the yawing moment, N, resulting from changes in the sideslip velocity, v, rate of roll, p, and rate of yaw, r, from the steady flight condition are set out below.

$$Y_a = Y_v v + Y_\zeta \zeta \qquad (3.31)$$

$$L = L_v v + L_p p + L_r r + L_\xi \xi + L_\zeta \zeta \qquad (3.32)$$

$$N = N_v v + N_p p + N_r r + N_\xi \xi + N_\zeta \zeta \qquad (3.33)$$

Normalisation of Derivatives
It should be noted that the derivatives are sometimes normalised and non-dimensionalised by dividing by the appropriate quantities – air density, airspeed, mean aerodynamic chord, wing span, etc. The appropriate quantities depend on whether it is a force/velocity derivative or a moment/angular rate derivative, etc. The equations of motion are also sometimes non-dimensionalised by dividing by the aircraft mass, moments of inertia, etc.

This enables aircraft responses to be compared independently of speed, air density, size, mass, inertia, etc. This stage has been omitted in this book for simplicity.

Readers wishing to find out more about non-dimensionalised derivatives and equations are referred to 'Further Reading' at the end of this chapter.

3.4.5   Equations of Longitudinal and Lateral Motion

Equations (3.28), (3.29) and (3.30) for X_a, Z_a and M can be substituted in the equations of motion (3.22), (3.24) and (3.26), respectively.

The equations of longitudinal motion for small disturbances then become:

$$X_u u + X_w w - mg\theta = m(\dot{u} + W_0 q) \qquad (3.34)$$

$$Z_u u + Z_w w + Z_{\dot{w}} \dot{w} + Z_q q + Z_\eta \eta = m(\dot{w} - U_0 q)) \qquad (3.35)$$

$$M_w w + M_q q + M_{\dot{w}} \dot{w} + M_\eta \eta = I_y \dot{q} \qquad (3.36)$$

Equations (3.31), (3.32) and (3.33) for Y_a, L and N can similarly be substituted in the equations of motion (3.23), (3.25) and (3.27), respectively. The equations of lateral motion for small disturbances then become:

$$Y_v v + Y_\zeta \zeta + mg\Phi = m(\dot{v} - W_0 p + U_0 r) \qquad (3.37)$$

$$L_v v + L_p p + L_r r + L_\zeta \zeta + L_\xi \xi = I_x \dot{p} \qquad (3.38)$$

$$N_v v + N_p p + N_r r + N_\zeta \zeta + N_\xi \xi = I_z \dot{r} \qquad (3.39)$$

These equations can be re-arranged as a set of first-order differential equations to express the first derivative of each variable as a linear equation relating all the variables in the set. The longitudinal equations have been simplified for clarity by omitting the $Z_q q$, $Z_{\dot{w}} \dot{w}$ and $M_{\dot{w}} \dot{w}$ terms. Hence, from Eqs. (3.34), (3.35) and (3.36) and noting that $q = \dot{\theta}$ for small disturbances,

$$\left.\begin{aligned}
\dot{u} &= \frac{X_u}{m} u + \frac{X_w}{m} w - W_0 q - g\theta \\
\dot{w} &= \frac{Z_u}{m} u + \frac{Z_w}{m} w + U_0 q \quad + \frac{Z_\eta}{m} \eta \\
\dot{q} &= \frac{M_w}{I_y} w + \frac{M_q}{I_y} q \quad + \frac{M_\eta}{I_y} \eta \\
\dot{\theta} &= q
\end{aligned}\right\} \qquad (3.40)$$

Similarly, from Eqs. (3.37), (3.38) and (3.39) and noting that $p = \dot{\Phi}$ for small disturbances,

$$\left.\begin{aligned}
\dot{v} &= \frac{Y_v}{m} v + W_0 p - U_0 r + g\Phi + \frac{Y_\zeta}{m} \zeta \\
\dot{p} &= \frac{L_v}{I_x} v + \frac{L_p}{I_x} p + \frac{L_r}{I_x} r \quad + \frac{L_\zeta}{I_x} \zeta + \frac{L_\xi}{I_x} \xi \\
\dot{r} &= \frac{N_v}{I_z} v + \frac{N_p}{I_z} p + \frac{N_r}{I_z} r \quad + \frac{N_\zeta}{I_z} \zeta + \frac{N_\xi}{I_z} \xi \\
\dot{\Phi} &= p
\end{aligned}\right\} \qquad (3.41)$$

These equations can be solved in a step-by-step, or iterative, manner by continuously computing the first derivative at each increment of time and using it to derive the change in the variables over the time increment and hence update the value at the next time increment.

As a simplified example, suppose the first-order equations relating three variables x, y, z are as follows:

$$\dot{x} = a_1 x + b_1 y + c_1 z$$
$$\dot{y} = a_2 x + b_2 y + c_2 z$$
$$\dot{z} = a_3 x + b_3 y + c_3 z$$

Let Δt = time increment, value of x at time $n\Delta t$ be denoted by x_n and value of ⬚ at time $n\Delta t$ be denoted by ⬚$_n$.

Value of x at time $(n + 1)\Delta t$, that is $x_{(n + 1)}$, is given by

$$x_{(n+1)} = x_n + \dot{x}_n \cdot \Delta t$$

(assuming Δt is chosen as a suitably small time increment). Hence

$$x_{(n+1)} = x_n + (a_1 x_n + b_1 y_n + c_1 z_n)\Delta t$$

a form suitable for implementation in a digital computer.

These equations can be expressed in the matrix format

$$\begin{bmatrix} \dot{x} \\ \dot{y} \\ \dot{z} \end{bmatrix} = \begin{bmatrix} a_1 & b_1 & c_1 \\ a_2 & b_2 & c_2 \\ a_3 & b_3 & c_3 \end{bmatrix} \begin{bmatrix} x \\ y \\ z \end{bmatrix}$$

Similarly, Eqs. (3.40) and (3.41) can be expressed as follows:

$$\begin{bmatrix} \dot{u} \\ \dot{w} \\ \dot{q} \\ \dot{\theta} \end{bmatrix} = \begin{bmatrix} \dfrac{X_u}{m} & \dfrac{X_w}{m} & -W_0 & -g \\ \dfrac{Z_u}{m} & \dfrac{Z_w}{m} & U_0 & 0 \\ 0 & \dfrac{M_w}{I_y} & \dfrac{M_q}{I_y} & 0 \\ 0 & 0 & 1 & 0 \end{bmatrix} \begin{bmatrix} u \\ w \\ q \\ \theta \end{bmatrix} + \begin{bmatrix} 0 \\ \dfrac{z_\eta}{m} \\ \dfrac{M_\eta}{I_y} \\ 0 \end{bmatrix} [\eta] \quad (3.42)$$

$$\begin{bmatrix} \dot{v} \\ \dot{p} \\ \dot{r} \\ \dot{\Phi} \end{bmatrix} = \begin{bmatrix} \dfrac{Y_v}{m} & W_0 & -U_0 & g \\ \dfrac{L_v}{I_x} & \dfrac{L_p}{I_x} & \dfrac{L_r}{I_x} & 0 \\ \dfrac{N_v}{I_z} & \dfrac{N_p}{I_z} & \dfrac{N_r}{I_z} & 0 \\ 0 & 1 & 0 & 0 \end{bmatrix} \begin{bmatrix} v \\ p \\ r \\ \Phi \end{bmatrix} + \begin{bmatrix} \dfrac{Y_\zeta}{m} & 0 \\ \dfrac{L_\zeta}{I_x} & \dfrac{L_\xi}{I_x} \\ \dfrac{N_\zeta}{I_z} & \dfrac{N_\xi}{I_z} \\ 0 & 0 \end{bmatrix} \begin{bmatrix} \zeta \\ \xi \end{bmatrix} \quad (3.43)$$

The use of a matrix format for the equations of motion gives much more compact expressions and enables the equations to be manipulated using matrix algebra.

The equations can be expressed in the general form

$$\dot{\mathbf{X}} = \mathbf{AX} + \mathbf{BU}$$

The bold letters indicate that $\mathbf{X}$, $\mathbf{A}$, $\mathbf{B}$, $\mathbf{U}$ are matrices. $\mathbf{X}$ is the state vector matrix, the elements comprising the state variables; $\mathbf{A}$ is the state coefficient matrix; $\mathbf{B}$ is the driving matrix; $\mathbf{U}$ is the control input vector matrix, the elements comprising the control input variables.

For example, in Eq. (3.43), the state variables v, p, r, Φ form the state vector

$$\mathbf{X} = \begin{bmatrix} v \\ p \\ r \\ \Phi \end{bmatrix}$$

The state coefficient matrix

$$\mathbf{A} = \begin{bmatrix} \dfrac{Y_v}{m} & W_0 & -U_0 & g \\ \dfrac{L_v}{I_x} & \dfrac{L_p}{I_x} & \dfrac{L_r}{I_x} & 0 \\ \dfrac{N_v}{I_z} & \dfrac{N_p}{I_z} & \dfrac{N_r}{I_z} & 0 \\ 0 & 1 & 0 & 0 \end{bmatrix}$$

The driving matrix

$$\mathbf{B} = \begin{bmatrix} \dfrac{Y_\zeta}{m} & 0 \\ \dfrac{L_\zeta}{I_x} & \dfrac{L_\xi}{I_x} \\ \dfrac{N_\zeta}{I_z} & \dfrac{N_\xi}{I_z} \\ 0 & 0 \end{bmatrix}$$

The control input vector

$$\mathbf{U} = \begin{bmatrix} \zeta \\ \xi \end{bmatrix}$$

Further treatment of state variable matrix equations is beyond the scope of this chapter, the main objective being to introduce the reader who is not familiar with the methods and terminology to their use. Appropriate textbooks are listed in the 'Further Reading' at the end of this chapter.

3.5 Longitudinal Control and Response

3.5.1 Longitudinal Control

In conventional (i.e. non-fly-by-wire) aircraft, the pilot controls the angular movement of the tailplane/elevators directly from the control column, or 'stick', which is mechanically coupled by rods and linkages to the tailplane/elevator servo actuator. (Fully powered controls are assumed.) To manoeuvre in the longitudinal (or pitch) plane, the pilot controls the tailplane/elevator angle and hence the pitching moment exerted about the CG by the tailplane lift. This enables the pilot to rotate the aircraft about its CG to change

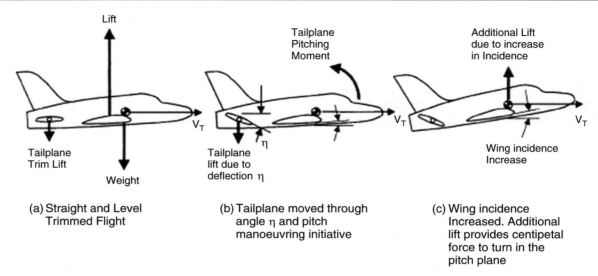

Fig. 3.24 Manoeuvring in the pitch plane

the wing incidence angle and hence control the wing lift to provide the necessary normal, or centripetal, force to change the direction of the aircraft's flight path (see Fig. 3.24).

The initial response of the aircraft on application of a steady tailplane/elevator angular movement from the trim position is as follows.

The resulting pitching moment accelerates the aircraft's inertia about the pitch axis causing the aircraft to rotate about its CG so that the wing incidence angle increases. The wing lift increases accordingly and causes the aircraft to turn in the pitch plane and the rate of pitch to build up. This rotation about the CG is opposed by the pitching moment due to incidence which increases as the incidence increases and by the pitch rate damping moment exerted by the tailplane. A steady condition is reached and the aircraft settles down to a new steady wing incidence angle and a steady rate of pitch which is proportional to the tailplane/elevator angular movement from the trim position.

The aircraft's inertia about the pitch axis generally results in some transient overshoot before a steady wing incidence is achieved, the amount of overshoot being mainly dependent on the pitch rate damping generated by the tailplane.

The normal acceleration is equal to the product of the forward speed and the rate of pitch and is directly proportional to the increase in wing lift resulting from the increase in wing incidence angle. For a given constant forward speed, the rate of pitch is thus proportional to the tailplane/elevator angular movement from the trim position, which in turn is proportional to the stick deflection. The rate of pitch is reduced to zero by returning, that is 'centralising', the stick to its trimmed flight position.

3.5.2 Stick Force/g

The wing lift is, however, proportional to both the wing incidence and the dynamic pressure, $\frac{1}{2}\rho V_T^2$, so the required wing incidence for a given normal acceleration will vary with height and speed over the flight envelope. Only a small change in wing incidence is required at the high dynamic pressures resulting from high-speed low-level flight, and vice versa.

The tailplane/elevator angular movement required per g normal acceleration, and hence stick displacement/g, will thus vary with height and speed: the variation can be as high as 40:1 over the flight envelope for a high-performance aircraft.

The aerodynamic forces and moments are proportional to the dynamic pressure and with manually operated flying controls (i.e. no hydraulically powered controls), the pilot has a direct feedback from the stick of the forces being exerted as a result of moving the elevator. Thus, at higher speeds and dynamic pressures only a relatively small elevator movement is required per g normal acceleration. The high dynamic pressure experienced, however, requires a relatively high stick force although the stick movement is small. At lower speeds and dynamic pressures, a larger elevator angle is required per g and so the stick force is still relatively high. The stick force/g thus tends to remain constant for a well-designed aeroplane operating over its normal flight envelope.

However, with the fully powered flying controls required for high-speed aircraft, there is no direct feedback at the stick of the control moments being applied by moving the tailplane/elevators.

Stick displacement is an insensitive control for the pilot to apply, stick force being a much more natural and effective control, and matches with the pilot's experience and training flying aircraft with manually operated controls.

The stick force is made proportional to stick deflection by either a simple spring loading system or by an 'artificial feel unit' connected directly to the stick mechanism. The artificial feel unit varies the stick stiffness (or 'feel') as a function of dynamic pressure $\frac{1}{2}\rho V_T^2$ so as to provide the required stick force/g characteristics. The rate of pitch at a given forward speed (i.e. normal acceleration) is thus made proportional to the stick force applied by the pilot. However, as already mentioned, the effectiveness of the tailplane/elevator can vary over a wide range over the height and speed envelope. Automatic control may be required to improve the aircraft response and damping over the flight envelope and give acceptable stick force per g characteristics.

3.5.3 Pitch Rate Response to Tailplane/Elevator Angle

The key to analysing the response of the aircraft and the modifying control required from the automatic flight control system is to determine the aircraft's basic control transfer function. This relates the aircraft's pitch rate, q, to the tailplane/elevator angular movement, η, from the trimmed position.

This can be derived from Eqs. (3.34), (3.35) and (3.36) together with the relationship $q = d\theta/dt$ by eliminating u and w to obtain an equation in q and η only. This results in a transfer function of the form below:

$$\frac{q}{\eta} = \frac{K(D^3 + b_2 D^2 + b_1 D + b_0)}{(D^4 + a_3 D^3 + a_2 D^2 + a_1 D + a_0)}$$

where $D = d/dt$.

K, b_2, b_1, b_0, a_3, a_2, a_1, a_0 are constant coefficients comprising the various derivatives. The transient response (and hence the stability) is determined by the solution of the differential equation $(D^4 + a_3 D^3 + a_2 D^2 + a_1 D + a_0)q = 0$.

$q = Ce^{\lambda t}$, where C and λ are constants, is a solution of this type of linear differential equation with constant coefficients. Substituting $Ce^{\lambda t}$ for q yields

$$(\lambda^4 + a_3 \lambda^3 + a_2 \lambda^2 + a_1 \lambda + a_0)Ce^{\lambda t} = 0$$

That is,

$$(\lambda^4 + a_3 \lambda^3 + a_2 \lambda^2 + a_1 \lambda + a_0) = 0$$

This equation is referred to as the characteristic equation. This quartic equation can be split into two quadratic factors,

which can be further factorised into pairs of conjugate complex factors as shown below:

$$(\lambda + \alpha_1 + j\omega_1)(\lambda + \alpha_1 - j\omega_1)(\lambda + \alpha_2 + j\omega_2) \\ \times (\lambda + \alpha_2 - j\omega_2) = 0$$

The solution is as follows:

$$q = \underbrace{A_1 e^{-\alpha_1 t} \sin(\omega_1 t + \phi_1)}_{\text{Short period motion}} + \underbrace{A_2 e^{-\alpha_2 t} \sin(\omega_2 t + \phi_2)}_{\text{Long period motion}}$$

A_1, ϕ_1, A_2, ϕ_2 are constants determined by the initial conditions, that is, value of $\ldots q$, $\ddot{q}$, $\dot{q}$, q at time $t = 0$.

For the system to be stable, the exponents must be negative so that the exponential terms decay to zero with time. (Positive exponents result in terms which diverge exponentially with time, i.e. an unstable response.)

The solution thus comprises the sum of two exponentially damped sinusoids corresponding to the short-period and long-period responses, respectively.

The initial basic response of the aircraft is thus a damped oscillatory response known as the short-period response. The period of this motion is in the region of 1 to 10 seconds, depending on the type of aircraft and its forward speed, being inversely proportional to the forward speed, for example, typical fighter aircraft would be about 1-second period whereas a large transport aircraft would be about 5- to 10-second period. This short-period response is generally fairly well-damped for a well-behaved (stable) aircraft but may need augmenting with an auto-stabilisation system over parts of the flight envelope of height and speed combinations.

The second stage comprises a slow lightly damped oscillation with a period ranging from 40 seconds to minutes and is known as the long period/motion. It is basically similar to the *phugoid* motion and is again inversely proportional to forward speed. The phugoid motion consists of a lightly damped oscillation in height and airspeed whilst the angle of incidence remains virtually unchanged and is due to the interchange of kinetic energy and potential energy as the aircraft's height and speed change. The damping of the phugoid motion is basically a task for the autopilot, the long period making it very difficult for the pilot to control.

Figure 3.25 illustrates the two types of motion in the response.

A simpler method of obtaining a good approximation to the q/n transfer function, which accurately represents the aircraft's short-period response, can be obtained by assuming the forward speed remains constant. This is a reasonable assumption as the change in forward speed is slow compared with the other variables.

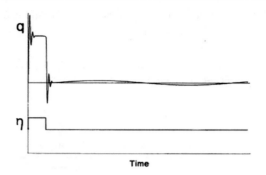

Fig. 3.25 Pitch response

This simpler method is set out in the next section as it gives a good 'picture' in control engineering terms of the behaviour of the basic aircraft.

3.5.4 Pitch Response Assuming Constant Forward Speed

The transfer functions relating pitch rate and wing incidence to tailplane (or elevator) angle, that is q/η and α/η, are derived below from first principles making the assumption of constant forward speed and small perturbations from steady straight and level, trimmed flight.

The derivative terms due to the rate of change of vertical velocity, $\dot{w}$, that is $Z_{\dot{w}}\dot{w}$ and $M_{\dot{w}}\dot{w}$, are neglected and also the $Z_q q$ term.

Referring to Fig. 3.26, an orthogonal set of axes OX, OY, OZ moving with the aircraft with the centre O at the aircraft's centre of gravity is used to define the aircraft motion. OX is aligned with the flight path vector, OY with the aircraft's pitch axis and OZ normal to the aircraft's flight path vector (positive direction downwards). These axes are often referred to as 'stability axes' and enable some simplification to be achieved in the equations of motion.

Velocity along OX axis is U (constant); velocity increment along OZ axis is w; velocity along OY axis is 0; rate of rotation about pitch axis, OY, is q; change in angle of incidence from trim value is α; aircraft mass is m; aircraft moment of inertia about pitch axis, OY, is I_y.

Considering forces acting in OZ direction:

1. Change in lift force acting on wing due to change in angle of incidence, α, from trim value is $Z_\alpha \alpha$.
2. Change in lift force acting on tailplane due to change in tailplane angle, η, from trim value is $Z_\eta \eta$.

Normal acceleration along OZ axis is $\dot{w} - Uq$, where $\dot{w} = dw/dt$ (refer to Sect. 3.4.3, Eq. (3.21)).
Equation of motion along OZ axis is

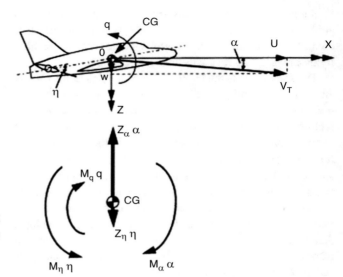

Fig. 3.26 Forces and moments – longitudinal plane (stability axes)

$$Z_\alpha \alpha + Z_\eta \eta = m\left(\dot{w} - U_q\right) \qquad (3.44)$$

Considering moments acting about CG:

1. Pitching moment due to change in tailplane angle, η, from trim value is $M_\eta \eta$.
2. Pitching moment due to change in angle of incidence, α, from trim value is $M_\alpha \alpha$.
3. Pitching moment due to angular rate of rotation, q, about the pitch axis is $M_q q$.

Equation of angular motion about pitch axis, OY, is

$$M_\eta \eta + M_\alpha \alpha + M_q q = I_y \dot{q} \qquad (3.45)$$

where $\dot{q} = dq/dt$ is angular acceleration and change in angle of incidence is

$$\alpha = \frac{w}{U} \qquad (3.46)$$

These simultaneous differential equations can be combined to give equations in terms of q and η only by eliminating w, or, α and η only by eliminating q. However, it is considered more instructive to carry out this process using block diagram algebra as this gives a better physical picture of the aircraft's dynamic behaviour and the inherent feedback mechanisms relating q and α.

Equation (3.45) can be written as

Fig. 3.27 Block diagram representation of Eq. (3.47)

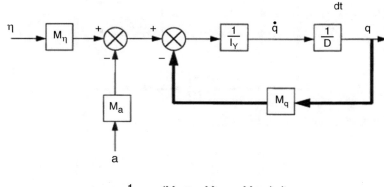

$$q = \frac{1}{I_Y} \int (M_a\, a + M_q\, q + M_\eta\, \eta)\, dt$$

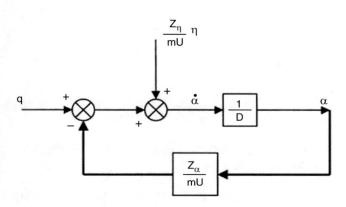

Fig. 3.28 Block diagram representation of Eq. (3.48)

$$q = \frac{1}{I_y} \int \left(M_\alpha \alpha + M_q q + M_\eta \eta\right) dt \qquad (3.47)$$

Substituting $U\dot{\alpha}$ for w in Eq. (3.44) and re-arranging yields

$$\alpha = \int \left(q + \frac{Z_\alpha}{mU}\alpha + \frac{Z_\eta}{mU}\eta\right) dt \qquad (3.48)$$

Figure 3.27 represents Eq. (3.47) in block diagram form. It should be noted that the signs of the derivatives have been indicated at the summation points in the block diagram. M_q is negative and an aerodynamically stable aircraft is assumed so that M_α is also negative. The $M_q q$ and $M_\alpha \alpha$ terms are thus negative feedback terms. (Positive M_α can be allowed for by the appropriate sign change.) Figure 3.28 is a block diagram representation of Eq. (3.48).

The q and α subsidiary feedback loops can be simplified using the relationship between output, θ_o, and input, θ_i, in a generalised negative feedback process of the type shown in Fig. 3.29.

That is,

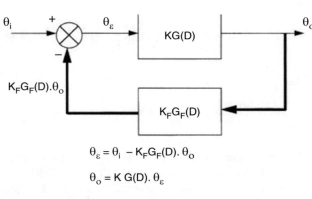

$$\theta_\varepsilon = \theta_i - K_F G_F(D)\cdot \theta_o$$

$$\theta_o = K\,G(D)\cdot \theta_\varepsilon$$

Hence $\quad \dfrac{\theta_o}{\theta_i} = \dfrac{KG(D)}{1 + KG(D)\cdot K_F G_F(D)}$

Fig. 3.29 Generalised negative feedback system

$$\frac{\theta_o}{\theta_i} = \frac{KG(D)}{1 + KG(D)\cdot K_F G_F(D)} \qquad (3.49)$$

where $KG(D)$ and $K_F G_F(D)$ are the transfer functions of the forward path and feedback path, respectively.

Thus, considering the $M_q q$ feedback loop

$$\frac{q}{(M_\eta \eta - M_\alpha \alpha)} = \frac{\frac{1}{I_y D}}{1 + \frac{1}{I_y D}\cdot M_q}$$

as $KG(D) = 1/I_y D$ and $K_F G_F(D) = M_q$.

$$\frac{q}{(M_\eta \eta - M_\alpha \alpha)} = \frac{1}{M_q}\cdot \frac{1}{(1 + T_1 D)} \qquad (3.50)$$

where

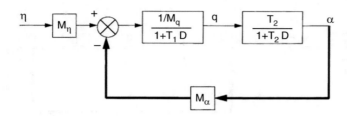

Fig. 3.30 Simplified overall block diagram

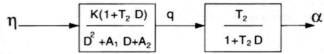

Fig. 3.31 Further simplified overall block diagram

$$T_1 = \frac{I_y}{M_q} \tag{3.51}$$

This represents a simple first-order lag transfer function.

Referring to Fig. 3.28, the α/q inner loop can be similarly simplified to

$$\frac{\alpha}{q} = T_2 \cdot \frac{1}{(1 + T_2 D)} \tag{3.52}$$

where

$$T_2 = \frac{mU}{Z_\alpha} \tag{3.53}$$

This represents another first-order lag transfer function. The overall block diagram thus simplifies to Fig. 3.30.

It should be noted that the $Z_\eta \eta / mU$ input term has been omitted for simplicity as it is small in comparison with the other terms. Also, because it is an external input to the loop, it does not affect the feedback loop stability and transient response to a disturbance.

Referring to Fig. 3.30 and applying Eq. (3.49) yields

$$\frac{q}{\eta} = M_\eta \cdot \frac{\frac{\frac{1}{M_q}}{(1 + T_1 D)}}{1 + \frac{\frac{1}{M_q}}{(1 + T_1 D)} \cdot \frac{T_2}{(1 + T_2 D)} M_\alpha} \tag{3.54}$$

This simplifies to

$$\frac{q}{\eta} = \frac{M_\eta}{I_y T_2} \cdot \frac{(1 + T_2 D)}{D^2 + \left(\frac{1}{T_1} + \frac{1}{T_2}\right)D + \frac{M_\alpha}{I_y} + \frac{1}{T_1 T_2}} \tag{3.55}$$

$$\frac{q}{\eta} = \frac{K(1 + T_2 D)}{(D^2 + A_1 D + A_2)} \tag{3.56}$$

where

$$K = \frac{M_\eta}{I_y T_2} = \frac{M_\eta Z_\alpha}{I_y mU} \tag{3.57}$$

$$A_1 = \frac{1}{T_1} + \frac{1}{T_2} = \frac{M_q}{I_y} + \frac{Z_\alpha}{mU} \tag{3.58}$$

$$A_2 = \frac{M_\alpha}{I_y} + \frac{1}{T_1 T_2} = \frac{M_\alpha}{I_y} + \frac{M_q Z_\alpha}{I_y mU} \tag{3.59}$$

Substituting

$$\alpha = \frac{T_2}{1 + T_2 D} \cdot q$$

in Eq. (3.56) yields

$$\frac{\alpha}{\eta} = \frac{KT_2}{(D^2 + A_1 D + A_2)} \tag{3.60}$$

The resulting overall block diagram is shown in Fig. 3.31. From Eq. (3.56)

$$(D^2 + A_1 D + A_2)q = K(1 + T_2 D)\eta \tag{3.61}$$

The transient response is given by the solution of the differential equation

$$(D^2 + A_1 D + A_2)q = 0 \tag{3.62}$$

The solution is determined by the roots of the characteristic equation

$$(\lambda^2 + A_1 \lambda + A_2) = 0$$

which are $\lambda = -\frac{1}{2}A_1 \pm \sqrt{A_1^2 - 4A_2}/2$.

In a conventional aircraft these are complex roots of the form $-\alpha_1 \pm j\omega$, where $\alpha_1 = A_1/2$ $\omega = \sqrt{4A_2 - A_1^2}/2$.

The solution in this case is an exponentially damped sinusoid

$$q = Ae^{-\alpha_1 t} \sin(\omega t + \phi)$$

when A and ϕ are constants determined from the initial conditions (value of $\dot{q}$ and q at time $t = 0$). The coefficient A_1 determines the value of α_1 ($= A_1/2$) and this determines the degree of damping and overshoot in the transient

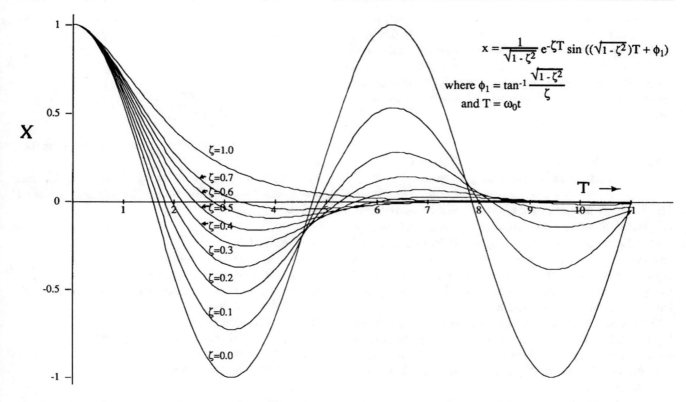

Fig. 3.32 Generalised second-order system transient response

response to a disturbance. The coefficient A_2 largely determines the frequency, ω, of the damped oscillation $\omega = \sqrt{4A_2 - A_1{}^2}/2$ and hence the speed of the response.

In order for the system to be stable the roots of the characteristic equation must be negative or have a negative real part, so that the exponents are negative and exponential terms decay to zero with time. This condition is met provided A_1 and A_2 are both positive.

Conversely, an aerodynamically unstable aircraft with a positive M_α would result in A_2 being negative thereby giving a positive exponent and a solution which diverges exponentially with time.

It is useful to express the quadratic factor $(D^2 + A_1D + A_2)$ in terms of two generalised parameters ω_0 and ζ, thus $(D^2 + 2\zeta\omega_0D + \omega_0^2)$ where ω_0 is the undamped natural frequency $= \sqrt{A_2}$.

$$\zeta = \text{Damping ratio} = \frac{\text{Coefficient of } D}{\text{Coefficient of } D \text{ for critical damping}}$$

That is,

$$\zeta = \frac{A_1}{2\sqrt{A_2}} \tag{3.63}$$

(Critical damping results in a non-oscillatory response and is the condition for equal roots, which occurs when $A_1^2 = 4A_2$.)

The transient response to an initial disturbance to a general second-order system of the type $(D^2 + 2\zeta\omega_0D + \omega_0^2)x = 0$ when x is any variable is plotted out in Fig. 3.32 for a range of values of damping ratio, ζ, and non-dimensional time $T = \omega_0 t$. These graphs can be used for any second-order system.

The steady-state values on application of a given tailplane (or elevator) angle can be obtained for both pitch rate and incidence by equating all the terms involving rates of change to zero. Thus in Eq. (3.61), $\ddot{q}$, $\dot{q}$ and $\dot{\eta}$ are all zero in the steady state. The suffix ss is used to denote steady-state values.

Whence $A_2q_{ss} = K\eta_{ss}$

$$q_{ss} = \frac{K}{A_2}\eta_{ss} \tag{3.64}$$

The steady-state change in wing incidence, α_{ss}, can be obtained in a similar manner from Eq. (3.60) and is given by

$$\alpha_{ss} = \frac{KT_2}{A_2}\eta_{ss} \tag{3.65}$$

It can be seen that changes in the aerodynamic derivatives over the flight envelope will alter both the speed and damping of the transient response to a tailplane movement or a disturbance. The steady-state wing incidence and pitch rate for a given tailplane (or elevator) movement will also change and hence the *Stick Force per g* will change over the flight envelope.

3.5.5　Worked Example on q/η Transfer Function and Pitch Response

The aerodynamic data below are very broadly representative of a conventional combat aircraft of transonic performance. These data have been derived by a 'reverse engineering' process from a knowledge of typical control 'gearings' in flight control systems (e.g. control surface movement/*g*) and actual aircraft short-period responses in terms of undamped natural frequency and damping ratio.

The objective of the worked example is to try to bring together the basic aerodynamic and aircraft control theory presented so far and show how this can be applied to give an appreciation of the aircraft's pitch response and also the need for auto-stabilisation.

The data and calculations are rounded off to match the accuracy of the estimates, and the aerodynamic behaviour has been greatly simplified and compressibility effects, etc., ignored.

In practice, the necessary aerodynamic derivative data would normally be available. The worked example below has been structured to bring out some of the aerodynamic relationships and principles.

Aircraft Data

Aircraft mass, $m = 25{,}000$ kg (55,000 lb)
Forward speed, $U = 250$ m/s (500 knots approx.)
Moment of inertia about pitch axis, $I_y = 6 \times 10^5$ kg m^2
Pitching moment due to incidence, $M_\alpha = 2 \times 10^7$ Nm/radian
Tailplane moment arm, $l_t = 8$ m
Wing incidence/*g*: 1.6° increase in wing incidence at 500 knots produces 1 *g* normal acceleration
Tailplane angle/*g*: 1° tailplane angular movement produces 1 *g* normal acceleration at 500 knots

Questions

(a) Derive an approximate *q/n* transfer function, assuming forward speed is constant.
(b) What is the undamped natural frequency and damping ratio of the aircraft's pitch response?

(c) What is the percentage overshoot of the transient response to a disturbance or control input?

$$\frac{q}{\eta} = \frac{K(1 + T_2 D)}{D^2 + A_1 D + A_2}$$

The problem is to derive K, T_2, A_1 and A_2 from the data given.

The undamped natural frequency, $\omega_0 = \sqrt{A_2}$, and damping ratio $\zeta = A_1/2\sqrt{A_2}$ can then be calculated knowing A_1 and A_2.

- *Derivation of T_2*
  Increase in wing lift is $Z_\alpha \alpha = m(ng)$
  where (ng) is normal acceleration
  1.6° change in wing incidence produces 1*g* normal acceleration at 500 knots
  Hence $Z_\alpha\ 1.6/60 = m\ 10$
  (Taking $1° \approx 1/60$ radian and $g \approx 10$ m/s^2)
  That is, $Z_\alpha = 375$ m
  $T_2 = mU/Z_\alpha = m\ 250/375$ m
  That is, $T_2 = 0.67$ s.

- *Relationship between M_q and M_η*
  Assuming an all moving tailplane
  Pitching moment due to tailplane angle change, $\eta = M_\eta \eta$
  Change in tailplane incidence due to pitch rate, $q = l_t\ q/U$
  Pitching moment due to pitch rate is $M_\eta l_t q/U = M_q q$
  Hence for an all moving tailplane, $M_q = l_t/U \cdot M_\eta$
  $l_t = 8$ m, $U = 250$ m/s
  $M_q = 8/250 \cdot M_\eta = 0.032 M_\eta$.

- *Derivation of M_η and M_q*
  1 g normal acceleration at 250 m/s forward speed corresponds to a rate of pitch equal to 10/250 rad/s, that is 2.4°/s approx. (taking $g = 10$ m/s^2).
  1.6° change in wing incidence produces 1*g* normal acceleration and requires 1° tailplane angular movement.
  Thus in the steady state (denoted by suffix *ss*),
  $M_\eta \eta_{ss} = M_\alpha \alpha_{ss} + M_q q_{ss}$
  Working in degrees, M_η
  $1 = (2 \times 10^7 \times 1.6) + (0.032 M_\eta \times 2.4)$
  Hence, $M_\eta = 3.5 \times 10^7$ Nm/rad and $M_q = 0.032 \times 3.5 \times 10^7$ Nm/rad/s, that is, $M_q = 1.1 \times 10^6$ Nm/rad/s.

- *Derivation of T_1*

$$T_1 = \frac{6 \times 10^5}{1.1 \times 10^6}$$

That is, $T_1 = 0.55$ s.

- *Derivation of coefficient A_1*

$$A_1 = \frac{1}{T_1} + \frac{1}{T_2} + \frac{1}{0.55} + \frac{1}{0.67}$$

That is, $A_1 = 2.3$.

- *Derivation of coefficient A_2*

$$A_2 = \frac{M_\alpha}{I_y} + \frac{1}{T_1 T_2} = \frac{2 \times 10^7}{6 \times 10^5} + \frac{1}{0.55 \times 0.67}$$

That is, $A_2 = 36$.

- *Derivation of K*

$$K = \frac{M_\eta}{I_y T_2} = \frac{3.5 \times 10^7}{6 \times 10^5 \times 0.67}$$

That is, $K = 87$.
Hence

$$\frac{q}{\eta} = \frac{87(1 + 0.67D)}{\left(D^2 + 2.3D + 36\right)}$$

By inspection $\omega_0^2 = 36$

- That is, $\omega_0 = 6$ rad/s (0.95 Hz)

$$\zeta = 2.3/(2 \times 6)$$

- That is, $\zeta = 0.2$ approx.

From examination of second-order system responses in Fig. 3.32.

Percentage overshoot of response to a disturbance or control input is just over 50%.

- *A candidate for auto-stabilisation*

These data will be used in a worked example on auto-stabilisation in Sect. 3.8.

3.6 Lateral Control

3.6.1 Aileron Control and Bank to Turn

The primary means of control of the aircraft in the lateral plane are the ailerons (see Fig. 3.15). These are moved differentially to increase the lift on one wing and reduce it on the other thereby creating a rolling moment so that the aircraft can be banked to turn.

Aircraft bank to turn so that a component of the wing lift can provide the essential centripetal force towards the centre of the turn in order to change the aircraft's flight path (see Fig. 3.33). The resulting centripetal acceleration is equal to (aircraft velocity) × (rate of turn), that is, $V_T \dot{\Psi}$. This centripetal acceleration causes an inertia force to be experienced in the reverse direction, that is, a centrifugal force.

In a steady banked turn with no sideslip the resultant force the pilot experiences is the vector sum of the gravitational force and the centrifugal force and is in the direction normal to the wings. The pilot thus experiences no lateral forces.

Banking to turn is thus fundamental to effective control and manoeuvring in the lateral plane because of the large lift forces that can be generated by the wings. The lift force can be up to nine times the aircraft weight in the case of a modern fighter aircraft. The resulting large centripetal component thus enables small radius, high g turns to be executed.

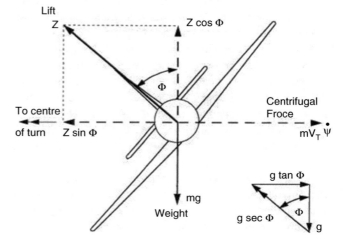

Fig. 3.33 Forces acting in a turn

Referring to Fig. 3.33: Horizontal component of the lift force is $Z \sin \Phi$. Equating this to the centrifugal acceleration gives

$$Z \sin \Phi = mV_T\dot{\Psi}$$

Vertical component of the lift force is $Z \cos \Phi$. Equating this to the aircraft weight gives

$$Z \cos \Phi = mg$$

from which

$$\tan \Phi = \frac{V_T\dot{\Psi}}{g} \qquad (3.66)$$

Thus the acceleration towards the centre of the turn is $g \tan \Phi$.

Referring to the inset vector diagram in Fig. 3.33, the normal acceleration component is thus equal to $g \sec \Phi$. Thus, a 60° banked turn produces a centripetal acceleration of 1.73 g and a normal acceleration of 2 g. At a forward speed of 100 m/s (200 knots approx.) the corresponding rate of turn would be 10.4°/s.

The lift required from the wings increases with the normal acceleration and the accompanying increase in drag requires additional engine thrust if the forward speed is to be maintained in the turn. The ability to execute a high g turn thus requires a high engine thrust/aircraft weight ratio.

To execute a coordinated turn with no sideslip requires the operation of all three sets of control surfaces, that is the ailerons and the tailplane (or elevator) and to a lesser extent the rudder. It is also necessary to operate the engine throttle (s) to control the engine thrust. The pilot first pushes the stick sideways to move the ailerons so that the aircraft rolls, the rate of roll being dependent on the stick movement. The rate of roll is arrested by centralising the stick when the desired bank angle for the rate of turn has been achieved. The pilot also pulls back gently on the stick to pitch the aircraft up to increase the wing incidence and hence the wing lift to stop loss of height and provide the necessary centripetal force to turn the aircraft. A gentle pressure is also applied to the rudder pedals as needed to counteract the yawing moment created by the differential drag of the ailerons. The throttles are also moved as necessary to increase the engine thrust to counteract the increase in drag resulting from the increase in lift and hence maintain airspeed.

The dynamics of the aircraft response to roll commands are complex and are covered later in Sect. 3.6.4. However, an initial appreciation can be gained by making the simplifying assumption of pure rolling motion, as follows.

Angular movement of the ailerons from the trim position, ξ, produces a rolling moment equal to $L_\xi\xi$ where L_ξ is the rolling moment derivative due to aileron angle movement. This is opposed by a rolling moment due to the rate of roll (refer to Sect. 3.4.4.2 and Fig. 3.23), which is equal to $L_p p$. The equation of motion is thus

$$L_\xi\xi + L_p p = I_x\dot{p} \qquad (3.67)$$

where I_x is moment of inertia of the aircraft about the roll axis.

This can be expressed in the form

$$(1 + T_R D)p = \frac{L_\xi}{L_p} \cdot \xi \qquad (3.68)$$

where T_R is the roll response time constant $= I_x / - L_p$ (note L_p is negative).

This is a classic first-order system, the transient solution being $p = Ae^{-t/\text{TR}}$, where A is a constant determined by the initial conditions.

For a step input of aileron angle, ξ_i, the response is given by

$$p = \frac{L_\xi}{L_p}\left(1 - e^{-1/T_R}\right)\xi \qquad (3.69)$$

This is illustrated in Fig. 3.34.

It can be seen that variations in the derivatives L_p and L_ξ over the flight envelope will affect both the speed of response and the steady-state rate of roll for a given aileron movement. It should be noted that aerodynamicists also refer to the roll response as a subsidence as this describes the exponential decay following a roll rate disturbance.

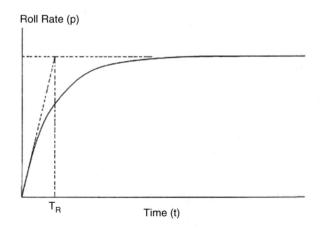

Fig. 3.34 Roll rate response

3.6.2 Rudder Control

Movement of the rudder creates both a lateral force and a yawing moment. The control action exerted by the rudder is thus used to:

(a) Counteract the yawing moment due to movement of the ailerons to bank the aircraft to turn as already explained.
(b) Counteract sideslipping motion.
(c) Counteract asymmetrical yawing moments resulting, say, from loss of engine power in the case of multi-engine aircraft or carrying asymmetrical stores/weapons in the case of a combat aircraft (or both).
(d) Deliberate execution of a sideslipping manoeuvre. The yawing moment created by the rudder movement will produce a sideslip incidence angle, β, and this will result in a side force from the fuselage and fin (see Fig. 3.21). This side force enables a flat sideslipping turn to be made. However, the sideways accelerations experienced in flat turns are not comfortable for the pilot (or crew and passengers for that matter). In any case, the amount of sideways lift that can be generated by the fin and fuselage body will only enable very wide flat turns to be made in general.
(e) To 'kick-off' the drift angle just prior to touch down when carrying out a crosswind landing. This is to arrest any sideways motion relative to the runway and so avoid side forces acting on the undercarriage at touch down.

Directional stability in the lateral plane is provided by the fin and rudder. Referring to Fig. 3.21, a sideslip velocity, v, with the aircraft's fore and aft axis at an incidence angle, β, to the aircraft velocity vector, V_T, (or relative wind) results in the fin developing a side force. As explained earlier, this side force tends to align the aircraft with the relative wind in a similar manner to a weathercock.

3.6.3 Short-Period Yawing Motion

This weathercock action can result in a lightly damped oscillatory motion under certain flight conditions because of the aircraft's inertia, if the yaw rate damping moment is small. The analysis of this motion can be simplified by assuming pure yawing motion and neglecting the cross-coupling of yawing motion with rolling motion (and vice versa). The direction of the aircraft's velocity vector is also assumed to be unchanged.

(It should be noted that the lateral response allowing for the effects of the roll/yaw cross-coupling is covered later in Sect. 3.6.4.) Referring to Fig. 3.21, it can be seen that:

Sideslip velocity, $v = V_T \sin\beta = V_T\beta$ (as β is a small angle)
$$= V_T\psi$$

as $\beta = \psi$, the incremental change in heading (or yaw) angle.
Thus yawing moment due to sideslip is $N_v V_T \psi$.

It should be noted that in general the yaw angle ψ is not the same as the sideslip incidence angle β, the yaw angle being the angle in the lateral plane between the aircraft's fore and aft axis and the original direction of motion. The two angles are only the same when the direction of the aircraft's velocity vector is unchanged. The moments acting about the CG are as follows:

1. Yawing moment due to sideslip is $N_v V_T \psi$.
2. Yawing moment due to yaw rate is $N_r r = N_r \dot{\Psi}$.

The equation of motion is as follows:

$$N_v V_T \Psi + N_r \dot{\Psi} = I_z \ddot{\Psi} \qquad (3.70)$$

where I_z is moment of inertia about yaw axis.
That is,

$$\left(D^2 + \frac{-N_r}{I_z}D + \frac{-N}{I_z}V_T \right)\Psi = 0 \qquad (3.71)$$

Note that derivatives N_r and N_v are both negative.
This is a second-order system with an undamped natural frequency

$$\omega_0 = \sqrt{\frac{N_v V_T}{I_z}}$$

and damping ratio

$$\zeta = \frac{1}{2}\frac{N_r}{\sqrt{N_v V_T I_z}}$$

Both the period of the oscillation $(2\pi/\omega_0)$ and the damping ratio are inversely proportional to the square root of the forward speed, that is, $\sqrt{V_T}$. Thus increasing the speed shortens the period but also reduces the damping. The period of the oscillation is of the order of 3 to 10 seconds.

This short-period yawing motion can be effectively damped out automatically by a yaw axis stability augmentation system (SAS) (or yaw damper) and will be covered in Sect. 3.7.

In practice there can be a significant cross-coupling between yawing and rolling motion and the yawing oscillatory motion can induce a corresponding oscillatory motion about the roll axis 90° out of phase known as '*Dutch roll*'. (Reputedly after the rolling–yawing gait of a drunken Dutchman.) The ratio of the amplitude of the rolling motion to that of the yawing motion is known as the '*Dutch roll ratio*' and can exceed unity, particularly with aircraft with large wing sweepback angles.

3.6.4 Combined Roll–Yaw–Sideslip Motion

The cross-coupling and interactions between rolling, yawing and sideslip motion and the resulting moments and forces have already been described in Sect. 3.4.4.2. The interactions between rudder and aileron controls have also been mentioned.

The equations of motion in the lateral plane following small disturbances from straight and level flight are set out in Sect. 3.4.5, Eqs. (3.37), (3.38) and (3.39), respectively.

The solution to this simplified lateral set of equations in terms of the transient response to a disturbance and resulting stability is a fifth-order differential equation. This can usually be factorised into two complex roots and three real roots as shown below.

$$\left(D^2 + 2\zeta\omega_0 D + \omega_0^2\right)\left(D + \frac{1}{T_R}\right)\left(D + \frac{1}{T_Y}\right)\left(D + \frac{1}{T_{Sp}}\right)x$$
$$= 0$$

where x denotes any of the variables v, p, r.

The transient solution is of the form

$$x = A_1 e^{-\zeta\omega_0 t}\sin\left(\sqrt{1 - \zeta^2}\cdot\omega_0 t + \phi\right) + A_2 e^{-t/T_R}$$
$$+ A_3 e^{-t/T_Y} + A_4 e^{-t/T_{Sp}}$$

where A_1, ϕ, A_2, A_3 and A_4 are constants determined by the initial conditions.

The quadratic term describes the Dutch roll motion with undamped natural frequency ω_0 and damping ratio ζ. T_R describes the roll motion subsidence and T_Y the yaw motion subsidence.

T_{Sp} is usually negative and is the time constant of a slow spiral divergence. It is a measure of the speed with which the aircraft, if perturbed, would roll and yaw into a spiral dive because of the rolling moment created by the rate of yaw. The time constant of this spiral divergence is of the order of 0.5–1 minute and is easily corrected by the pilot (or autopilot).

3.7 Powered Flying Controls

3.7.1 Introduction

It is appropriate at this point to introduce the subject of mechanically signalled powered flying controls as this is relevant to Sect. 3.8 'Stability Augmentation Systems' and also to Chap. 4.

The forces needed to move the control surfaces of a jet aircraft become very large at high speeds and are too high to be exerted manually. The control surfaces are thus fully power operated with no mechanical reversion; the actuators must also be irreversible to avoid flutter problems.

The hydraulic servo actuation systems which move the main control surfaces are known as *Power Control Units* (PCUs). The PCU is also an essential part of an FBW control system; there are, however, a number of modifications required to adapt the PCU for use in an FBW system as will be explained in the next chapter.

Figure 3.35 shows the mechanically signalled flying control system of the BAE Systems Hawk aircraft to give an appreciation of a typical system.

Figure 3.36 is a schematic illustration of the operation of a PCU. Movement of the control column (or rudder bar) is transmitted through mechanical rods and linkages to the PCU servo control valve. The resulting displacement of the control valve allows high-pressure hydraulic fluid to act on one side of the actuator ram whilst the other side of the ram is connected to the exhaust (or tank) pressure of the hydraulic supply.

The ram moves and this movement is fed back through a mechanical linkage so as to close the valve. The actuator ram thus follows the control column movement with little lag – a simple follow-up type of servo mechanism. A more detailed explanation together with the derivation of the PCU transfer function is given in the next subsection.

Typical hydraulic pressure for a PCU is 200 bar or 20 MPa (3000 psi), so that the actuator can exert a force of several tonnes, depending on the ram cross-sectional area. The forces required to move the PCU valve are small – of the order of a kilogram or less. The pilot, however, can exert a force of 50 kg on the servo control valve, if necessary, to overcome a jammed valve. Although this is a very unlikely event, the airworthiness authorities cite the case of a piece of locking wire getting past the oil filter and jamming the control valve. A force of 50 kg is deemed sufficient to shear through any such obstruction; hence, the relatively hefty yokes, or control columns, and the mechanical advantage built into the linkage/control rod system to enable the pilot to exert such forces.

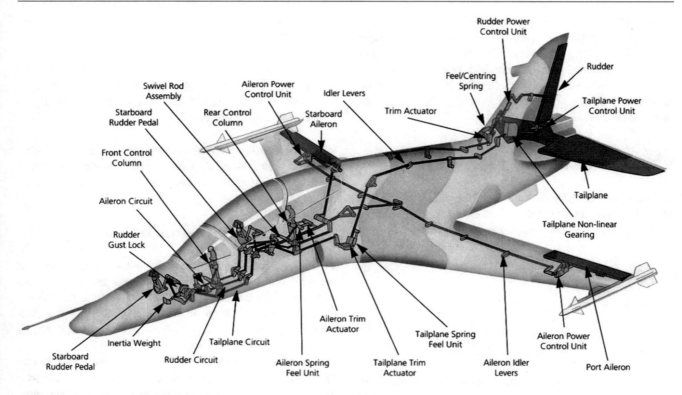

Fig. 3.35 Mechanically signalled flight control system. (Courtesy of BAE Systems)

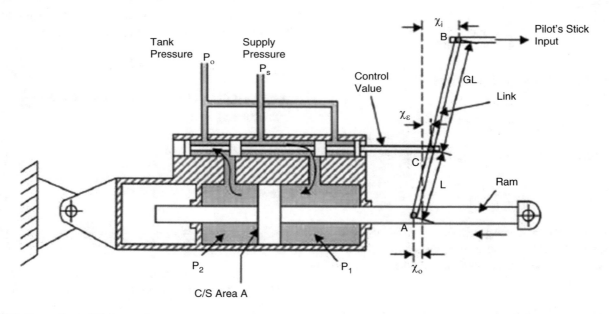

Fig. 3.36 Power Control Unit operation

The PCU uses a tandem arrangement of actuators and servo control valves operated from two totally independent hydraulic supplies to meet the safety and integrity requirements. Mechanically signalled PCUs are extremely reliable systems which have accumulated a vast operating experience, as they are universally installed in modern military and civil jet aircraft.

3.7.2 PCU Transfer Functions

Referring to Fig. 3.36, movement of the pilot's control column, x_i, causes the link AB to pivot about A.

This produces a valve movement which is equal to

$$\frac{l}{Gl + l} \cdot x_i$$

That is

$$\frac{1}{G + 1} \cdot x_i$$

Once the spool valve ports have opened, flow passes to the ram, which starts to move in the opposite direction. This causes the link to pivot about B and produces a valve movement equal to

$$\frac{Gl}{Gl + l} \cdot x_0$$

where x_0 is the actuator displacement, that is

$$\frac{G}{G + 1} \cdot x_0$$

Hence, the net valve opening, x_ε, is given by

$$x_\varepsilon = \frac{1}{G + l}(x_i - Gx_0) \tag{3.72}$$

(G is the stick-to-actuator gearing).

The relationship between ram velocity, $\dot{x}_0$, and valve opening, x_ε, can be derived as follows, by making the following assumptions:

(i) The spool valve has square ports of equal area so that the uncovered valve port area, a, is directly proportional to the spool valve displacement, x_ε, from the closed position, that is $a = K_1 x_\varepsilon$, where K_1 is a constant which is dependent on the valve dimensions.
(ii) The hydraulic fluid is assumed to be incompressible.
(iii) The external forces acting on the ram are small, that is the ram is unloaded, and control surface inertia can be

ignored. (The load inertia torques are generally fairly small in comparison with the actuator torque, which can be exerted.)
(iv) The exhaust (or tank) pressure, P_0, of the hydraulic supply is normally atmospheric pressure, which is small in comparison with the supply pressure, P_s, so that P_0 can be neglected.

It can be shown by Bernoulli's equation that the velocity, V, of the fluid flowing through an orifice is directly proportional to the square root of the pressure difference across the orifice, that is,

$$V = K_2\sqrt{\text{Pressure difference}}$$

where K_2 is a constant which is dependent on the density of the fluid and which also takes into account losses at the orifice (Discharge Coefficient). Flow through the orifice, $Q = aV$. Hence, flow to actuator is $K_1 K_2 x_\varepsilon \sqrt{P_s - P_1}$. Flow from actuator to exhaust is $K_1 K_2 x_\varepsilon \sqrt{P_2 - P_0} = K_1 K_2 x_\varepsilon \sqrt{P_2}$ as P_0 is assumed to be zero.

The flows into and out of the actuator are equal so that $P_2 = (P_s - P_1)$, and, neglecting external loads and the load inertia, $P_1 A = P_2 A$, where A is the ram cross-sectional area. Hence, $P_1 = P_2 = P_s/2$.

Flow through the valve orifices, Q, is equal to the displacement flow created by the ram movement, that is $Q = A\dot{x}_0$. Whence $A\dot{x}_0 = K_1 K_2\sqrt{P_s/2} \cdot x_\varepsilon$, that is,

$$\dot{x}_0 = K_v x_\varepsilon \tag{3.73}$$

where K_v is the actuator velocity constant $K_1 K_2 \frac{\sqrt{P_s}}{A\sqrt{2}}$.

Substituting x_ε from Eq. (3.72) yields

$$\dot{x}_0 = K_v \frac{1}{G + 1}(x_i - Gx_0) \tag{3.74}$$

from which

$$\frac{x_0}{x_i} = \frac{1}{G} \cdot \frac{1}{(1 + T_{Act}D)} \tag{3.75}$$

where T_{Act} is the actuator time constant $= (G + 1)/GK_v$.

Hence, the actuator transfer function is a simple first-order lag.

Typical values for a PCU time constant are around 0.1 s. This corresponds to a bandwidth (−3 dB down) of 1.6 Hz approx. (10 rads/s).

The phase lag at 1 Hz is often used as a measure of the actuator performance and would be 32° in the above case.

3.8 Stability Augmentation Systems

3.8.1 Limited Authority Stability Augmentation Systems

The possible need for improved damping and stability about all three axes has been referred to in the preceding sections. This can be achieved by an auto-stabilisation system, or, as it is sometimes referred to, a stability augmentation system.

Yaw stability augmentation systems are required in most jet aircraft to suppress the lightly damped short-period yawing motion and the accompanying oscillatory roll motion due to yaw/roll cross-coupling known as Dutch roll motion, which can occur over parts of the flight envelope (refer to Sect. 3.6.4). In the case of military aircraft, the yaw damper system may be essential to give a steady weapon aiming platform as the pilot is generally unable to control the short-period yawing motion and can in fact get out of phase and make the situation worse.

A yaw damper system is an essential system in most civil jet aircraft as the undamped short-period motion could cause considerable passenger discomfort.

As mentioned earlier, a yaw damper system may be insufficient with some aircraft with large wing sweepback to suppress the effects of the yaw/roll cross-coupling and a roll damper (or roll stability augmentation) system may also be necessary. The possible low damping of the short-period pitch response (refer to Sect. 3.5) can also require the installation of a pitch damper (or pitch stability augmentation) system. Hence, three-axis stability augmentation systems are installed in most high-performance military jet aircraft and very many civil jet aircraft.

A single channel, limited authority stability augmentation system is used in many aircraft, which have an acceptable though degraded and somewhat lightly damped response over parts of the flight envelope without stability augmentation. 'Single channel' means that there is no back-up or parallel system to give a failure survival capability.

The degree of control, or authority, exerted by the stability augmenter is limited so that in the event of a failure in the stability augmentation system, the pilot can over-ride and disengage it. Typical authority of the stability augmentation system is limited to a control surface movement corresponding to ± 0.5 g.

A simple stability augmentation system is shown schematically in Fig. 3.37. It comprises a rate gyroscope which senses the angular rate of rotation of the aircraft about the auto-stabilised axis and which is coupled to an electronic unit which controls the stability augmentation actuator. The stability augmentation actuator controls the main servo actuator (driving the control surface) in parallel with the pilot's control by means of a differential mechanism – typically a lever linkage differential.

The control surface angular movement is thus the sum of the pilot's stick input and the stability augmentation output. The stability augmentation system produces a control surface movement which is proportional to the aircraft's rate of rotation about the axis being stabilised and hence a damping moment proportional to the angular rate of rotation of the aircraft about that axis. In effect a synthetic $N_r r$ or $L_p p$ or $M_q q$ term. In practice, the stability augmentation damping term is more complicated than a simple (constant) $\times$ (aircraft rate of rotation) term because of factors such as stability augmentation servo performance and the need to avoid exciting structural resonances, that is, 'tail wagging the dog' effect. This latter effect is due to the force exerted by the control surface actuator reacting back on the aircraft structure, which has a finite structural stiffness.

Fig. 3.37 Simple stability augmentation system

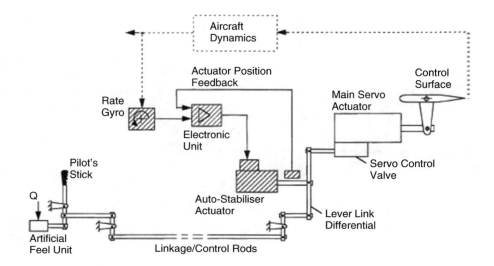

The resulting bending of the aircraft structure may be sensed by the stability augmentation rate gyro sensor and fed back to the actuator and so could excite a structural mode of oscillation, if the gain at the structural mode frequencies was sufficiently high. In the case of stability augmentation systems with a relatively low-stability augmentation gearing, it is generally sufficient to attenuate the high-frequency response of the stability augmentation with a high-frequency cut-off filter to avoid exciting the structural modes. The term 'stability augmentation gearing' is defined as the control surface angle/degree per second rate of rotation of the aircraft.

With higher-stability augmentation gearings, it may be necessary to incorporate '*notch filters*', which provide high attenuation at the specific body mode flexural frequencies. (This is briefly covered again in Chap. 4.)

Most simple, limited authority yaw dampers feed the angular rate sensor signal to the stability augmentation actuator through a 'band pass filter' with a high-frequency cut-off (see Fig. 3.38). The band pass filter attenuation increases as the frequency decreases and becomes infinite at zero frequency (i.e. DC) and so prevents the stability augmentation actuator from opposing the pilot's rudder commands during a steady rate of turn. (This characteristic is often referred to as 'DC blocking'.) The yawing motion frequency is in the centre of the band pass range of frequencies. The angular rate sensor output signal resulting from the short-period yawing motion thus passes through the filter without

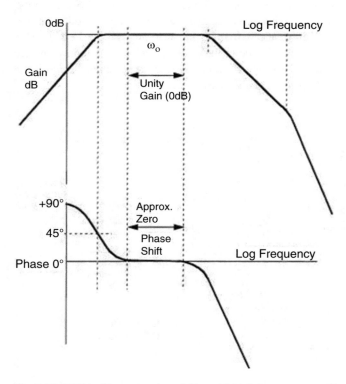

Fig. 3.38 DC blocking or 'wash-out' filter with high-frequency cut-off

attenuation or phase shift and so provides the required signal to the stability augmentation actuator to damp the oscillatory yawing motion. (It should be noted that this type of filter in flight control systems is often referred to as a 'wash-out filter', particularly in the USA, as it effectively 'washes out' the DC response.)

The variation in the aircraft response and control effectiveness over the flight envelope may require the stability augmentation gearings to be switched as a function of height and airspeed to possibly two or three different values.

3.8.1.1 'Worked Example' of Simple Pitch Stability Augmentation

An example of a typical representative simple pitch stability augmentation is given below to show how initial calculations can be made to obtain an order of magnitude estimate of the stability augmentation 'gearing' required using the transfer functions derived in Sect. 3.5 (stability augmentation 'gearing' being the tailplane angle in degrees per degree/s rate of pitch). It should be pointed out that much more complex computer modelling and simulation is necessary to establish the optimum values and allow for a more complex transfer function. For example:

(a)   Certain derivatives have been ignored.
(b)   Lags in the actuator response have been neglected together with any non-linear behaviour which might be present.
(c)   The transfer functions of any filters required to attenuate rate gyro noise and avoid excitation of structural resonances have been omitted.

Referring to Sect. 3.5.4, pitch rate, q, to tailplane angle, η, transfer function is

$$\frac{q}{\eta} = \frac{K(1 + T_2 D)}{\left(D^2 + 2\zeta\omega_0 D + {\omega_0}^2\right)}$$

where $T_2 = mU/Z_\alpha$ and $K = M_\eta/I_y T_2$.

Referring to Fig. 3.39 (ignoring lags in auto-stab actuator, main actuator, gyro and filters, etc.):

Stability augmentation output is $G_q q$

$$\text{Tailplane angle}, \eta = \delta_i - G_q q \qquad (3.76)$$

where G_q is stability augmentation gearing and δ_i is the pilot's stick input. Whence

$$q = \frac{K(1 + T_2 D)}{\left(D^2 + 2\zeta\omega_0 D + \omega_0^2\right)} \cdot \left(\delta_i - G_q q\right)$$

from which

Fig. 3.39 Pitch stability augmentation loop

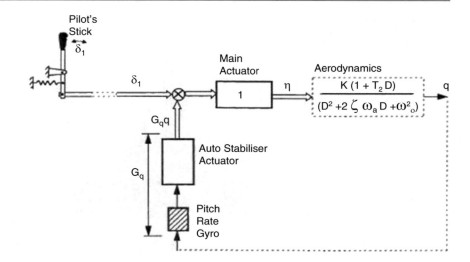

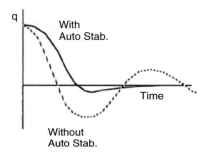

Fig. 3.40 Pitch response with and without stability augmentation

$$\left[D^2 + \left(2\zeta\omega_0 + KT_2G_q\right)D + \left(\omega_0{}^2 + KG_q\right)\right]q$$
$$= K(1 + T_2D)\delta_i$$

The worked example in Sect. 3.5.2 provides the data for the basic aircraft response with $K = 87$, $T_2 = 0.67$, $\omega_0 = 6$ rad/s, $\zeta = 0.2$.

The required value of G_q to improve the damping of the aircraft's pitch response to correspond to a damping ratio of 0.7 is derived below.

Let $\omega_{0\ stab}$ and ζ_{stab} denote the undamped natural frequency and damping ratio of the quadratic factor in the transfer function q/δ_i of the aircraft with pitch stability augmentation.

Hence

$$\omega_{0\ stab}^2 = \omega_0^2 + KG_q$$
$$2\zeta_{stab}\omega_{0\ stab} = 2\zeta\omega_0 + KT_2G_q$$
$$\zeta_{stab} = 0.7$$

$\therefore 2 \times 0.7 \times \omega_{0\ stab} = (2 \times 0.2 \times 6) + (87 \times 0.67 \times G_q)$, whence $\omega_{0\ stab} = 1.71 + 41.6G_q$, viz. $1734G_q^2 + 55G_q - 33 = 0$, from which $G_q = 0.12°$ tailplane/degree per second

rate of pitch. Whence $\zeta_{stab} = 0.7$ and $\omega_{0\ stab} = 6.8$ rad/s (1.1 Hz) approx. The q/δ_i transfer function with stability augmentation is thus

$$\frac{q}{\delta_i} = \frac{87(1 + 0.67D)}{D^2 + 9.5D + 46}$$

The improvement in damping is shown in Fig. 3.39.

It should be noted that the loop gain is only 0.29 – much lower than would be used in an FBW pitch rate command loop (Fig. 3.40).

3.8.2 Full Authority Stability Augmentation Systems

In many cases, aerodynamic performance and manoeuvrability do not always match with stability and damping over the whole of the flight envelope and can lead to restricting compromises. There may thus be a need for full authority stability augmentation over part of the flight envelope (as opposed to the limited authority systems just described). Because the system has full authority, a failure could be catastrophic and so a failure survival system of triplex level redundancy (or higher) is essential.

Triplex level redundancy means three totally independent parallel systems of sensors, electronic units (or controllers) and actuation units with independent electrical and hydraulic supplies.

Basically the triplex system operates on a majority voting of the 'odd man out' principle. Because the three parallel systems or 'channels' are totally independent, the probability of two systems failing together at the same instant in time is negligibly low (of the order typically of 10^{-6}/hour failure probability). A failure in a channel is thus detected by cross-comparison with the two 'good' channels and the failed

Fig. 3.41 Lift generation: fixed-wing aircraft compared with a helicopter

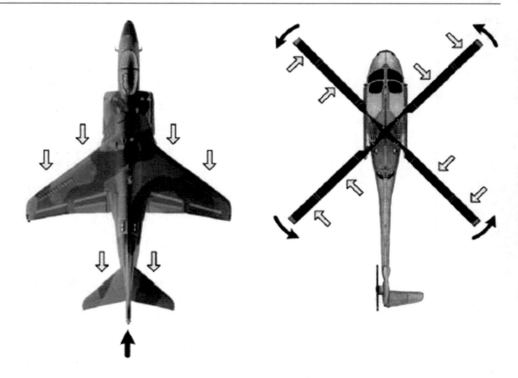

channel can be disconnected leaving the system at duplex level redundancy. Failure survival redundant configurations are discussed in some detail in Chap. 4, and at this stage it is sufficient to appreciate the basic triplex concept.

A second failure results in a stalemate situation with the failed stability augmentation actuator being counteracted by the 'good' channel. The result is a passive failure situation and the control surface does not move.

The operational philosophy with such a system after the first failure would generally be to minimise the time at risk at duplex level by, say, decelerating and possibly descending until a speed and height condition is reached where stability augmentation is not essential.

The relatively high stability augmentation gearings that may be required together with the failure survival system requirements lead to what is known as a 'manoeuvre command fly-by-wire' flight control system. This is discussed in Chap. 4.

3.9 Helicopter Flight Control

3.9.1 Introduction

The aim of this section is to explain the basic principles of helicopter flight control including stability augmentation. This provides the lead-in to helicopter fly-by-wire control systems, which are covered in Chap. 4.

Like fixed-wing aircraft, helicopters depend on the aerodynamic lift forces resulting from the flow of air over the wing surfaces, in this case, the rotor blades. The airflow over the wings of a fixed-wing aircraft is a result of the forward speed of the aircraft as it is propelled by the thrust exerted by the jet engines, or propellers.

In the case of the helicopter, however, the airflow over the wings (rotor blades) is caused by the rotation of the whole rotor system; thus, the aircraft need not move.

Figure 3.41 illustrates the fundamental difference between fixed-wing aircraft and rotary wing aircraft (helicopters) in the generation of aerodynamic lift.

This gives the helicopter the major advantages of being able to take-off and land vertically, move forward or backwards, move sideways, and the ability to hover over a fixed point.

These capabilities are widely exploited in both civil and military helicopter usages. For example, civil applications such as air ambulances, police surveillance, air–sea rescue, and transporting passengers to and from locations with restricted space.

Military applications include operations by special forces in inaccessible terrain, transport of troops and equipment where roads are poor, possibly mined and subject to ambush, evacuation of the wounded, and ground attack (helicopter gunships). Naval applications include anti-submarine operations, airborne early warning, shipping strikes using air-launched guided missiles, and ship-to-ship or ship-to-shore transport of men and equipment.

There are a number of basic rotor configurations which are used in helicopters. These are described briefly below.

Fig. 3.42 Single main rotor configuration – AgustaWestland AW 159 'Lynx Wildcat' helicopter. (Courtesy of AgustaWestland)

Single Main Rotor This is the most common and widely used rotor configuration. An essential feature of this configuration is a means of applying a torque about the yaw axis to counteract the reaction torque acting on the fuselage to the turning of the main rotor. This is usually achieved with a tail rotor driven from the engine(s), the yawing moment being controlled by changing the pitch of the tail rotor blades. An alternative method is the 'NOTAR'[1] system, which uses the reaction torque from a jet of air supplied from the engine compressor.

The rotor can comprise two blades, three blades, four blades or even five blades depending on the size and roles of the particular helicopter.

Figure 3.42 shows an example of a single main rotor helicopter configuration and is an illustration of the new AgustaWestland AW 159 'Lynx Wildcat' army/naval helicopter, which made its first flight in November 2009 at Yeovil, Somerset. This new helicopter is a development of the well-proven Westland WG 13 'Lynx' helicopter, which first entered service with the Royal Navy and the British Army over 30 years ago and has since been exported worldwide.

Co-axial Rotor Configuration with Contra-Rotating Rotors This configuration eliminates the rotor reaction torque problem and dispenses with the need for a tail rotor. The new Sikorsky S 97 'Raider' light assault helicopter, currently under construction, exploits the co-axial contra-rotating rotor configuration together with a pusher propeller providing forward thrust. The basic configuration is shown in Fig. 3.43. This configuration is used in the earlier Sikorsky 'X-2 Technology Demonstrator', which has demonstrated a level speed of over 250 knots.

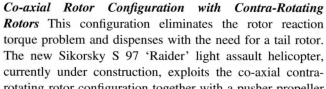

Fig. 3.43 Co-axial contra-rotating rotors configuration – Sikorsky S 97 helicopter

Yaw control is achieved by increasing the collective pitch of one rotor and reducing it on the other rotor to produce a torque about the yaw axis.

Contra-rotating rotors reduce the effects of dissymmetry of lift resulting from the forward motion of the helicopter, which increases the lift force on the advancing rotor blades and at the same time decreases the lift on the retreating rotor blades. The increase, or decrease, in lift is at a maximum when the rotor blade is at 90 to the direction of forward motion. The effects are reduced because the two rotors are turning in opposite directions causing the blades to advance on either side at the same time.

Other benefits include increased payload for the same engine power as all the engine power is available for lift and thrust, whereas some of the engine power is required to drive the anti-torque rotor on a single rotor design. The configuration can also be more compact overall. The principal disadvantage is the increased mechanical complexity.

Contra-Rotating Tandem Rotor Configuration Another configuration which eliminates the rotor reaction torque problem is the contra-rotating tandem rotor configuration. The well-proven Boeing Chinook helicopter illustrated in Fig. 3.44 is a good example of this configuration.

Yaw control is achieved by differential main rotor torque.

---

[1]NOTAR is an acronym for 'No Tail Rotor'.

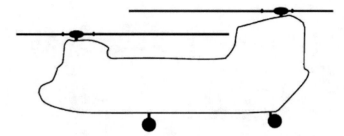

Fig. 3.44 Tandem rotor configuration – Boeing Chinook helicopter

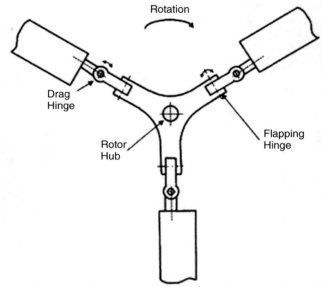

Fig. 3.46 Drag-and-flapping hinges on an articulated rotor

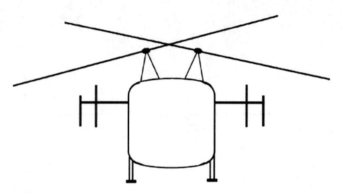

Fig. 3.45 Intermeshing rotor configuration – Kaman helicopter

Intermeshing Contra-Rotating Rotor Configuration This configuration also has zero net rotor reaction torque. Figure 3.45 shows the intermeshing rotor system developed by the Kaman helicopter company in the USA. The new Kaman K-Max helicopter exploits this configuration and features an impressive lifting capability. Yaw control is achieved by differential rotor torque.

The methods adopted for attaching the rotor blades to the rotor head have a direct influence on the dynamic response of the helicopter and can be divided into three basic categories, comprising:

* Articulated rotors
* Semi-rigid rotors
* Rigid rotors

Articulated Rotor Head Fig. 3.46 illustrates the drag-and-flapping hinge design of the rotor head.

The flapping hinges give the rotor blades limited freedom about the flapping hinges to move up or down as the lift increases on the advancing rotor blade and decreases on the retreating rotor blade.

It should be appreciated that there are large centrifugal forces exerted on the rotor blades due to their rotational velocity. These centrifugal forces oppose the deflection of the rotor blades from the plane of rotation normal to the rotor drive shaft.

The rotor blades thus deflect from this normal plane until the moment about the flapping hinge due to the rotor lift force is balanced by the moment due to the centrifugal force component normal to the rotor blade.

The drag hinges, also known as 'lead-lag' hinges, allow the rotor blades to move freely backward and forward by a limited amount in the plane of rotation relative to the other blades (or the rotor head). The flapping-and-drag hinges enable a lighter mechanical rotor head to be used; the down side is the increased mechanical complexity and wear and maintenance of the hinge bearings.

Semi-Rigid Rotors The semi-rigid rotor is illustrated schematically in Fig. 3.47, and is an integral rotor unit, without blade articulation.

There are no flapping or drag hinges but it has, however, limited flexibility in the flapping-and-drag planes. As in all helicopters, the flexibility of the long, relatively thin rotor blades enables them to bend up or down under the changing lift forces.

This configuration eliminates the mechanical complexity of flapping-and-drag hinges and reduces the bending moments on the rotor shaft. It also has a much faster pitch attitude response and enables high manoeuvrability to be achieved (in conjunction with the command/stability augmentation system). There are many examples of helicopters in service with semi-rigid rotors, for example the 'Lynx' helicopter.

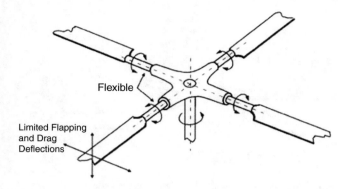

Flexible

Limited Flapping
and Drag
Deflections

Fig. 3.47 Semi-rigid rotor

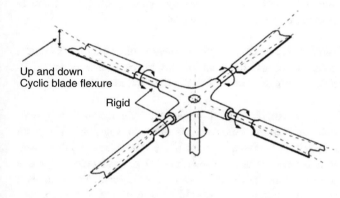

Up and down
Cyclic blade flexure

Rigid

Fig. 3.48 Rigid rotor

Fig. 3.49 Schematic diagram of
the swash plate mechanism

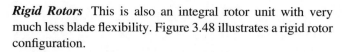

Rigid Rotors This is also an integral rotor unit with very much less blade flexibility. Figure 3.48 illustrates a rigid rotor configuration.

3.9.2 Control of the Helicopter in Flight

Control of the flight path of a helicopter is achieved by controlling the angle of incidence of the individual rotor blades to the airstream, that is, their *pitch angle*. This is achieved by means of a swash plate mechanism.

The basic swash plate mechanism is illustrated schematically in Fig. 3.49.

The lower half of the swash plate mechanism comprises a co-axial collar which can move axially up or down with respect to the rotor head. The co-axial collar, in turn, incorporates a spherical bearing which enables the lower swash plate to tilt about the centre of rotation of this bearing. The lower swash plate can thus move axially up or down and can also be independently tilted at an angle to the helicopter fore and aft axis, or the helicopter lateral axis. Actuators control the up or down axial movement and the fore and aft, and lateral, tilt of the lower swash plate. The upper swash plate, which is coupled to the rotor shaft by a scissor linkage, in turn controls the individual rotor blade pitch angles through the pitch linkage arms.

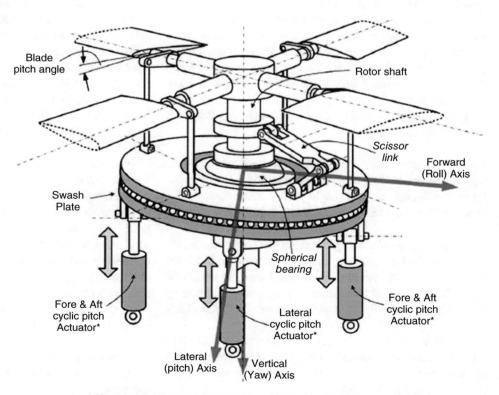

** Note that the three cyclic actuatros all move up and down
together for collective control and differentially for cyclic control*

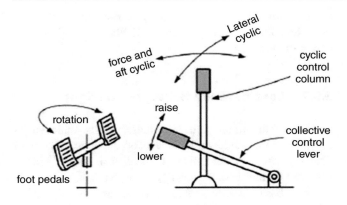

Fig. 3.50 Pilot's controls

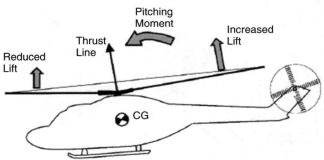

Fig. 3.51 Forces and moments acting on the helicopter

Pilot's Controls   Fig. 3.50 illustrates the layout of the pilot's controls.

The pilot has a *collective* control lever, usually located on the left side, to control the vertical motion of the helicopter by changing the rotor collective pitch angle.

There is a *cyclic* control column to control the fore and aft and lateral motion of the helicopter by changing the cyclic pitch angles of the rotor blades.

Collective and cyclic pitch control will be explained in the next sections.

There is also a *foot pedal* control for the pilot to control the pitch angle of the variable pitch tail rotor and hence the yawing moment (and side force) it exerts.

It should be noted that the rotor speed is controlled automatically, in modern helicopters, and is maintained at a sensibly constant value in a particular flight mode. This is governed by the limits set by the advancing rotor blade tip velocity not becoming supersonic and the retreating rotor blades not stalling.

The rotor lift is thus mainly determined by the rotor collective pitch angle.

Collective Pitch Control   Operating the pitch control lever raises or lowers the collar of the lower part of the swash plate thereby changing the pitch angle of each rotor blade by the same amount, simultaneously. Refer to Fig. 3.49.

Increasing the collective pitch angle increases the rotor lift and the helicopter climbs; similarly, decreasing the rotor pitch angle reduces the rotor lift so the helicopter descends. The helicopter can be maintained in a hovering condition by adjusting the rotor lift to exactly balance the helicopter's weight.

It is, however, also necessary to adjust the tail rotor pitch angle to counter the change in the rotor reaction torque acting on the helicopter about the yaw axis, when the rotor collective pitch angle is changed.

Cyclic Pitch Control   Forward flight (and also rearward or sideways flight) is achieved by cyclic pitch control and is explained in the next subsections.

Pitch Axis Control   Forward movement of the pilot's control column causes the front of the lower swash plate to tilt downwards with respect to the fore and aft axis.

The rotor blade pitch angles thus change cyclically over each revolution of the rotor, as shown in Fig. 3.49. The pitch angle of each advancing rotor blade is progressively decreased as it approaches the fore and aft position, so that the lift force is reduced at the front of the rotor disc. At the same time, the pitch angle of each retreating rotor blade is increased as it approaches the rear of the fore and aft axis thereby increasing the lift force on the rear half of the rotor disc.

The rotor disc is thus deflected about the helicopter pitch axis. In simple terms the deflected rotor disc results in both a horizontal fore and aft force and a pitching moment. The resulting attitude response is very much influenced by the rotor design, which can vary between very rigid and very flexible, as explained earlier.

The more rigid the rotor attachment, the more moment is imparted to the aircraft, so this results in a rapid change in pitch attitude.

Figure 3.51 illustrates the forces and moments acting on the helicopter.

The longitudinal force is dominant in the case of the more flexible rotor attachment, and this is applied to the rotor mast head. The attitude change develops as the helicopter accelerates forward to a constant speed when the forces and moments at the mast head cancel out, as shown in Fig. 3.52.

In general, at a given forward velocity there will be a corresponding pitch angle, that is, the greater the speed, the more the nose-down attitude. In equilibrium, the forward component of rotor force is balanced by the aerodynamic drag force.

It should be noted that the pitch control of a helicopter is more complex than that of a fixed-wing aircraft, where the short-period dynamics can be represented as a simple second-order system, as explained earlier in this chapter.

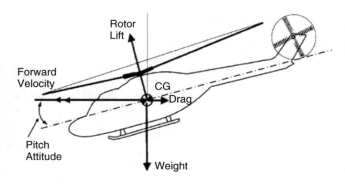

Fig. 3.52 Steady-state forward motion

In contrast, the pitch axis control depends on the rotor dynamics, the helicopter rigid body dynamics and the resultant effect of the airspeed. The pitch axis motion of a helicopter has been described as the equivalent to a fixed-wing aircraft with an unstable short-period phugoid.

Roll axis This is similar to the pitch axis (at least at low speed around the hover) in that roll stick deflection results in a lateral tilt of the rotor disc, followed by a bank angle and a lateral acceleration. At higher speed, the bank angle results in a side force and hence (ideally) a coordinated turn, although this involves appropriate control of the tail rotor and the collective axes of operation.

Yaw Axis The pilot's pedals provide a yawing moment (also a side force) via the variable pitch tail rotor, so that the helicopter rotates about a vertical axis. As the airspeed increases, the natural weathercock stability of the airframe comes into effect, and the function of the pedals and the tail rotor is to minimise sideslip, hence helping turn coordination during banked turns.

3.9.3 Stability Augmentation

Controlling the flight path of a basic helicopter (without stability augmentation) can present the pilot with a relatively high workload in certain flight conditions. It can involve the simultaneous coordination of the following:

Fore and aft and lateral motion of the helicopter through the control column cyclic pitch control system

Lift control through the collective pitch control lever

Control of yawing motion through the foot pedals

The pitch attitude response can have neutral stability, or even be divergently unstable without the intervention of the pilot in the loop. A basic Stability Augmentation System (SAS) may thus be required to give the helicopter good control and handling characteristics. The helicopter is also basically asymmetric and this requires cross-axis coupling in

various areas to suppress these effects; for instance, collective to yaw and collective to pitch.

Less sophisticated helicopters have a single channel, or simplex, limited authority three-axis stability augmentation system providing pitch, roll, yaw rate damping about the respective pitch, roll, yaw axes. The system is basically similar to the fixed-wing stability augmentation system shown in Fig. 3.37; the stability augmentation actuators in this case coupling into the fore and aft, and lateral, cyclic pitch and tail rotor pitch main actuators (or Power Control Units [PCUs] as they are generally called). Washed-out, or DC-blocked, pitch attitude and roll attitude signals are frequently included in the pitch and roll channels.

The actuator authority, in the case of a simplex system, is set at a value which limits the magnitude of the disengagement transient to a safe acceptable level, following a failure in the stability augmentation system.

This inevitably limits their capability, and more sophisticated systems which require larger actuator authority employ fault tolerant system architectures.

Figure 3.53 (**a**, **b**, **c**) illustrates the various fault tolerant SAS architectures comprising duplex, dual–dual and triplex systems.

A duplex system is shown in Fig. 3.53a. All axes have two entirely independent lanes operating in such a manner that the resultant blade angle change produced in response to sensor outputs is the average of the two individual lanes. Under normal operating conditions the two lanes are so matched as to be practically identical. The outputs of the Computer and the lane actuator position in each lane are continually monitored and cross-compared. Any significant disparities are displayed to the pilot on the Controller and Monitor Unit and the pilot can disengage the stability augmentation system.

The duplex architecture ensures that a lane failure results in only a 'passive' failure of the system; that is, the summed output of the two lane actuators remains at the position it was in at the time of the failure. This minimises the disengagement transient when a failure occurs and the system reverts to manual control.

The dual–dual system (Fig. 3.53b) is basically a dual version of the duplex system. It can survive the first failure and reverts to manual control with minimal transient in the event of a second failure. This is a low-probability event, and very rarely will the pilot be without the handling and control benefits of the system.

The triplex system shown in Fig. 3.53c is able to survive the first failure, and reverts to a duplex system. The second failure is a passive failure, with the system reverting to manual operation with minimal transient.

Modern stability augmentation systems are implemented digitally and have a very much higher reliability than the older analogue systems. They also incorporate efficient and effective self-monitoring and self-test facilities.

Fig. 3.53 (**a**) Duplex stability augmentation system; (**b**) Dual–dual stability augmentation system; (**c**) Triplex stability augmentation system

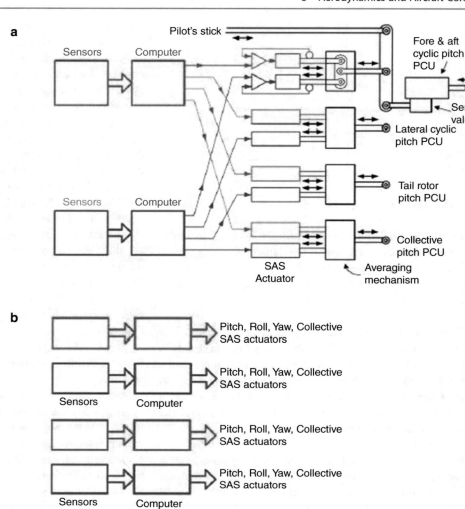

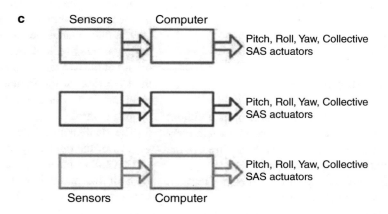

Angular momentum gyro-based sensor systems have been replaced with strap-down Attitude/Heading Reference Systems (AHRS) using solid state rate gyros and accelerometers. They provide pitch, roll attitude and heading data together with pitch, roll, yaw rates and forward, lateral, normal acceleration components from the one unit (refer Chap. 5). The MTBF of a strap-down AHRS is in the region of 50,000 hours, compared with an angular momentum gyro-based sensor system with an MTBF of around 2500 hours.

3.9.3.1 Fault Tolerant Pitch Axis Stability Augmentation

The helicopter load can result in the centre of gravity (CG) being offset from the rotor thrust line. This requires an increase in the cyclic pitch angle to provide a balancing trim moment and hence increased stick force. (Stick force is proportional to stick deflection and hence cyclic pitch angle.) The CG offset can also change in flight; for example, hovering over a fixed spot and lowering a crew member and

Fig. 3.54 Pitch axis stability augmentation system

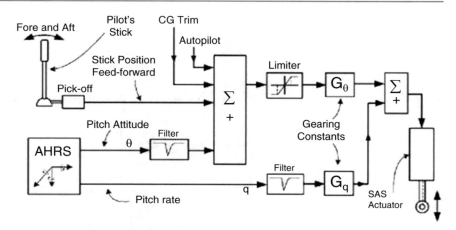

Note: Only one lane shown

winching up a casualty in a rescue operation. The stick deflection to pitch attitude relationship thus changes when the CG offset changes.

Fault tolerant, pitch axis stability augmentation systems augment the characteristic stick deflection to attitude change. Figure 3.54 illustrates a single channel of a typical fault tolerant pitch system. Command augmentation is provided by stick position feed forward (or 'stick cancellation'), whilst stability augmentation and damping is provided by pitch attitude and pitch rate feedback. The command augmentation system is normally set up to match the normal stick to attitude relationship to minimise disengagement transients.

The augmentation system provides damping and reduces the effects of turbulence. The stability augmentation system opposes and reduces changes in the pitch attitude from that set by the stick position when the helicopter is subjected to a disturbance. The pilot workload to counter a change in CG offset is thus reduced.

A typical system will maintain a constant datum pitch attitude over an approximate ±18° range. Over this range the attitude of the helicopter is proportionally related to the fore and aft cyclic stick position. The stabiliser actuator stroke is limited, so the gearing of blade angle to attitude or stick position has two slopes, with higher gearing over an attitude range of around ±5°.

The AHRS pitch attitude and pitch rate outputs are filtered by structural 'notch' filters to attenuate the harmonic vibration frequencies generated by the rotor rotation, which are sensed by the gyros. ('Notch' filters are covered in Chap. 4.)

An input from the trim wheel enables the pilot to set the null (mid-stroke) position of the stability augmentation actuator.

3.9.3.2 Roll Axis Stability Augmentation

A typical roll axis stability augmentation system is shown in Fig. 3.55, which is similar in many aspects to the pitch axis system. 'Wings'-level attitude is maintained using roll attitude and roll rate information from the AHRS.

Roll attitude is typically proportional to control stick movement for up to about 7° of bank angle and is limited when it reaches that value. Control then reverts substantially to a roll rate system. Large bank angles are achieved and maintained for executing higher g turns, by a combination of roll rate and 'washed-out' roll angle. This enables a demanded roll attitude to be maintained.

Because the signal is steady-state DC blocked, it does not permanently saturate the system by demanding a steady-state full actuator stroke. (The stabiliser actuator authority in roll is typically less than ±2° of blade angle.)

The roll angle and roll rate signals need to be filtered by notch filters to attenuate the vibration harmonic frequencies. A further structural 'notch' filter is used to suppress excitation of the rotor lag plane resonance.

A modest amount of phase advance may also be required to achieve acceptable damping. This is because of the cumulative effect of the lags introduced by the notch filters and the PCU actuator response (typically around 1 Hz bandwidth).

The control stick cancellation signals from the stick position pick-offs are typically fed through a simple lag filter. This improves the 'matching' of stick input to attitude and rate attained giving a better response to pilot input.

3.9.3.3 Yaw Axis Stability Augmentation

Figure 3.56 shows a single lane of a basic yaw axis stability augmentation system.

The control signals are passed through notch filters to avoid exciting structural resonance modes. Lateral acceleration signals can be added to improve yaw stability.

3.9.3.4 Collective

Normal acceleration signals from the strap-down AHRS can be used to augment the helicopter pitch stability at high forward speeds and assist autopilot performance.

The normal acceleration signals are typically filtered by a simple lag filter to attenuate higher-frequency noise.

Fig. 3.55 Roll axis auto-stabilisation system

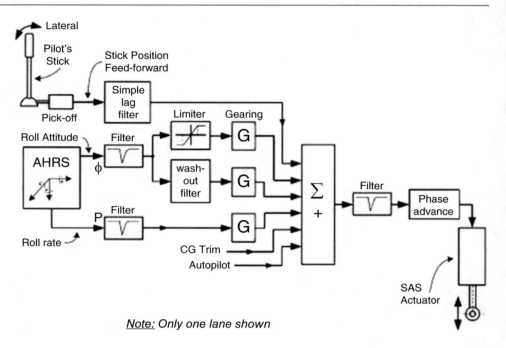

Note: Only one lane shown

Fig. 3.56 Basic yaw rate stability augmentation system

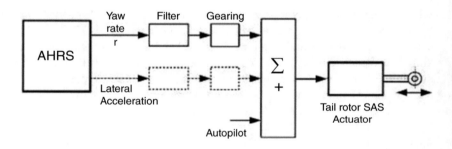

Fault tolerant stability augmentation systems give a very significant improvement in the control and handling characteristics of the basic helicopter. The improvement, however, is constrained by their limited authority and the need to be able to revert safely to manual control in the event of the loss of the augmentation system.

The full benefits of automatic control require a full authority system which is able to survive two failures in the system from any cause. Replace all mechanical links with electrical links and the system becomes an FBW system similar to those used in fixed-wing aircraft. The all-round improvement in helicopter control and handling characteristics it can confer are covered in the next chapter.

Further Reading

Babister, A.W.: Aircraft Dynamic Stability and Response. Pergammon Press

Bertin, J.J. and Smith, M.L.: Aerodynamics for Engineers, Prentice Hall

Blakelock, J.H.: Automatic Control of Aircraft and Missiles, Wiley

Cook, M.V.: Flight Dynamics Principles. Arnold, Hodder Headline Group (1997)

Dommasch, D.D., Sherby, S.S., Connolly, T.F.: Airplane Aerodynamics. Pitman

Etkin, B.: Dynamics of Atmospheric Flight. Wiley

McLean, D.: Automatic Flight Control Systems. Prentice Hall

McRuer, D., Ashkenas, I., Graham, D.: Aircraft Dynamics and Automatic Control. Princeton University Press

Nelson, R.C.: Flight Stability and Automatic Control. McGraw-Hill

Pratt, R.W.: Flight Control Systems Practical Issues in Design and Implementation, IEE Control Engineering Series, vol. 57. The Institution of Electrical Engineers

4.1 Introduction

The introduction of fly-by-wire (FBW) flight control systems has been a watershed development in aircraft evolution as it has enabled technical advances to be made, which were not possible before. One of the unique benefits of an FBW system is the ability to exploit aircraft configurations, which provide increased aerodynamic efficiency, like more lift and lower drag, but at a cost of reduced natural stability. This can include negative stability, that is, the aircraft is unstable over part of the range of speed and height conditions (or flight envelope [FE]).

The FBW system provides high-integrity automatic stabilisation of the aircraft to compensate for the loss of natural stability and thus enables a lighter aircraft with a better overall performance to be produced compared with a conventional design. It also provides the pilot with very good control and handling characteristics which are more or less constant over the whole flight envelope and under all loading conditions. Other benefits an FBW system can provide are manoeuvre command control, 'carefree manoeuvring' and not least the elimination of the bulk and mechanical complexity of the control rods and linkages connecting the pilot's stick to the control surface power control units (PCUs) and consequent weight saving.

Aircraft with FBW flight control systems first came into service in the late 1970s using analogue implementation. Digital FBW systems have been in service since the late 1980s. The concepts are not new; in fact, all guided missiles use this type of control. What has taken the time has been the development of the failure survival technologies to enable a high-integrity system to be implemented economically with the required safety levels, reliability and availability. A major factor has been the development of failure survival digital flight control systems and their implementation in VLSI microcircuits. There are other technologies where development has been essential for FBW control, such as failure survival actuation systems to operate the control surfaces.

All new fighter designs exploit FBW control. Figure 1.2 (Chap. 1) illustrates the Eurofighter Typhoon as a typical example.

A recent development in military aircraft is the emergence of 'stealth' technology where the aircraft configuration and shape are specifically designed to reduce its radar cross-section. In general, the stealth features reduce the aircraft's natural stability and damping, and FBW control is essential to achieve good handling and control characteristics.

The current generation of civil airliners exploit FBW control. Examples are the Airbus A319, A320, A330, A340, A350, A380, and the Boeing 777 and 787.

Very many tens of millions of flying hours have now been accumulated by aircraft with digital FBW flight control systems, and their safety and integrity have been established.

4.2 FBW Flight Control Features and Advantages

4.2.1 FBW System Basic Concepts and Feature

Figure 4.1 shows the basic elements of an FBW flight control system.

Note:

- The total elimination of all the complex mechanical control runs and linkages – all commands and signals are transmitted electrically along wires, hence the name fly-by-wire.
- The interposition of a computer between the pilot's commands and the control surface actuators.
- The aircraft motion sensors which feedback the components of the aircraft's angular and linear motion to the computer.
- The air data sensors which supply height and airspeed information to the computer.

R. P. G. Collinson, *Introduction to Avionics Systems*, https://doi.org/10.1007/978-3-031-29215-6_4

Fig. 4.1 Basic elements of an
FBW flight control system

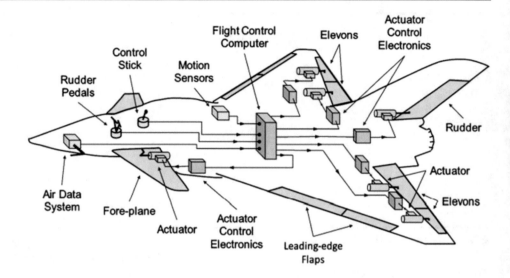

- Not shown in the figure is the redundancy to enable failures in the system to be absorbed. The flaps can also be continually controlled by the flight control computer. (Command links and actuators omitted for clarity.)

The pilot thus controls the aircraft through the flight control computer, and the computer determines the control surface movement for the aircraft to respond in the best way to the pilot's commands and achieve a fast, well-damped response throughout the flight envelope.

The key features are described in more detail below.

4.2.1.1 Electrical Data Transmission

Electrical transmission of signals and commands is a key element in an FBW system. Modern systems use a serial digital data transmission system with time division multiplexing. The signals can then be transmitted along a network or 'highway' comprising two wires only, as only one set of data is being transmitted at any particular time.

Figure 4.2 shows how a digital flight control system is interconnected using a digital data bus.

Military FBW systems generally use the well-established MIL-STD-1553 data bus system. The links and bus use a screened twisted pair of wires with connection to the bus through isolating transformers. This is a command/response system with the bus controller function embedded in the flight control computers. It has a data rate of 1 Mbit/s and a word length of 20 bits to encode clock, data and address and so can receive or transmit up to 50,000 data words a second.

The Boeing 777 uses the ARINC 629 data bus system. This is an autonomous system and operates at 2 Mbit/s. The links and bus use an unscreened twisted pair of wires with connection to the bus through demountable current transformer couplers.

The electronic complexity required in these systems to code the data and transmit it, or to receive data and decode it, is encapsulated in one or two integrated microcircuits. Data bus systems are covered in Chap. 9.

4.2.1.2 FBW Control Surface Actuation

The actuation systems which control the movements of the control surfaces are vital elements in an FBW system. They must be able to survive any two failures and carry on operating satisfactorily in order to meet the aircraft safety and integrity requirements (discussed later in the chapter). The servo actuation systems driving the control surfaces comprise a two-stage servo system with the FBW servo actuators driving the duplex control valves of the main power control actuators. Both electrohydraulic and electrical first-stage actuation systems are used, although the trend is now towards direct drive electric motors.

Linear and rotary electro-magnetic (EM) actuators are used with multiple independent windings; three windings in the case of a triplex system or four windings in a quadruplex system.

A typical quadruplex actuation system comprises four totally independent first-stage actuators which force-add their outputs to drive the power control unit (PCU) servo control valve. Figure 4.3 shows a quadruplex actuation system schematically with electrical first-stage actuators driving the PCU servo control valve. There is no mechanical feedback from the PCU actuator to the servo control valve as there is in a conventional non-FBW system. Instead, the position of the control surface is fed back electrically to the input of the actuator control electronics; four independent position sensors are used to maintain the required integrity. The overall feedback improves the speed of response of the actuation system by a factor of about 10 compared with a conventional PCU. Fast response is absolutely essential in an

Fig. 4.2 Flight control system bus configuration

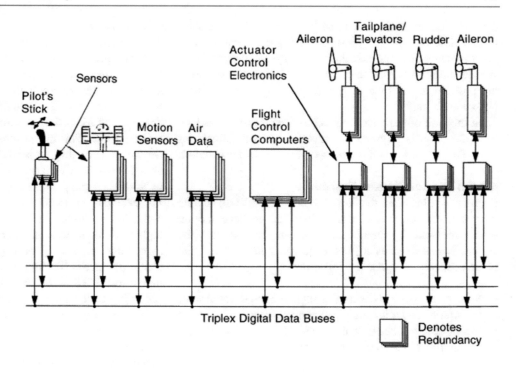

Fig. 4.3 Quadruplex actuation system

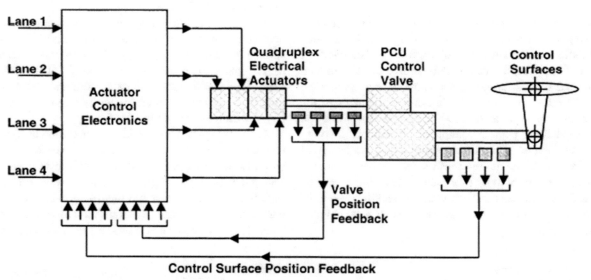

FBW actuation system in order to minimise the lags in the FBW loop. A typical agile fighter which is aerodynamically unstable would diverge exponentially, with the divergence doubling every 0.2 s in the absence of FBW control. The response of the FBW system to correct any divergence must thus be very fast. The actuator response requirements for a modern agile fighter correspond to a phase lag of less than 12° at 1 Hz.

The failure survival philosophy of a quadruplex actuation system is that if one actuator fails the three good ones can override it. The failed actuator is identified by comparing its control signals with the other three on the assumption that the probability of more than one failing at precisely the same instant is extremely remote. (All four actuators are totally independent in terms of separate power supplies, control electronics, etc.). The failed actuator is then disconnected (or, in the case of an electrohydraulic first-stage actuator, is hydraulically bypassed) leaving three good actuators in command. A second subsequent actuator failure is detected by a similar process; two actuators are in agreement, one differs, therefore it must be the failed one. The second failed actuator is then disconnected (or bypassed) leaving the remaining two good actuators in control. In the extremely unlikely event of a subsequent third failure, the control surface would remain in the position at the time of failure with one good actuator opposing the failed one.

In the case of a triplex architecture, some form of in-lane fault detection is required to survive a second failure, for example, comparison with a computer model of the actuator.

4.2.1.3 Motion Sensor Feedback

Figure 4.4 illustrates the key features of FBW in simple diagrammatic form. An FBW system has to have motion sensor feedback by definition – without these sensors the system is classified as a 'direct electric link' system. The motion sensors comprise the following:

- Rate gyros which measure the angular rates of rotation of the aircraft about its pitch, roll and yaw axes.
- Linear accelerometers which measure the components of the aircraft's acceleration along these axes.

The feedback action of these sensors in automatically stabilising the aircraft can be seen from Fig. 4.4. Any change in the motion of the aircraft resulting from a disturbance of any sort (e.g. gust) is immediately sensed by the motion sensors and causes the computer to move the appropriate control surfaces so as to apply forces and moments to the aircraft to correct and suppress the deviation from the commanded flight path. An automatic 'hands-off' stability is achieved with the aircraft rock steady if the pilot lets go of the control stick. The motion sensors also enable a manoeuvre command control to be exercised by the pilot, as will be explained later. Because of their vital role and the need to be able to survive failures, they are typically at a quadruplex level of redundancy.

4.2.1.4 Air Data

The need for air data information on the airspeed and height is to compensate for the very wide variation in the control surface effectiveness over the aircraft's flight envelope of height and speed combinations; for example, low speed at low altitude during take-off and landing, high speeds approaching Mach 1 at low altitude in the case of a military strike aircraft, cruising subsonic flight at high altitude where the air is very thin, supersonic flight at medium to high altitude, etc. The variation in control surface effectiveness, or 'stick force per g' as it is referred to, can be as high as 40:1. For example, at 45,000 ft. it may require 20° of tailplane deflection to produce a normal acceleration of 1 g. At very high subsonic speeds of around 600 knots at very low altitude, however, it may only need 0.5° deflection, and 20° would produce sufficient g to break up the aircraft. It is thus necessary to adjust or scale the control surface deflection according to the aircraft's airspeed and height as it is not possible to achieve a stable closed-loop control system with such a wide variation in the open-loop gain. The aircraft response and controllability are also dependent on its Mach number, that is, the ratio of the true airspeed of the aircraft to the local speed of sound.

The FBW system is thus supplied with airspeed, height and Mach number in order to adjust or scale the control surface deflections accordingly. (This process is referred to as 'air data gain scheduling' in the USA). Totally independent, redundant sources of air data information are required in order to meet the safety and integrity requirements. Generally, quadruplex sources are used.

The FBW system also requires information on the aircraft incidence angles, that is, the local flow angles in the pitch and yaw planes between the airstream and the fuselage datum. The pitch incidence angle controls the wing lift and it is essential to monitor that the incidence angle is below the maximum value to ensure that a stall condition is not reached. (A stall results when the airflow starts to break away from the

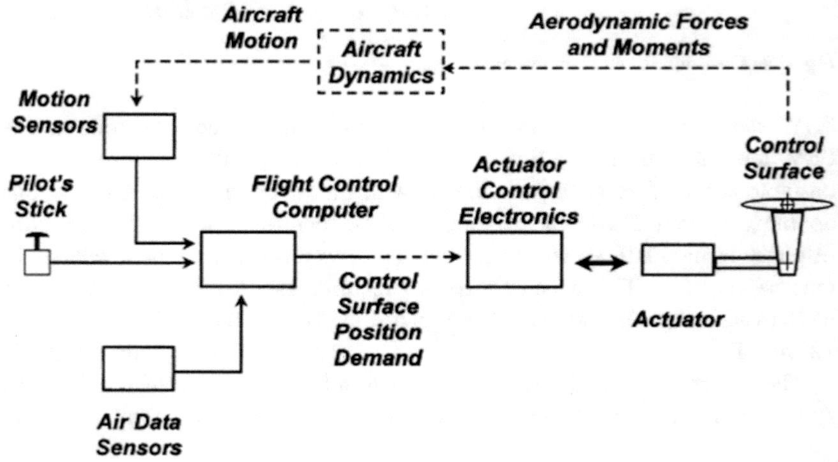

Fig. 4.4 Fly-by-wire flight control system

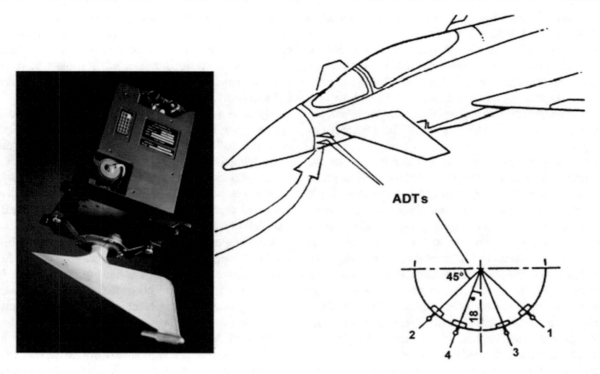

Fig. 4.5 Integrated air data transducer system. (Courtesy of BAE Systems)

upper surface of the wing with consequent sudden loss of lift and control.) The incidence angle in the pitch plane (or angle of attack as it is often called) is used as a control term in the pitch FBW system. The incidence angle in the yaw plane is known as the angle of sideslip, and is used as a control term in the FBW rudder control system.

Figure 4.5 illustrates an integrated air data sensor unit, which combines the incidence vane and the pitot-static probe; the vane aligns itself with the airstream in the same manner as a weather vane. The unit contains the total pressure and static pressure sensors together with the associated electronics including a microprocessor to carry out the air data computations. It provides height, calibrated airspeed, Mach number and local flow angle information to the FBW system. Four of these air data transducer systems as they are known are installed on the Eurofighter Typhoon to meet the failure survival and integrity requirements.

4.2.1.5 High Integrity, Failure Survival Computing System

The flight control computing system must be of very high integrity and have the failure survival capability to meet the flight safety requirements. The tasks carried out by the computing system comprise the following:

(a) Failure detection
(b) Fault isolation and system reconfiguration in the event of a failure
(c) Computation of the required control surface angles

(d) Monitoring
(e) Built-in test

4.2.1.6 Very High Overall System Integrity

The overall system integrity must be as high as the mechanical control system it replaces. The probability of a catastrophic failure must not exceed 10^{-9}/hour for a civil aircraft and 10^{-7}/hour for a military aircraft.

4.2.2   Advantages of FBW Control

The main advantages of a well-designed FBW flight control system are:

4.2.2.1 Increased Performance

FBW enables a smaller tailplane, fin and rudder to be used, thereby reducing both aircraft weight and drag, active control of the tailplane and rudder making up for the reduction in natural stability.

For a civil airliner, reducing the stability margins and compensating for the reduction with an FBW system thus result in a lighter aircraft with a better performance and better operating economics and flexibility than a conventional design, for example, the ability to carry additional freight. Conventional airliners may have the space for additional freight containers, but the resulting rearward shift of the centre of gravity (CG) would give the aircraft unacceptably marginal handling characteristics. It should be noted that the carriage of containerised freight as well as passengers forms a very significant part of an airline's revenue.

A civil FBW airliner can be configured so that its control and handling characteristics are very similar to those of comparable mechanically signalled aircraft. This is a considerable advantage in civil airline operation where pilots may be interchanging with existing mechanically signalled aircraft in the airline fleet. (The Boeing 777 incorporates this philosophy.)

For a military aircraft, such as an air superiority fighter, the FBW system enables aircraft configurations with negative stability to be used. These give more lift, as the trim lift is positive, so that a lighter, more agile fighter can be produced – agility being defined as the ability to change the direction of the aircraft's velocity vector. An increase in instantaneous turn rate of 35% is claimed for some of the new agile fighters. It should be noted that the aerodynamic centre moves aft at supersonic speeds increasing the longitudinal stability. In fact, most agile fighters which are longitudinally unstable at subsonic speeds become stable at supersonic speeds. In the case of a conventional aircraft, the increase in longitudinal stability at supersonic speeds requires larger pitch control moments; a highly stable aircraft resists changing its incidence whereas manoeuvrability requires rapid changes in incidence.

4.2.2.2 Reduced Weight
Electrically signalled controls are lighter than mechanically signalled controls. FBW eliminates the bulk and mechanical complexity of mechanically signalled controls with their disadvantages of friction, backlash (mechanical lost motion), structure flexure problems, periodic rigging and adjustment. The control gearings are also implemented through software, which gives greater flexibility.

4.2.3 FBW Control Sticks/Passive FBW Inceptors

FBW flight control enables a small, compact pilot's control stick to be used allowing more flexibility in the cockpit layout. The displays are un-obscured by large control columns; the cockpit flight deck is very valuable 'real estate'.

Figures 1.3 and 1.8 in Chap. 1 show the Eurofighter Typhoon cockpit and the Airbus A350 flight deck, respectively, with their FBW control stick installations. The FBW control stick is often referred to as an FBW 'inceptor'.

An FBW inceptor is defined as a device which translates the pilot's control inputs into electrical signals. They can be divided into two basic types – passive and active.

The passive type of FBW inceptor provides a fixed stick force–stick displacement relationship by means of a mechanical spring-box arrangement.

The active type of inceptor, in contrast, can provide a wide range of stick force–stick displacement characteristics by computer control of force motors, which back drive the control stick. Active inceptors are a subject in their own right and are covered in Sect. 4.7 to maintain the balance of this section.

Figure 4.6 illustrates the features of a passive FBW inceptor. The illustration shows a centrally mounted control stick as installed in the Eurofighter Typhoon (right) and the side mounted stick (left) in the Lockheed F22 Raptor fighter.

The diagram shows the construction schematically. The stick is a two-axis one and provides pitch and roll electrical command signals; typically, four independent electrical position pick-offs are provided for each axis of control.

The damper is an essential element; without it the stick would be considerably underdamped because of the low friction in the mechanism and the significant mass of the hand grip. It provides a smooth feel to the stick movement; the spring–mass–damper combination acting as a low-pass filter on the stick movement.

The stick force characteristics are shown in Fig. 4.6. A small break-out force is required to displace the stick from the central position in pitch and roll. Roll control is a simple linear spring characteristic. Pitch, however, requires a step increase in force at larger stick displacements and the spring rate increases so that larger forces have to be exerted when commanding high g manoeuvres. These characteristics are carefully tailored to meet the consensus of pilot approval.

4.2.3.1 Automatic Stabilisation
Hands-off stability as explained earlier.

4.2.3.2 Carefree Manoeuvring
The FBW computer continually monitors the aircraft's state to assess how close it is to its manoeuvre boundaries. It automatically limits the pilot's command inputs to ensure that the aircraft does not enter an unacceptable attitude or approach too near its limiting incidence angle (approaching the stall) or carry out manoeuvres which would exceed the structural limits of the aircraft.

A number of aircraft are lost each year due to flying too close to their manoeuvre limits and the very high workload in the event of a subsequent emergency. The FBW system can thus make a significant contribution to flight safety.

It should be noted that some FBW systems (e.g. Boeing 777) allow the pilots to retain ultimate control authority of the aircraft and break through or override the bank angle and stall limits if they are concerned about the aircraft's behaviour in extreme conditions.

4.2.3.3 Ability to Integrate Additional Controls
These controls need to be integrated automatically to avoid an excessive pilot workload – too many things to do at once:

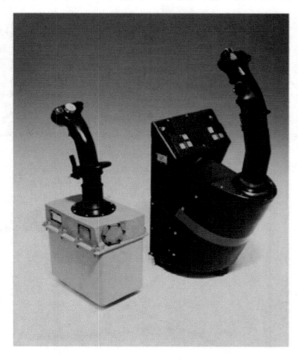

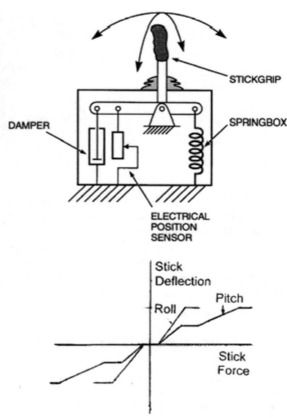

Fig. 4.6 FBW control stick/passive FBW inceptor features

- Leading and trailing edge flaps for manoeuvring and not just for take-off and landing.
- Variable wing sweep.
- Thrust vectoring.

4.2.3.4 Ease of Integration of the Autopilot

The electrical interface and the manoeuvre command control of the FBW system greatly ease the autopilot integration task. The autopilot provides steering commands as pitch rate or roll rate commands to the FBW system. The relatively high bandwidth manoeuvre command 'inner loop' FBW system ensures that response to the outer loop autopilot commands is fast and well damped, ensuring good control of the aircraft flight path in the autopilot modes. A demanding autopilot performance is required for applications such as automatic landing, or automatic terrain following at 100–200 ft. above the ground at over 600 knots where the excursions from the demanded flight path must be kept small. This was one of the 'drivers', in fact, for the FBW system installed in the Tornado strike aircraft, together with the need to maintain good handling and control when carrying a large load of external 'stores'.

4.2.3.5 Closed-Loop Manoeuvre Command Control

A closed-loop manoeuvre command control is achieved by increasing the gain of the motion sensor feedback loops. The control surface actuators are thus controlled by the difference, or error, between the pilot's command signals and the measured aircraft motion from the appropriate sensors, for example, pitch rate in the case of a pitch rate command system and roll rate in the case of a roll rate command system. Other control terms may also be included such as airstream incidence angles and possibly normal and lateral accelerations.

The flight control computer derives the required control surface movements for the aircraft to follow the pilot's commands in a fast, well-damped manner.

It should be stressed that the achievement of such a response requires extensive design and testing and a well-integrated combination of aircraft and FBW control system. For instance, adequate control power must be available from the control surfaces and non-linear behaviour taken into account (e.g. rate limiting in the actuation system).

A well-designed closed-loop control system offers the following advantages over an open-loop control system:

Fig. 4.7 Roll rate command system

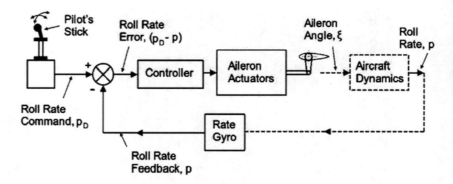

- The steady-state output to input relationship is substantially independent of changes in the loop gain provided this remains sufficiently high.
- The system bandwidth is improved and the phase lag when following a dynamically varying input is reduced.
- A fast well-damped response, which is little affected by normal changes in the loop gain, can generally be achieved by suitable design of the control loop.

Figure 4.7 is a block diagram of a closed-loop roll rate command system.

Consider what happens when the pilot pushes the stick to command a roll rate. At the instant the command is applied the roll rate is zero, so that the roll rate error produces a large aileron deflection. This creates a relatively large rolling movement on the aircraft so that the roll rate builds up rapidly. The roll rate error is rapidly reduced until the roll rate error is near zero, and the aircraft roll rate is effectively equal to the commanded roll rate. Because the roll rate creates an aerodynamic damping moment which opposes the rate of roll, the aileron deflection cannot be reduced to zero but is reduced to a value where the rolling moment produced is equal and opposite to the aerodynamic damping moment. The controller gain is sufficiently high, however, to keep the steady-state roll rate error to a small value.

A much faster roll response can be obtained compared with a conventional open-loop system, as can be seen in Fig. 4.8; the variation in response across the flight envelope is also much less.

Aircraft need to bank to turn, so a fast, precise roll response is required. Push the stick sideways and a roll rate directly proportional to the force exerted on the stick is obtained. Return the stick to the centre when the desired bank angle is reached and the aircraft stops rolling, without any overshoot.

The improvement brought by closed-loop control can be analysed by making certain simplifying assumptions. The roll rate/aileron angle relationship has been derived in Chap. 3, Sect. 3.6.1, Eq. (3.68).

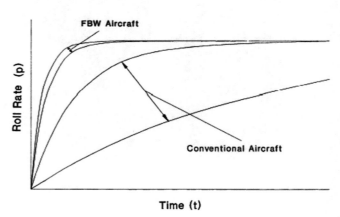

Fig. 4.8 Roll rate response

Assuming pure rolling motion and the actuators have a transfer function of unity, the open-loop transfer function is

$$\frac{p}{(p_D - p)} = K_c \cdot \frac{L_\xi}{L_p} \cdot \frac{1}{1 + T_R D} \quad (4.1)$$

where p_D = demanded roll rate; K_c = controller gain = aileron angle in degrees/degree per second roll rate error; ξ = aileron angle; L_ξ = rolling moment derivative due to aileron angle; lL_p = rolling moment derivative due to roll rate, p; I_x = moment of inertia about roll axis; and T_R = roll time constant = I_x/L_p.

Whence the closed-loop response is given by

$$\frac{p}{p_D} = \frac{\frac{K_c L_\xi / L_p}{(1 + T_R D)}}{1 + \frac{K_c L_\xi / L_p}{(1 + T_R D)}} \quad (4.2)$$

From this

$$\frac{p}{p_D} = \frac{K}{(1 + K)} \cdot \frac{1}{1 + \frac{T_R}{(1+K)} D} \quad (4.3)$$

where

$$K = \frac{K_c L_\xi}{L_p} = \text{open-loop gain}.$$

If

$$K >> 1, \text{then} \quad \frac{p}{p_D} \approx \frac{1}{1 + \frac{T_R}{K}D} \quad (4.4)$$

The steady-state value of the roll rate for a given stick input is thus substantially constant over the flight envelope provided K is sufficiently large. The time constant of the roll response is also greatly reduced, being divided by the open-loop gain K.

It should be stressed again that the above treatment has been greatly simplified by assuming pure rolling motion and no roll/yaw/sideslip cross-coupling, and also ignoring lags in the actuator and filters, etc. The object is to explain basic principles without resorting to a mathematically complex model.

Figure 4.9 shows an FBW pitch rate command system. Consider now what happens when the pilot exerts a force on the stick to command a pitch rate. The aircraft pitch rate is initially zero, so the resultant pitch rate error causes the computer to demand an appropriate deflection of the tailplane from the trim position. The ensuing lift force acting on the tailplane exerts a pitching moment on the aircraft about its CG causing the pitch attitude to change and the wing incidence to increase. The resulting lift force from the wings provides the necessary force at right angles to the aircraft's velocity vector to change the direction of the aircraft's flight path so that the aircraft turns in the pitch plane. The increasing pitch rate is fed back to the computer, reducing the tailplane angle until a condition is reached when the aircraft pitch rate is equal to the commanded pitch rate. The pitch rate error is thus brought to near zero and maintained near zero by the automatic control loop.

Pitch rate command enables precise 'fingertip' control to be achieved. For example, to change the pitch attitude to climb, gentle pressure back on the stick produces a pitch rate of a few degrees per second; let the stick go back to the central position and the pitch rate stops in less than a second with negligible overshoot with the aircraft at the desired attitude. Increasing the stick force produces a proportionate increase in pitch rate. The normal acceleration, or g, is equal to the aircraft velocity multiplied by the pitch rate, so for a given speed the g is directly proportional to the rate of pitch.

4.2.3.6 Aerodynamics Versus 'Stealth'

The concept of reducing the radar cross-section of an aircraft so that it is virtually undetectable (except at very close range) has been given the name 'stealth' in the USA. Radar reflection returns are minimised by faceted surfaces which reflect radar energy away from the direction of the source, engine intake design and the extensive use of radar energy absorbing materials in the structure. An example of a stealth aircraft is the Northrop B2 shown in Fig. 4.10.

Stealth considerations and requirements can conflict with aerodynamics requirements, and FBW flight control is essential to give acceptable, safe handling across the flight envelope.

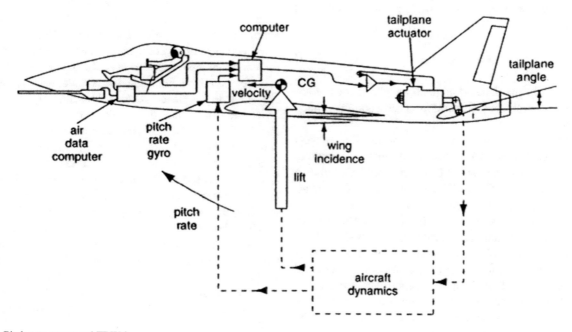

Fig. 4.9 Pitch rate command FBW loop

Fig. 4.10 Northrop B2 'Spirit' strategic bomber. (Courtesy of the Royal Aeronautical Society Library). Note 'flying wing' configuration and absence of vertical surfaces; yaw control being achieved by the operation of split ailerons

4.3 Control Laws

The term 'control laws' is used to define the algorithms relating the control surface demand to the pilot's stick command and the various motion sensor signals and the aircraft height, speed and Mach number.

As an example, a very simple basic pitch rate command law would be:

$$\eta_D = K(\theta_i - G_q q) \tag{4.5}$$

where η_D = tailplane demand angle; θ_i = pilot's input command; q = pitch rate; K = forward loop gain; and G_q = pitch rate gearing.

In practice, additional control terms from other sensors (e.g. incidence and normal acceleration) may be required. The value of K and the sensor gearings would also almost certainly need to be varied with height and airspeed by an air data gain scheduling system, as already mentioned.

Control terms proportional to the derivative or rate of change of error and integral of error are also used to shape and improve the closed-loop response. A brief explanation of the action of these control terms is given below in view of their importance in closed-loop flight control systems.

Proportional plus derivative of error control provides a phase advance characteristic to compensate for the lags in the system, for instance actuator response, and hence improve the loop stability. This increases the damping of the aircraft response and reduces the overshoot to a minimum when responding to an input or disturbance.

The control term is filtered to limit the increase in gain at high frequencies and smooth the differentiation process, which amplifies any noise present in the error signal. Figure 4.11 illustrates the gain and phase versus frequency characteristics of a simple filtered 'one plus derivative' error control (usually known as phase advance control), with a transfer function:

$$\frac{(1 \pm TD)}{(1 + T/n\,D)}$$

the characteristics being plotted for $n = 3, 5, 7, 10$. The amount of phase advance is determined by the value of n and is generally limited by the resulting increase in gain at high frequencies and avoiding exciting structural resonance modes. Typical values used for n are between 4 and 5 giving a maximum phase lead of about 40°.

A more physical explanation of the damping action of a phase advance element is shown in Fig. 4.12. This shows how a control moment, which is proportional to error plus rate of change of error (suitably smoothed), changes sign before the error reaches zero and hence applies a retarding moment to decelerate and progressively reduce the velocity before the error reaches zero and hence minimises the overshoot.

Proportional plus integral of error control eliminates steady-state errors and reduces the following lag. The integral of error term increases the loop gain at low frequencies up to theoretically infinite gain at dc so that there are zero steady-state errors due to out of trim external moments or forces acting on the aircraft. The error when following a low-frequency input command is also reduced and is zero for a constant input rate. The need for an integral term control can be seen by examining the simple system shown in Fig. 4.13. In the steady state, if θ_i is constant, θ_o must also be constant and thus the resultant moment acting on the aircraft must also be zero and the control surface moment is equal and opposite to the out of trim moment.

However, to generate a control surface moment requires an error signal so that θ_o cannot equal θ_i exactly. The magnitude of the steady-state error is dependent on the loop gain K, which in turn is limited by the loop stability.

The effect of 'proportional plus integral' error control can be seen in Fig. 4.14. The steady-state error is zero; however, the integral of the error is not zero and reaches a value sufficient to generate the necessary control moment to balance the trim moment. The value of an integral term control can thus be seen in achieving automatic trimming of the control surfaces.

Fig. 4.11 Phase advance
function frequency response

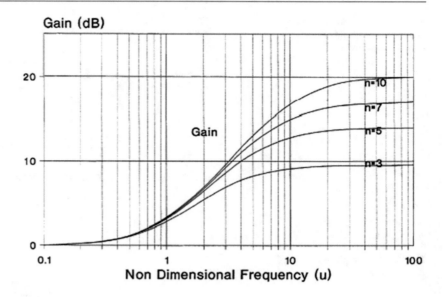

Figure 4.15 illustrates the gain and phase versus frequency characteristics for proportional plus integral control. The transfer function being:

$$\left(1 + \frac{1}{TD}\right), \text{that is } \frac{(1 + TD)}{TD}$$

It can be seen that the gain approaches unity (0 dB) at high frequencies, with zero phase shift so that the loop stability is not degraded, providing the integral time constant, T, is suitably chosen.

It should be noted that integral terms need to be initially synchronised to the initial conditions defined at start up to avoid large transients on switching on.

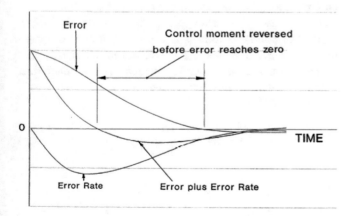

Fig. 4.12 Damping action – 'one plus derivative' of error

Fig. 4.13 Simple closed-loop control system subjected to external moment

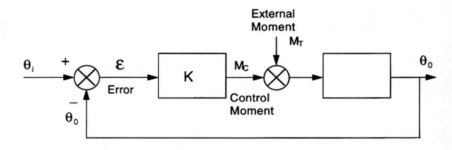

In the steady state, $M_C = M_T$

$$\text{Steady State Error} = \frac{M_T}{K}$$

Fig. 4.14 'One plus integral' of error control

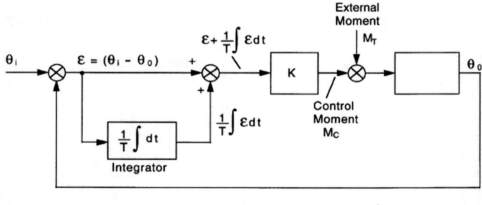

In the steady state $\quad M_C = M_T$

Steady State Error $= 0$

$$\frac{1}{T}\int \varepsilon \, dt = \frac{M_T}{K}$$

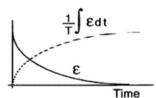

Mention has already been made of the 'carefree manoeuvring' characteristic that can be incorporated in an FBW system. This is achieved by continually monitoring the aircraft's state and automatically limiting the authority of the pilot's command input by means of a suitable control law which takes into account the aircraft's manoeuvre boundaries and control limits. This prevents the pilot from attempting to manoeuvre the aircraft into an unacceptable attitude, or approach too near to the limiting angle of incidence, or exceed the structural limits of the aircraft in a manoeuvre.

4.3.1 Pitch Rate Command Control

A block diagram of a pitch rate command FBW flight control system is shown in Fig. 4.16 (the essential redundancy has been omitted for clarity).

The primary feedback of pitch rate for the pitch rate command loop is provided by the pitch rate gyro(s).

In the case of an aerodynamically unstable aircraft, an integral of pitch rate term is also used to provide a 'quasi-incidence term' to counteract a positive (unstable) M_α.

The transfer function relating incidence, α, and pitch rate, q, has been derived in Chap. 3, Sect. 3.5.4.

From Eq. (3.52)

$$\alpha = T_2 \frac{1}{1 + T_2 D} q \tag{4.6}$$

Compared with

Fig. 4.15 'One plus integral'
function frequency response

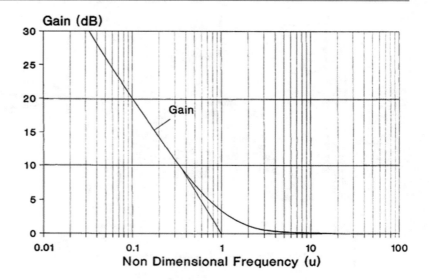

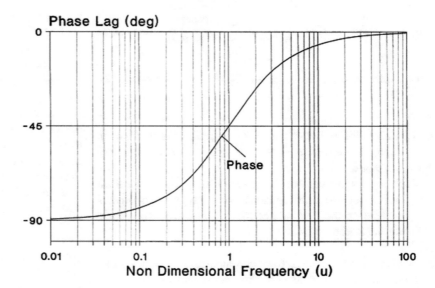

$$\int q\,dt = \frac{1}{D}q \qquad (4.7)$$

The relationship between α and $\int q\,dt$ can be seen from the two equations. The frequency response of the α/q transfer function, in fact, approaches that of a simple integrator at frequencies where $\omega T_2 \gg 1$, the gain approaching $1/\omega$ and the phase $-90°$.

Hence $\int q\,dt$ H α for values of w such that $\omega T2 \gg 1.\alpha$

The proportional plus integral control is generally applied to the pitch rate error, $q_E = (q_D - q)$, where $q_D =$ demanded pitch rate.

This has the advantages of eliminating steady-state following errors and enabling the pilot to change the pitch attitude easily. (It should be noted that as far as loop stability is concerned, there is no difference in placing the stabilising

transfer function in the pitch rate error path as opposed to the pitch rate feedback path.)

In the case of an aerodynamically unstable aircraft, a well-designed *one plus integral of pitch rate error* control can provide an acceptable stable loop, although not an optimum one, without any other sensor terms such as airstream incidence or normal acceleration. This has the advantage that the key motion feedback sensors, namely the pitch rate gyros, are very rugged and reliable devices.

Solid state rate gyros are very reliable sensors with an MTBF in the region of 50,000–100,000 hours – 'fit and forget' devices.

It should be noted that while airstream incidence sensors can also provide the stabilising term to counter an unstable M_α, they have two inherent shortcomings as far as a 'core' stabilising function is concerned.

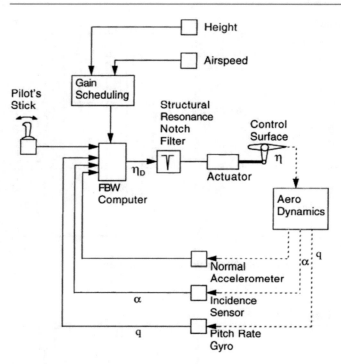

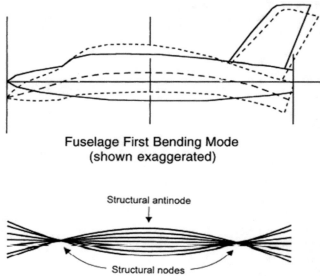

Fig. 4.17 First body bending mode

Fig. 4.16 Pitch rate command FBW loop (redundancy omitted for clarity)

1. Vulnerability to damage on the ground, or bird strikes in flight, as the sensors have to be mounted externally in the airstream.
2. Problems in matching due to the differences in the local airflow at different locations, for example, two sensors located on the port side of the fuselage and two sensors on the starboard side.

The incidence terms from the airstream sensors can be blended with the pitch rate error terms and the 'gearings' (or gain coefficients) of these terms adjusted as higher angles of attack are approached. This enables the angle of incidence to be kept within the safe limits while manoeuvring at high angles of incidence, that is, angle of incidence limiting.

Normal acceleration terms from the normal accelerometers are also blended with the other control terms and the gearings of the terms adjusted as high normal accelerations are demanded so as to give normal acceleration limiting.

The blending of normal acceleration and pitch rate is sometimes referred to as C_N^* ('C star') control. The blending of pitch rate and normal acceleration to meet pilot handling criteria is, however, a more complex subject and beyond the scope of this chapter. Sufficient to say that optimum handling requires a blend of control terms, the mixing of which needs to be adjusted over the flight envelope.

Air data gain scheduling is necessary to adjust the gearings of the control terms to cope with the changes in control effectiveness over the flight envelope as explained earlier.

So far, the aircraft has been treated as a rigid body in deriving the dynamic response. The aircraft structure, however, is flexible and has a variety of flexural and torsional modes. Figure 4.17 illustrates the first body bending mode in the aircraft's longitudinal plane. The frequency of this structural mode is typically between 8 and 15 Hz for a fighter/strike aircraft and between 2 and 4 Hz for a large transport aircraft. There are also second and third harmonic modes at twice and three times the first body mode frequency and possibly significant at even higher modes. The wings, tailplane (or foreplane) and rudder have structural (flexural and torsional) vibration modes of their own, which couple with the aircraft structure.

The FCS sensors that sense the aircraft motion also sense the structural deflections and hence couple the structure modes into the control loops. The location of the motion sensors with respect to the nodes and anti-nodes of these structural modes is important. In the example shown in Fig. 4.17, ideally, the pitch rate gyros would be located on an anti-node (direction of the gyro input axis is unchanged) and the normal accelerometer on a node (zero linear displacement). In this context, the best location for a normal accelerometer is the forward node because it reduces the non-minimum phase effect of the control loop.

The ideal locations, however, may not be practical and acceptable compromises have to be made on the sensor locations. For instance, modern strap-down inertial measuring units require the gyros and accelerometers to be co-located in a rigid block. The compromise position is usually between the forward structural node and the central anti-node.

The structural modes can be excited in two ways.

1. *Inertially*: Through the inertial forces resulting from the offset of the centre of mass of the control surface from the hinge line. These forces are directly proportional to the angular acceleration of the control surface.
2. *Aerodynamically*: Through the changes in the aerodynamic pressure distribution over the control surface and adjacent surfaces when the control surface is deflected from its datum position. These forces are proportional to the deflection of the control surface and the dynamic pressure.

It should be noted that the inertial excitation forces are in anti-phase with the aerodynamic excitation forces (as the angular acceleration of the vibrating control surface is 180° out of phase with the angular deflection). Both components are taken into account in assessing the structural coupling effects. Inertially induced excitation forces are predominant at low airspeeds in the case of all moving control surfaces such as tailplanes. The aerodynamic excitation forces are large and dominant at high airspeeds in the case of relatively light weight, but aerodynamically powerful, wing trailing edge control surfaces such as elevons and ailerons and can excite the wing structural modes.

It is thus necessary to attenuate the FCS loop gain at the structural mode frequencies to avoid driving the surfaces at these frequencies and exciting the flexible modes. Figure 4.18 illustrates the frequency response of a typical 'notch filter', so called because the gain versus frequency graph has a notch shape. The carriage of external stores and the fuel state produces significant changes in the modal frequencies and must be taken into account in the notch filter and FCS loop design. The phase lag introduced by the notch filter at frequencies below the notch frequency (Fig. 4.18) has a de-stabilising effect on the FCS loop.

4.3.1.1 Agile Fighter Pitch Rate Command Loop Example

A worked example for a pitch rate command loop for a hypothetical FBW agile fighter with a 12% negative (unstable) static margin is set out below in the diagram and Aircraft data. It is stressed that the aircraft is a hypothetical one based on published data on several modern fighter aircraft and order of magnitude estimates.

The transfer functions in this example have been expressed in terms of the Laplace operator, *s*, by substituting *s* for the operator *D*(*d*/*dt*) to conform with normal control engineering practice. The use of Laplace transforms has many advantages in dealing with the initial conditions in higher-order differential equations. Readers not familiar with Laplace transforms will find them covered in most engineering mathematics textbooks.

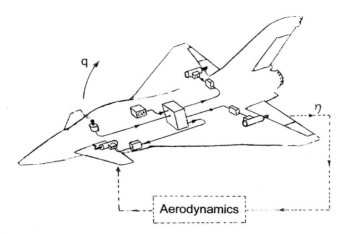

Diagram for Agile Fighter Pitch Rate Command Loop Example

Aircraft data

Mass, $m = 16,000$ kOverall length $= 14.5$ mWing span $= 11$ mWing area $S = 50$ m^2

Moment of inertia about pitch axis, $I_y = 2.5 \times 105$ kg m^2

Airspeed $VT = 300$ m/s (600 knots approximately)

- Wing incidence/g at 600 knots $= 2/3$ degree/g
- Static margin $= 12\%$ negative
- Pitching moment derivative due to control surface deflection, $M_\eta = 5 \times 106$ Nm/radian
- Pitching moment derivative due to pitch rate, $M_q = 5 \times 105$ Nm/radian/s

4.3.1.2 Problem

From the data above derive a suitable pitch control law, neglecting lags in the actuator response, non-linear effects, structural resonance notch filters, etc.

The stages in the proposed solution are set out below.

Stage 1 Derivation of q/\ transfer function of basic aircraft
- Derivation of T_1 and T_2

$$T_1 = \frac{I_Y}{M_q} = \frac{2.5 \times 10^5}{5 \times 10^5}$$

Fig. 4.18 Frequency response of typical notch filter

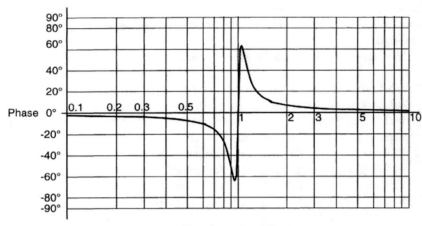

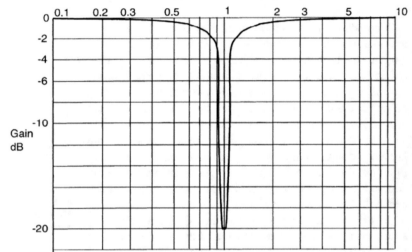

That is, $T_1 = 0.5$ s.

Given 2/3 degree wing incidence/g at 600 knots

$$Z_\alpha \frac{2}{3} \times \frac{1}{60} = 16,000 \times 10$$

($1° = 1/60$ rad and $g = 10$ m/s² approximately). Hence $Z_{\langle} = 1.44 \times 10^7$ N/rad

$$T_2 = \frac{mU}{Z_\alpha} = \frac{16,000 \times 300}{1.44 \times 10^7}$$

$T_2 = 0.33$ s.

- Derivation of M_α

Aerodynamic mean chord,

$$c = \frac{\text{wing area}}{\text{wing span}} = \frac{50}{11}$$

That is, $c = 4.5$ m

$$\text{Static margin} = \frac{\text{distance between CG and aerodynamic centre}}{\text{aerodynamic mean chord}}$$

$$= 12\%$$

Distance between CG and aerodynamic centre

$$= -0.12 \times 4.5 = -0.54\,\text{m}$$

$$M_\alpha = Z_\alpha \times (\text{Distance CG to aerodynamic centre})$$

$$= 1.44 \times 10^7 \times 0.54$$

That is, $M_\alpha = 7.8 \times 10^6$ Nm/radian

The block diagram representation of the basic aircraft dynamics is shown in Fig. 4.19 with the numerical values of M_1, M_q, $M_{\langle}$, T_1, T_2 inserted (refer to Fig. 3.30 in Sect. 3.5.4, Chap. 3).

The q/l is obtained using block diagram algebra:

$$\frac{q}{\eta} = 5 \times 10^6 \cdot \frac{\frac{1/5 \times 10^5}{(1+0.5\,\text{s})}}{1 + \frac{1/5 \times 10^5}{(1+0.5\,\text{s})} \cdot \frac{0.33}{(1+0.33\,\text{s})} \cdot \left(-7.8 \times 10^6\right)}$$

This simplifies to

Fig. 4.19 Block diagram of basic
aircraft pitch transfer function

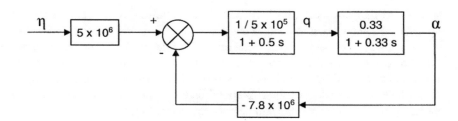

$$\frac{q}{\eta} = \frac{60(1 + 0.33\,\mathrm{s})}{(s^2 + 5s - 25)}$$

$$\frac{q}{\eta} = \frac{60(1 + 0.33\,\mathrm{s})}{(s + 8.1)(s - 3.1)}$$

The response of the basic aircraft to a disturbance or a control input is divergent with a divergent exponential component with a time constant of 1/3.1 s, that is, 0.32 s.

That is, $e^{t/0.32}$

Now $e^{0.7}$ H 2, so that the time to double amplitude $t_{\times 2}$ is given by

$$\frac{t_{\times 2}}{0.32} = 0.7$$

Hence time to double amplitude = 0.22 s.

Stage 2 Derivation of stabilising transfer function
The pitch rate demand loop can be stabilised by a controller with a transfer function

$$G_q = \left(1 + \frac{1}{Ts}\right) q_E$$

where G_q = controller gearing = degrees control surface movement per degree per second pitch rate error; T = integrator time constant; and q_E = pitch rate error.

The overall loop is shown in block diagram form in Fig. 4.20.

The open-loop transfer function is

$$\frac{q}{q_E} = \frac{60\ G_q(1 + Ts)(1 + 0.33\,\mathrm{s})}{Ts(s + 8.1)(s - 3.1)} = KG(s)$$

The closed-loop transfer function is

$$\frac{q}{q_D} = \frac{KG(s)}{1 + KG(s)}$$

The stability is determined by the roots of the equation

$$1 + KG(s) = 0$$

which must have a negative real part for stability.

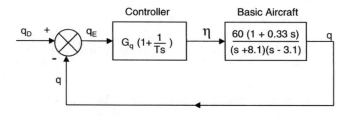

Fig. 4.20 Block diagram of overall loop

$$\left[1 + \frac{60\ G_q(1 + Ts)(1 + 0.33\,\mathrm{s})}{Ts(s + 8.1)(s - 3.1)}\right] q = 0$$

This simplifies to

$$\left[s^3 + (5 + 20\,G_q)\,s^2 + \left(60\,G_q + 20\frac{G_q}{T} - 25\right)s + 60\frac{G_q}{T}\right] = 0$$

$$(4.9)$$

This can be factorised:

$$(s + a)\left(s^2 + 2\zeta\omega_0 s + \omega_0^2\right) = 0 \qquad (4.10)$$

The desired value of $\omega 0$ is about 6.3 rad/s (1 Hz), which is well below the likely first body flexure mode frequency of, say, about 12 Hz (high g capability and aeroelastic deformation effects at high values of dynamic pressure require high stiffness structures). An $\omega 0$ of about 1 Hz also corresponds to that of many existing conventional fighters with good handling qualities. The value of damping ratio, ζ, should be about 0.6 to achieve a well-damped response and allow for the erosion of the loop phase margin due to lags in the actuator and structural resonance 'notch' filters, etc.

The desired quadratic factor is thus $(s^2 + 7.56s + 39.7)$ the roots of which are $(-3.8 + j5)$ and $(-3.8 - j5)$.

Having decided on these roots, the values of G_q and T can be obtained from the open-loop response using the root locus method. This method is considered to be beyond the scope of an introductory book. It is, however, well covered in most standard textbooks on control engineering and readers wishing to find out more about the method are referred to appropriate references at the end of this chapter.

It is, however, possible to derive the required values of G_q and T by equating coefficients of the two equations, although this is not a route which would normally be used in practice. Nevertheless, it does convey the concept of manipulating the system gains and time constants to achieve the desired roots for the closed-loop response.

Expanding Eq. (4.10) yields

$$\left[s^3 + (a + 2\zeta\omega_0)s^2 + \left(\omega_0^2 + 2\zeta\omega_0 a\right) s + a\omega_0^2 \right] = 0 \tag{4.11}$$

Equating the coefficients of s^2, s and the constant terms and substituting $\zeta = 0.6$ and $\omega_0 = 6.3$ yields:

$$5 + 20 G_q = a + 7.56 \tag{4.12}$$

$$60 G_q + 20 \frac{G_q}{T} - 25 = 39.7 + 7.56 a \tag{4.13}$$

$$60 \frac{G_q}{T} = 39.7 a \tag{4.14}$$

Solving these three simultaneous equations for the three unknowns G_q, T and a gives the following:

- $G_q = 0.46$ degree control surface/degree/s pitch rate error
- $T = 0.105$ s
- $a = 6.58 \text{ s}^{-1}$

The transient response to a disturbance or control input is thus made up of a subsidence with a time constant of $1/6.58$ s, that is 0.15 s, and a 0.6 critically damped sinusoid with an undamped natural frequency of 1 Hz (damped natural frequency is 5 rad/s or 0.8 Hz) – that is a fast, well-damped response.

The required stabilising transfer function is as follows:

$$0.46 \left[1 + \frac{1}{0.105 \, s} \right]$$

4.3.2 Lags in the Control Loop

The worked example has been very much simplified by ignoring the lags which are present in a real system and by assuming a rigid aircraft. The object is to illustrate basic principles, and in particular the use of a proportional plus integral of pitch rate error control law to stabilise an aerodynamically unstable aircraft.

The design of an FBW flight control system must take into account all the sources of lags which are present in the control loop and accurately model the structural coupling effects of the flexural modes.

Lags in any closed-loop system are de-stabilising and the significant sources of lags are as follows:

1. Actuators
2. Sensor dynamics and noise filters
3. Latency in the computational processes
4. Notch filters

These effects are briefly discussed, as follows.

Actuator response The response of the actuators at low frequencies is basically that of a low-pass (or first-order) filter, but as the frequency increases the lags in the first-stage actuation system become dominant. The output/input ratio falls at an increasing rate and the phase lag rapidly increases. The frequency response of a typical FBW actuator is shown in Fig. 4.21. Phase lag at 1 Hz is in the region of 10–12°, at 5 Hz around 50°. At 10 Hz, the lag is in the region of 90° and the output/input ratio has fallen to nearly −6 dB (approximately 0.6). It should be noted that a full dynamic model of a typical FBW actuator is around an 11th-order system when all the elements are taken into account, including the effects of compressibility in the hydraulic fluid.

The response can, however, change very suddenly if rate limiting occurs and the actuator can display a sudden large increase in phase lag, as shown in the dotted lines in Fig. 4.21. Rate limiting can occur when the control valve travel limits are reached under conditions of large-amplitude demands. The behaviour under rate-limiting conditions is non-linear and the onset and the effects are dependent on the amplitude and frequency of the input demand.

If rate limiting is encountered, say during extreme manoeuvres, the resulting additional phase lag in the actuator response can lead to a severe temporary reduction in the FBW loop stability margins. Potential handling difficulties can then occur as a consequence.

Particular care is therefore taken in the design stage to avoid rate-limiting effects in extreme manoeuvres. The control valve ports and valve travel need to be adequately sized.

Sensor dynamics The sensor dynamic response may need to be taken into account, depending on the types of sensors used. For example, angular momentum rate gyros of the two axis, dynamically tuned-type gyros (DTGs), have torque balance loops with a bandwidth in the region of 60–70 Hz. Solid state, micro-machined quartz tuning fork rate gyros also have a bandwidth in the region of 60 Hz. The phase

Fig. 4.21 Frequency response of typical FBW actuator

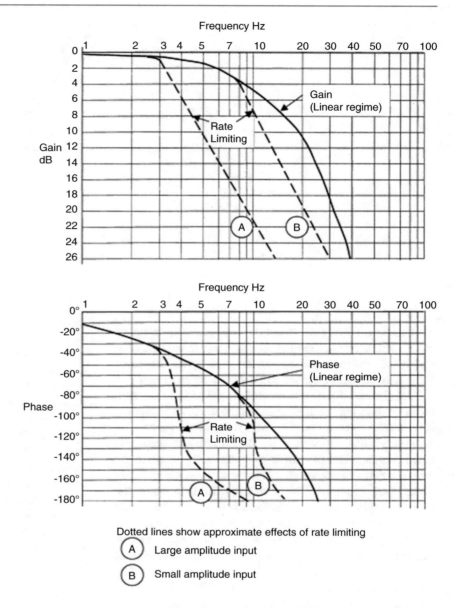

Dotted lines show approximate effects of rate limiting

(A) Large amplitude input

(B) Small amplitude input

lag at low frequencies with such sensors is fairly small (about 5° at 5 Hz) but may need to be taken into account if there are higher-frequency structural resonances present.

Noise filters may also be required on the sensor outputs; these filters also produce phase lags.

Latency This effect arises from the use of digital computers in the processing of the control laws and sensor data. It results from two effects, either of which causes a time delay and hence a phase lag. These effects are a sampling delay (the sensor data is sampled intermittently) and a transport delay (there is always some delay between when the computations are carried out and when the results of the computation are passed on to the actuators). Figure 4.22 illustrates the phase lag produced by a sampling delay; the mean time delay is half the sampling period. The phase lag produced by a time delay is directly proportional to frequency and is equal to the angular frequency multiplied by the time delay. For example, a time delay of 10 ms produces a phase lag of 7.2° at an input frequency of 2 Hz; at 10 Hz the lag is 36°.

The amplitude, however, is not attenuated as the frequency increases, as is the case with a simple lag filter, and so the de-stabilising effect is worse.

Notch filters The frequency response of a typical simple notch filter is shown in Fig. 4.18. Note the rapidly increasing phase lag as the notch frequency is approached. This is followed by a rapid change to a phase lead above the notch frequency, which reduces to zero phase shift as the frequency increases still further.

Fig. 4.22 Phase lag introduced by sampling delay

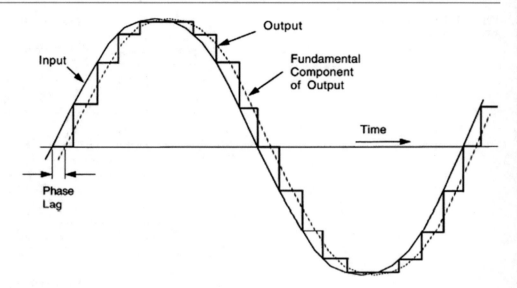

A modern FBW aircraft can have, perhaps, up to ten or more notch filters in the pitch control loop to cover a wide variety of structural resonances. Perhaps up to six notch filters may be required in the roll and yaw control loops.

Digital notch filters are generally used rather than analogue because the notch frequencies are very stable and the lane match errors are small. There may, however, be analogue notch filters as well as the digital filters. It should be noted that the effect of sampling and transport delays in the case of a digitally implemented filter is to cause a slight shift in the notch frequency. The normal practice is to 'warp' the filter by slightly adjusting the constants so that the actual frequency comes out right after allowing for sampling.

The lags contributed by successive notch filters at low frequencies erode the phase and gain margins of the control loops. Refer to Fig. 4.18, which shows phase versus frequency. Although the phase lag at the FBW loop response frequencies (e.g. 1–3 Hz) may only be a few degrees, the cumulative lag from several notch filters can be significant.

The introduction of a phase advance transfer function in the control law to restore the phase margin, however, increases the gain at the higher frequencies where the structural resonances are present. This limits the amount of phase advance which can be used. Referring to Fig. 4.11, it can be seen that a value of $n = 3$ will produce a maximum phase lead of 28° but increases the gain at high frequencies by nearly 10 dB (n is the ratio of the time constants in the numerator and denominator of the transfer function, respectively). This clearly reduces the effective depth of the notch filters so that there is a considerable optimisation task to be undertaken.

An appropriate example of the effect on the FCS loop stability from the phase lags produced by several notch filters is shown on a Nichols chart in Fig. 4.23, together with the effect of a phase advance element.

Nichols charts show the open-loop gain and phase at particular frequencies and enable the gain and phase margins to be read off directly.

The gain margin is the amount the loop gain can be increased before instability results because the open-loop gain at the frequency where there is 180° phase lag has reached 0 dB (unity). Typical gain margins are around 9 dB.

The phase margin is the additional phase lag that will cause instability by producing 180° phase lag at the frequency where the open-loop gain is 0 dB. Typical phase margins are around 45°.

It should be noted that the phase of a lightly damped structural resonance changes very rapidly from near zero just before the resonant frequency to −180° just above the resonant frequency, being −90° at the resonant frequency. The order of the system is greatly increased by the structural resonances and notch filters so that the phase exceeds multiples of 360° at higher frequencies. (The phase of an nth-order system reaches $n \times 90°$.) Modern FBW agile fighters can have up to 80th-order longitudinal control loops when all factors are taken into account.

In order for the loop to be stable at the structural resonance frequencies, the open-loop gain must not exceed 0 dB where the phase is −180° plus multiples of −360°; that is −540°, −900°, −1260°, etc. (as well as −180°). An adequate gain margin of, say, 9 dB is required at these frequencies.

Structural resonance frequencies where the open-loop phase lag is −360° or multiples of −360° can exceed 0 dB gain at these frequencies, as the action of the loop is to suppress these frequencies.

Clearly, the open-loop phase as well as the gain must be known at the structural resonance frequencies under all circumstances, if phase stabilisation is being relied on. The alternative is to require an adequate gain margin, say 9 dB, at all the resonant frequencies and not rely on an accurate knowledge of the phase.

Fig. 4.23 Nichols chart open-loop frequency responses

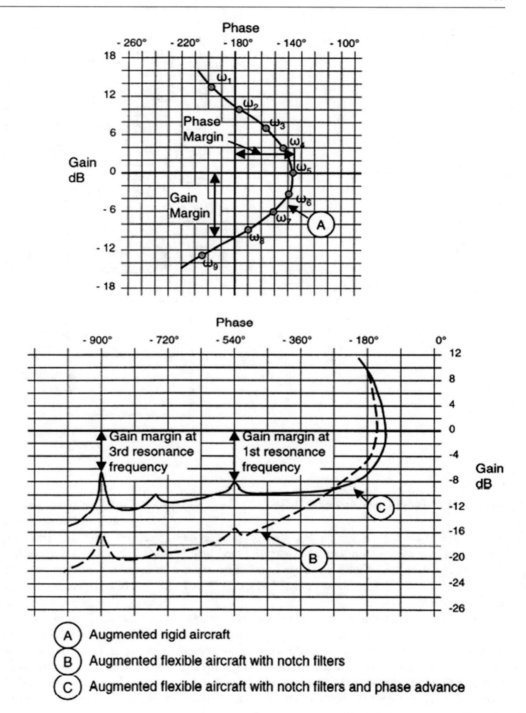

(A) Augmented rigid aircraft

(B) Augmented flexible aircraft with notch filters

(C) Augmented flexible aircraft with notch filters and phase advance

4.3.3   Roll Rate Command Control

A roll rate command system is shown in Fig. 4.24, the primary feedback term being roll rate from the roll rate gyros. Other sensors are also used to control the rudder automatically to counteract and suppress the lateral cross-coupling effects described earlier in Chap. 3 (Sect. 3.5). A yaw rate, r, feedback term derived from the yaw rate gyros

provides the yaw auto-stabilisation function. Other terms can include yaw incidence angle, β, from yaw incidence sensors and lateral acceleration from lateral accelerometers.

As in the pitch rate command loop, air data gain scheduling is used to adjust the gearings (or gains) of the control terms. Derivative and integral terms may also be used. Notch filters are also used to attenuate the loop gain at the structural resonance frequencies.

Fig. 4.24 Lateral FBW control loops – roll rate command

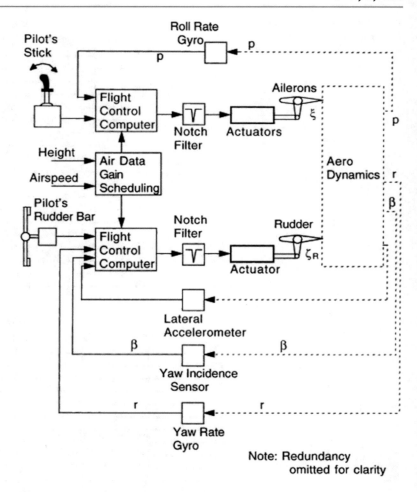

Note: Redundancy omitted for clarity

4.3.4 Handling Qualities and PIOs

The control of the aircraft's flight path by the pilot can be represented as a closed-loop process as shown in the block diagram, Fig. 4.25. The pilot is an integral part of the outer loop in series with an inner loop comprising the FCS and aircraft dynamics.

The pilot controls the pitch attitude and bank angle through the FCS so as to change the magnitude and spatial direction of the lift force vector. The lift force vector acting on the aircraft then produces the desired change in the aircraft's flight path through the flight path kinematics. The loop closure is through the pilot's senses. The matching of the aircraft's response to the pilot's response thus determines the aircraft handling (or flying) characteristics. Good handling qualities can be defined as those characteristics of the dynamic behaviour of the aircraft which enable the pilot to exercise precise control with low pilot workload. The achievement of good handling characteristics is a major objective in the FCS design.

The quality of the handling characteristics and the pilot workload are generally expressed in terms of the Cooper–Harper scale of ratings from 1 to 10; 1 being excellent and

10 the worst qualities possible. These ratings are derived from pilots' comments in carrying out a range of tasks, which are defined in terms of flight phase categories A, B and C. Category A consists of demanding tasks such as air to air or air to ground combat and in-flight refuelling. Category B covers less demanding tasks such as climb, cruise and descent. Category C covers terminal tasks such as landing and take-off. Table 4.1 shows the Cooper–Harper scale of gradations for measuring pilot opinion.

Deficiencies in the FCS design can result in what are referred to as 'pilot-induced oscillations', or 'PIOs' for short. These are sustained or uncontrollable oscillations resulting from the efforts of the pilot to control the aircraft. They can have catastrophic results in extreme cases.

The phenomenon of PIOs has been encountered in aircraft since the beginning of powered flight; the original Wright Flyer was susceptible to PIOs. PIOs in conventional un-augmented aircraft have generally been due to design deficiencies such as low damping of the short-period pitch mode or low control sensitivity (stick force/g).

With the advent of FBW aircraft with high gain feedback control systems, however, PIOs have generally been the result of some non-linear event such as rate limiting in the

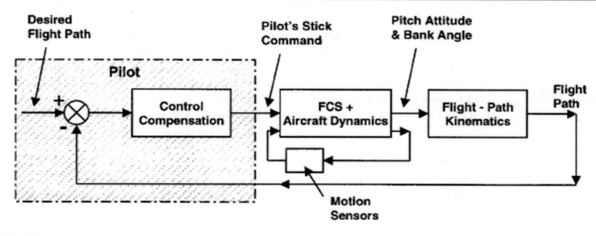

Fig. 4.25 Pilot in the loop

Table 4.1 Cooper–Harper scale

Level	Pilot rating	Aircraft characteristics	Demands on the pilot in selected tasks or required operations
1 Satisfactory without improvement	1	Excellent, highly desirable	Pilot compensation not a factor for desired performance
	2	Good, negligible deficiencies	Pilot compensation not a factor for desired performance
	3	Fair, some mildly unpleasant deficiencies	Minimal pilot compensation required for desired performance
2 Deficiencies warrant improvement	4	Minor but annoying deficiencies	Desired performance required moderate pilot compensation
	5	Moderately objectionable deficiencies	Adequate performance required considerable pilot compensation
	6	Very objectionable but tolerable deficiencies	Adequate performance required extensive pilot compensation
3 Deficiencies warrant improvement	7	Major deficiencies	Adequate performance not attainable with maximum tolerable pilot compensation; controllability not in question
	8	Major deficiencies	Considerable pilot compensation is required for control
	9	Major deficiencies	Intense pilot compensation is required to retain control
Improvement mandatory	10	Major deficiencies	Control will be lost during some portion of required operation

actuation system. Rate limiting can result in an abrupt increase in the actuator lag and consequent degradation and lag in the FCS response. PIOs were, in fact, encountered in early development flight testing on the Lockheed YF22 fighter and the prototype SAAB JAS 39 Gripen fighter and also during early flight testing of the C17 transport and Boeing 777. These have been attributed to rate-limiting effects. It should be noted, however, that rate limiting does not necessarily result in PIOs and the methodology of predicting PIO susceptibility, particularly following a non-linear event, is not yet at a mature state. Analysing the effects of non-linear behaviour in an element in a control loop is very much complicated by the fact that the response of the element is non-linear and is dependent on the frequency and amplitude of the pilot's inputs.

The subject of handling characteristics and PIOs involves very considerable background knowledge and it is only possible within the space constraints to give a very brief overview.

The control exercised by the pilot (refer to Fig. 4.24) is influenced by the following factors. There is an inherent time delay of the order of 0.3 second in the pilot's response. The pilot's control sensitivity, or gain, however, can be varied and the pilot is able to provide a phase lead to compensate for the aircraft's response, or a smoothing lag filter response if required. The pilot's bandwidth is of the order of 0.5 Hz. It should be noted that pilots dislike having to apply lead or lag because of the resulting increase in workload.

Consider now the pilot as part of a closed-loop control system. There are many lag elements in the outer control loop comprising the pilot and the FCS/airframe combination and the phase lag round the loop reaches and exceeds 180° as the frequency is increased. It can be seen that, as in any closed-loop control system, increasing the gain exercised by the pilot at the 180° lag frequency will result in the loop going unstable if the loop gain reaches unity.

Hence, if the pilot increases his gain in an attempt to get tighter control and overcome a lag in the aircraft response, the consequence can be that the aircraft becomes more out of control. The pilot gets out of phase and the harder the pilot works to control the aircraft, the worse the situation becomes and PIOs result.

References on the subject of handling characteristics are given at the end of this chapter.

4.3.5 Modern Control Theory

The treatment of closed-loop flight control systems in this book has used the classic linear control theory approach. The aircraft dynamics have been simplified where possible to enable a good basic understanding to be obtained – for example, simplifying the pitch dynamics to a second-order system transfer function by assuming the forward speed is constant over the short-period motion, deriving the roll rate response to aileron movement by assuming pure rolling motion, and neglecting roll/yaw cross-coupling effects.

Similarly, the short-period yawing motion is first derived by assuming pure yawing motion and the yaw/roll cross-coupling effects neglected. A more accurate representation of the dynamic behaviour has then been derived and the representation of the aircraft dynamics as a matrix array of first-order state equations has been explained in Chap. 3 (Sect. 3.4.5). The representation of the system dynamics and the control processes in the form of matrix arrays of first-order equations is a key feature of what is now referred to as modern control theory.

Flight control systems come into the category of multi-input/multi-output (MIMO) closed-loop control systems as control is exerted about three axes and there are six degrees of freedom. The classic control theory approach is very suitable for single input/single output (SISO) closed-loop control systems and some single input/multi-output (SIMO) systems (e.g. two outputs).

However, modern control theory has been specifically developed to deal with MIMO systems and is now widely used in the design of modern flight control systems. The techniques are very powerful and can deal with non-linear systems. The methods, however, require a good understanding of matrix algebra and not all readers are familiar with this subject. For this reason, the classic control theory approach has been considered more suitable for this book as it is intended to act as an introduction to what is a large multi-discipline subject. It should also be stressed that classic control theory methods are extensively used in parallel with modern control theory methods because of their robustness and visibility in particular areas.

Appropriate references on modern control theory methods are given at the end of this chapter for those readers wishing to know more about the subject.

4.4 Redundancy and Failure Survival

4.4.1 Safety and Integrity

Clearly, the FBW flight control system must be no less safe than the simple mechanical control systems which it replaces. The safety levels required are specified in terms of the probability of a catastrophic failure occurring in the system from any cause whatsoever which could result in loss of control of the aircraft. It is generally specified that the probability of a catastrophic failure in the flight control system must not exceed 1×10^{-7}/hour for a military aircraft or 1×10^{-9}/hour for a civil aircraft.

These very low-probability figures are difficult to appreciate and also impossible to verify statistically. To give some idea of their magnitude, a failure probability figure of 1×10^{-9}/hour means that a fleet of 3000 aircraft flying an average of 3000 hours per annum would experience one catastrophic failure of the FBW system in 100 years! Military aircraft utilisation per annum is much lower than civil transport aircraft, so a figure of 1×10^{-7}/hour for the FBW system is acceptable.

It should be noted that the statistical level of safety currently being achieved with civil aircraft transport corresponds to a figure of around 1×10^{-6}/hour. This figure is derived from the total number of civil aircraft crashes occurring in a year from all causes divided by the total number of aircraft flying and their annual operating hours.

The mean time between failures, or MTBF, of a single channel FBW system is in the region of 3000 hours. The FBW system must thus possess redundancy with multiple parallel channels so that it is able to survive at least two failures, if these very low failure probability figures are to be met.

There is clearly an economic limit to the number of parallel redundant channels. Apart from the cost of the additional channels, the overall system MTBF is reduced and hence the system availability is lowered. While it may be acceptable to fly with one failed channel, assuming sufficient redundancy, the impact on availability is obvious if the overall MTBF is too low.

Fig. 4.26 Quadruplex system configuration. (Note: position feedback from control surface actuators omitted for clarity)

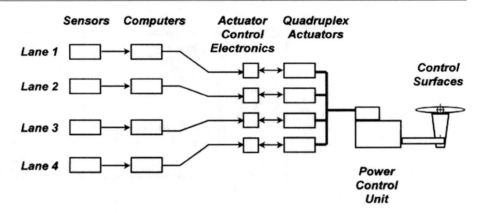

4.4.2 Redundant Configurations

A well-established redundant configuration comprises four totally independent channels of sensors and computers in a parallel arrangement to give the required failure survival capability – such a configuration is referred to as a quadruplex system. The four independent channels are then configured to drive a failure survival actuation system with sufficient redundancy such that the overall FBW system of interconnected sensors, computers and actuators can survive any two failures from whatever cause. Figure 4.26 illustrates the basic quadruplex configuration.

The integrity of the electrical power supplies and the hydraulic power supplies is absolutely vital and adequate redundancy must be provided so that the system can survive failures in both the electrical and hydraulic power supplies. In fact the starting point in the design of any FBW system is the safety and integrity and redundancy levels required of the electrical and hydraulic power supply systems as their integrity dominates the overall FBW system integrity.

With four totally independent channels, the assumption is made that the probability of three or four channels failing at the same instant in time is negligible. This assumption will be discussed in Sect. 4.4.5. Thus if one system fails 'hard over' (i.e. demanding maximum control surface movement), the other three parallel 'good' channels can override the failed channel. However, to survive a second failure it is necessary to disconnect the first failed channel, otherwise it would be stalemate – two good channels versus two failed hard over channels so that the control surface does not move – a 'fail passive' situation.

Failures are therefore detected by cross-comparison of the parallel channels and majority voting on the 'odd man out' principle. The quadruplex system is thus able to survive two failures by majority voting and disconnecting the failed channels, the system degrading to triplex redundancy after the first failure and duplex redundancy after the second failure. A third failure results in a fail passive situation, the 'good' channel counteracting the failed channel.

The incorporation of a monitoring system to check the correct functioning of a channel to a very high confidence level can also enable a failed channel to be identified and disconnected and this leads to an alternative failure survival configuration known as 'monitored triplex'. A monitored triplex configuration comprises three totally independent parallel channels with each channel monitored by a dissimilar system to detect a failure. Provided this monitoring is to a sufficiently high degree of integrity and confidence level, such a system can survive two failures.

The respective merits of the two systems are fairly evenly divided; the monitored triplex system has rather less hardware and so can be of lower cost. This is offset by the better 'visibility' in terms of failure survival confidence of the quadruplex system, particularly when it incorporates self-monitoring to a high confidence level.

Figure 4.27 illustrates the quadruplex and monitored triplex redundancy configurations schematically.

4.4.3 Voting and Consolidation

As already mentioned, failures are detected by cross-comparison and majority voting. However, it is necessary to allow for normal variations in the outputs of the parallel sensors measuring a particular quantity due to inherent errors in the sensors and their manufacturing tolerances. Normal errors in sensors include such parameters as follows:

(a) Scale factor errors
(b) Linearity errors
(c) Null or zero offset errors
(d) Hysteresis errors
(e) Variation of the above parameters with temperature

The absolute accuracy of the sensor is rather less important (within reason) than the variation, or 'tracking error', spread between individual sensors particularly under dynamic conditions.

Fig. 4.27 Redundancy configurations

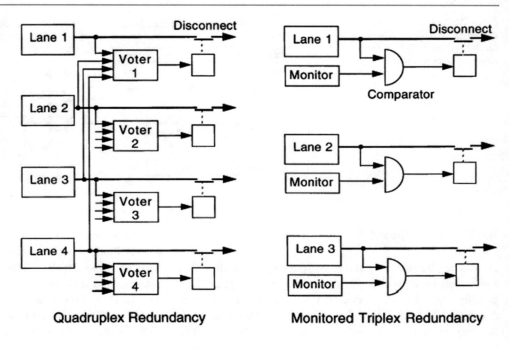

Quadruplex Redundancy

Monitored Triplex Redundancy

Fig. 4.28 Sensor tracking – effect of a small phase difference

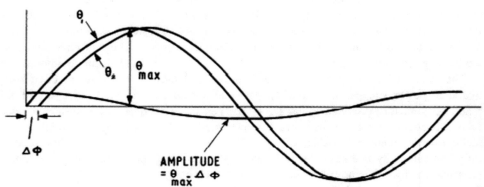

For instance, a situation where all the sensors were in error by nearly the same amount and the variations between the individual sensors were small would be preferable to a higher absolute accuracy but a larger variation. Variations in the dynamic response of the parallel sensors must be small. For instance, consider two sensors measuring a sinusoidally varying quantity but with a slightly different phase lag between the two sensor outputs. The difference between these two sensor outputs is a sinusoidally varying quantity 90° out of phase with the input quantity and with an amplitude directly proportional to this phase difference (Fig. 4.28). Thus, a phase difference of 2° (1/30 rad approximately) results in a tracking spread of over 3% of the peak value of the input quantity.

This result is arrived at as follows: let θ_1 and θ_2 be the two sensor outputs measuring an input quantity θ, the first sensor having a phase lag φ and the second sensor having a phase lag $(\varphi + \Delta\varphi)$ and assume for simplicity that both sensors measure the peak amplitude θ_{max} without error.

Thus difference between the two sensor outputs is

$$\theta_{max} \ \sin(\varphi t - \varphi) - \theta_{max} \ \sin[\varphi t - (\varphi + \Delta\varphi)]$$
$$\approx \Delta\varphi\theta_{max} \ \cos(\varphi t - \varphi)$$

The sensors should thus have a high bandwidth so that the phase lags at aircraft control frequencies are small anyway and the phase variations between the sensors will thus be very small. The variations between the filters used to filter sensor output noise are also minimised by using accurate, stable components in the filters.

The failure detection algorithms enable the detection and isolation of a sensor whose output departs by more than a specified amount from the normal error spread. The value of this disconnect threshold in conjunction with the failure detection and isolation algorithm determines the magnitude of the transient which the aircraft may experience as a result of the failed sensor being disconnected.

There are thus two conflicting requirements:

1. Low number of nuisance disconnections
2. Minimum transient on disconnecting a failed sensor

If the disconnect threshold is set too low, a sensor can be deemed to have failed and be disconnected when it is merely at the edge of its specification tolerances, that is, a nuisance disconnection. Conversely, if set too high, when a real failure does occur, the magnitude of the transient on disconnecting the failed sensor can be unacceptably large.

Failures can be divided into the following categories:

(i) *'Hard over'* failures whereby the failed sensor output is hard over full scale and would demand full authority control surface movement with catastrophic results.

(ii) *Zero output failures* whereby the sensor output is zero and thus no control action would take place.

(iii) *Slow over failures* whereby the sensor output is slowly increasing or drifting with time, although the input is stationary, and would eventually cause a full scale hard over.

(iv) *Oscillatory failures* whereby the sensor output is oscillating, the amplitude and frequency being dependent on the type of failure. Loss of feedback, for instance, can result in a stop-to-stop oscillation.

(v) *Soft failures* whereby the sensor is functioning but its output is outside the specification tolerances.

(vi) *Intermittent failures* whereby the sensor fails and then recovers intermittently.

There are very large numbers of practical voting algorithms. A detailed discussion of voting algorithms is beyond the scope of this chapter and only one commonly used algorithm is therefore described. The basis of this algorithm is to select the middle value of the parallel sensor signals and compare the other signals with it. In the case of a quadruplex system, the lower of the two middle values is selected. The differences from the selected middle value of the highest and lowest sensor values are continually monitored. The four sensor outputs are valid providing these differences are less than the disconnect threshold value. Conversely, if one sensor differs by more than the disconnect threshold it is deemed to have failed and its output is disconnected.

For example, suppose full scale sensor output is 100 and the normal sensor tolerance is ±0.5%. The disconnect threshold is generally taken as (1.5 × maximum signal variation), and in this example would thus be equal to 1.5. Suppose for example sensor outputs for a particular quantity were as follows:

$$\theta_1 = 60, \quad \theta_2 = 60.5, \quad \theta_3 = 60.7, \quad \theta_4 = 59.7$$

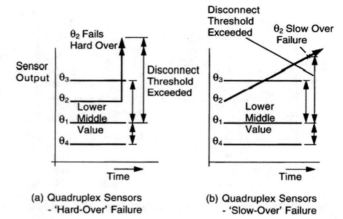

(a) Quadruplex Sensors - 'Hard-Over' Failure

(b) Quadruplex Sensors - 'Slow-Over' Failure

Fig. 4.29 Operation of failure detection algorithm – first failure

The lower middle value is 60 and the differences of the higher and lower value sensor outputs from this middle value are 0.5 and 0.3, respectively, so that all sensor outputs would thus be valid.

Consider now two failure examples: (a) a hard over failure and (b) a slow over failure, as shown in Fig. 4.29.

(a) The hard over failure results from, say, sensor θ_2 output going hard over to the full scale value of 100, that is, $\theta_2 = 100$, with the input quantity stationary so that sensor outputs θ_1, θ_3, θ_4 are unchanged. The difference between θ_2 and this lower middle value of 60 greatly exceeds the disconnect threshold and θ_2 would be detected as having failed and disconnected accordingly.

(b) The slow over failure results from say sensor θ_2 drifting with time from its original value of 60.5 and with the input quantity stationary so that θ_1, θ_3, θ_4 are unchanged. When θ_2 reaches a value of just over 61.5, the difference from the new middle value of 60 exceeds the disconnect threshold value of 1.5 and θ_2 is disconnected.

After either of the above failures, the system would be at triplex level and the same two failure cases are considered for a second sensor failure, and are shown in Fig. 4.30.

Suppose sensor values are $\theta_1 = 60$, $\theta_3 = 60.7$, $\theta_4 = 59.7$:

(a) Hard over failure results from, say, θ_3 failing hard over to full scale, that is, $\theta_3 = 100$, with the input quantity stationary. The difference of θ_3 from the middle value of 60 greatly exceeds the disconnect threshold so that the θ_3 failure is detected and isolated.

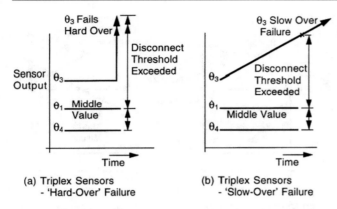

Fig. 4.30 Operation of failure detection algorithm – second failure

(b)   Slow over failure results from θ_3 slowly drifting off towards full scale from its original value of 60.7, with the input quantity stationary. θ_3 will be detected as having failed when its output exceeds a value of 61.5 and disconnected accordingly.

The adoption of digital computing technology enables more sophisticated failure isolation strategies to be adopted to reduce 'nuisance disconnections'. For instance, a sensor whose output differs from the other sensors during dynamic or transient conditions by an amount which just exceeds the disconnect threshold can be 'put on ice' and temporarily isolated. Its output can be continually compared with the other sensors and if it recovers within tolerance it can then be reconnected. The system can be designed to allow, say, up to ten reconnections before completely isolating the sensor.

The use of integral terms in the control laws makes it essential that the outputs of the individual parallel sensors are consolidated and only one signal is used for control purposes in each of the computing channels. This is because any differences between the sensor outputs would be integrated up with time so that the four computer outputs would diverge.

Thus if ε_1, ε_2, ε_3, ε_4 are the respective errors in the four sensor outputs, the four computer outputs would differ by $\int\varepsilon_1 dt$, $\int\varepsilon_2 dt$, $\int\varepsilon_3 dt$, $\int\varepsilon_4 dt$ and hence will diverge with time.

The lower middle value of the four sensor outputs (or middle of the three sensor outputs in the triplex case) is generally used as the consolidated output by the four computers.

4.4.4   Quadruplex System Architecture

The tasks carried out by each lane of a quadruplex system are shown in Fig. 4.31. A typical quadruplex system architecture is shown in Fig. 4.32 with voters and output consolidation.

The essential buffering and isolation of outputs (e.g. electro-optic isolators) is omitted for clarity, but is essential to prevent electrical faults in one channel propagating into another – common mode failures.

4.4.5   Common Mode Failures

As already mentioned, the basis of the fault detection and isolation technique described relies on the probability of a single event causing all the parallel channels to fail simultaneously as being negligibly small. The type of failure which can affect all systems at the same time is termed a Common Mode Failure.

Examples of common mode failures are as follows:

- Lightning strike
- Electro-magnetic interference
- Fire/explosion/battle damage
- Incorrect maintenance
- Common design errors – software

Every care is taken to minimise the probability of these failures occurring. For instance, very stringent electromagnetic (EM) shielding practices are used including screened cables, segregation of cables and units, enclosure of all electronic components in EM shielded boxes with incoming wires terminated with EM filters, etc. An attractive alternative is to transmit all the signals as coded light pulses suitably time division multiplexed along fibre optic cables, the optical transmission media being unaffected by any electro-magnetic interference (EMI). The fibre optic cable provides complete electrical isolation and eliminates the possibility of propagation of electrical faults between units. It also offers a much higher data rate transmission capability. The use of such technology is sometimes described as a 'fly-by-light' flight control system.

Hazards from fire, explosions or battle damage are minimised by the physical segregation and separation of the individual channels and is sometimes referred to as a 'brick wall' separation philosophy.

Very stringent control, inspection and maintenance disciplines are exerted to eliminate as far as is practical the possibility of common maintenance errors.

The common design error, which could affect all the independent channels, is a very major problem. One of the most difficult areas in the case of a digital FBW system is the possibility of an undetected error in the software, which could affect all the channels. Very great care is taken in software generation to eliminate software errors as far as possible by the adoption of very stringent design procedures: these are briefly covered in Sect. 4.5.3.4.

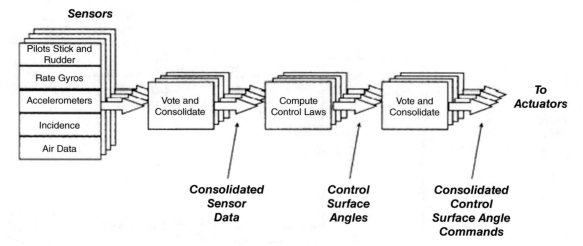

Fig. 4.31 Lane processing tasks

Fig. 4.32 Quadruplex system architecture

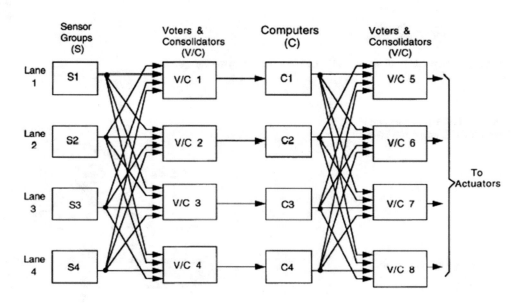

The problem of eliminating the possibility of a common mode failure from whatever cause has thus led to the use of what is known as dissimilar redundancy.

4.4.6 Dissimilar Redundancy

1. Use of two or more different types of microprocessors with dissimilar software.
2. Use of a backup analogue system in addition to the main digital system, which is at quadruplex or triplex level of redundancy.
3. Use of a backup system using different sensors, computing and control means, for example, separate control surfaces.
4. Combinations of Eqs. (4.1) to (4.3) above.

Figure 4.33 shows a generalised dissimilar redundant flight control system architecture.

Modern airliner design practice is to split the basic control surfaces into two or three sections with each section controlled independently by its own actuators. The moments exerted by each section of the control surface are thus summed aerodynamically and a failure of an individual section can be tolerated. This gives increased failure survivability to the flight control system. Most of the control surfaces are controlled by two actuators: in some cases dissimilar types of actuator.

Figure 4.34 shows the flight control surfaces of the Airbus A380 airliner, which is very similar to that of the new A350 airliner. On the A380, roll control is provided on each wing by three ailerons complemented by outboard spoilers. Pitch control is provided by the Trimmable horizontal Stabiliser

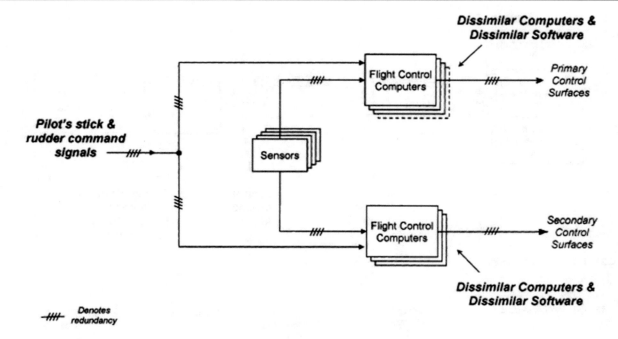

Fig. 4.33 Generalised dissimilar redundant flight control system architecture

Fig. 4.34 A380 Flight control surfaces. (Courtesy of Airbus)

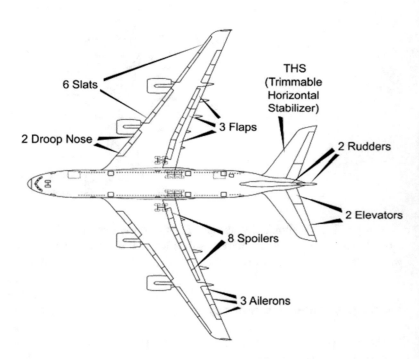

(THS) and four elevators hinged on the THS. Yaw control is provided by two rudders Whereas, on the A350, Roll control is provided on each wing by two ailerons complemented by inboard spoilers. Pitch control is provided by the Trimmable Horizontal Stabiliser (THS) and two elevators hinged on the THS; yaw control is provided by a single rudder. There are 12 slats, 4 adaptive dropped hinge flaps and 2 droop nose devices.

The Airbus A318, A319, A320, A330, A340, A350, A380 share a common flight control system architecture with similar control laws that provide similar flight handling characteristics enabling easy cross-qualification across the range. The A350 and A380 flight control system is shown in Fig. 4.35; the control architecture shown in Fig. 4.35 applies across the range.

Fig. 4.35 A380 flight control system. (Courtesy of Airbus)

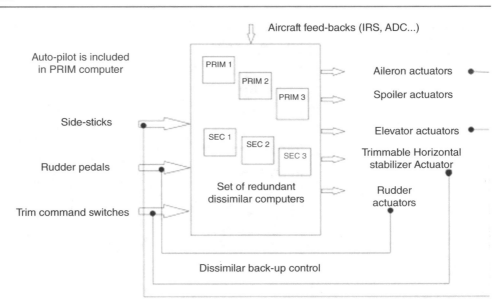

Three totally independent primary (PRIM) flight control computers control one second set of ailerons, spoilers, elevators and rudder. The autopilot is included in the PRIM.

Three totally independent secondary (SEC) flight control computers control a second set of ailerons, spoilers, elevators, rudder and the THS. Each computer is capable of controlling pitch, roll and yaw axes through its associated actuators.

Each computer comprises two independent processing channels with different internal architectures, the processors being procured from different manufacturers with their software generated by different teams. In the case of disagreement between the two processing channels, the computer is automatically disconnected.

The A380 and the A350 can be dispatched with any one of the six computers failed.

Under normal control laws, the pilot exercises a manoeuvre command control through the side-stick of load factor for longitudinal control and roll rate for lateral control.

(Load factor is directly related to the normal acceleration, ng, and is equal to $(1 + n)$g. A load factor of 1 g corresponds to straight and level flight.)

The A350 flight control system benefits from evolutions introduced on the A380:

- Integration of the flight guidance (FG) and flight envelope (FE) functions in the primary flight control computers (PRIMs)
- Replacement of all mechanical backup controls by electrical backup controls
- Addition of a new pitch trim switch, which replaces the trim wheels
- Introduction of active stability longitudinal and lateral axes

- Introduction of electro-hydrostatic actuators (EHAs) and electro-backup hydraulic actuators (refer to Actuators)

Inside the normal flight envelope of both the A350 and A380, the main features are:

- Neutral static stability and short-term attitude stability
- Automatic longitudinal trimming
- Lateral attitude hold and automatic elevator in turn
- Dutch roll damping
- Turn coordination

Flight control protection (load factor, angle of attack, speed, attitude) is also part of the flight control laws and provides the pilot with instinctive procedures to get maximum aircraft performance in emergency situations.

Aircraft feedback is provided by the Air Data Inertial Reference System (ADIRS) and dedicated sensors (accelerometers, rate gyros, etc.).

The flight control functions are exercised through the PRIM and SEC computers, as shown in Fig. 4.36.

One of the PRIMs is the master computer and carries out the computation function. It computes the control surface deflections for the aircraft to respond to the pilot's, or Autopilot, commands in an optimum manner from the appropriate control laws and the aircraft feedback from the ADIRS.

The three PRIMs and the three SECs each carry out the execution function of control and monitoring of their assigned servo actuators so that control surface deflections obey the computer demands from the master PRIM.

The master PRIM also carries out self-monitoring, and checks its commands are being executed by comparing them with the aircraft response feedback from the ADIRS.

Fig. 4.36 A350 flight control Backup Power Supply System. (Courtesy of Airbus)

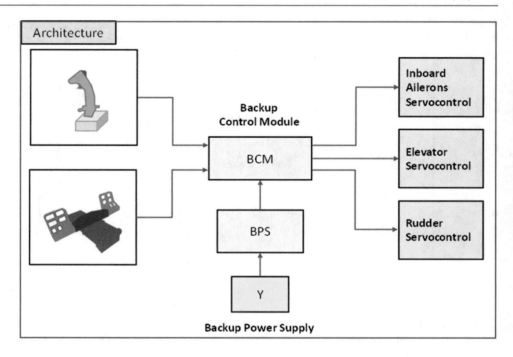

If a malfunction is detected, the master PRIM passes the computation function to another PRIM. The master PRIM may continue to perform the execution function depending on the malfunction.

If all the PRIMs are lost, each SEC performs the computation and execution function; there is no master SEC.

Progressive control law reconfigurations occur as a function of a system failure status and are indicated to the pilot through the following laws:

- *Normal law*: Auto-trim control laws and flight domain protections available.
- *Normal flight control law* is available after any single failure of sensors, electrical system, hydraulic systems of PRIM.
- *Alternate law*: Auto-trim control laws and flight domain warnings.
- *Direct law:* Manual trim, direct relationship between sidestick deflection and surface positions, longitudinal and lateral stabilisation and flight domain warnings.

An electrical backup system provides control of the aircraft in the extremely improbable case of loss of all the flight control computers, or electrical power supplies. The electrical backup system is totally segregated from the normal flight control system and uses dedicated sensors and transducers in the pilot controls.

The backup control module (BCM) controls and monitors the inboard elevators, the THS, the inboard ailerons and the rudder only. Specific control laws apply whenever the electrical backup is active, and provide pitch motion damping, yaw damping and roll control.

The THS can also be controlled from trim switches located on the pedestal.

An electrical backup system controls the aircraft in the case of the failure of the following:

- All the PRIMs and all SECs
- The electrical power supply of the PRIMs and the SECs

The electrical backup system is totally segregated from the normal flight control systems and has:

- A backup power supply (BPS). The BPS is an electrical generator that is activated in the case of computer or electrical generation failure. The yellow hydraulic circuit supplies the BPS.
- A backup control module (BCM). The BCM controls and monitors the following:
  - The inboard ailerons
  - The elevators
  - The rudder

The direct control laws apply whenever the electrical backup system is active, with the following features:

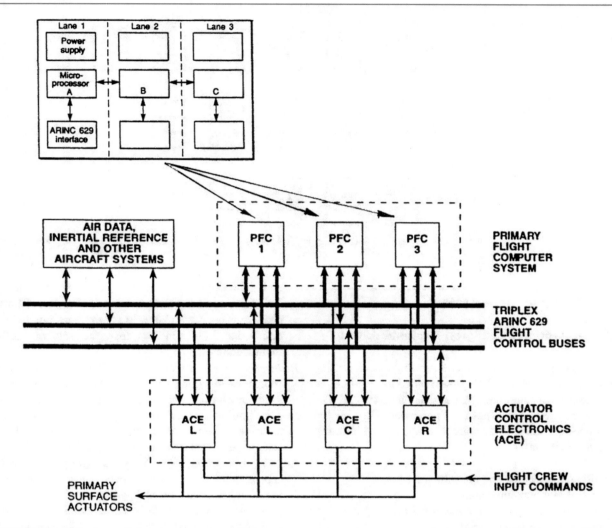

Fig. 4.37 Boeing 777 primary flight computer system architecture. (Note: Primary flight computers are located in different places – two in the E/E bay, one aft of the forward cargo door)

- Pitch motion damping
- Yaw damping
- Direct roll

The FBW flight control system of the Boeing 777 also makes extensive use of dissimilar redundancy.

Figure 4.37 is a simplified block diagram of the primary flight control system architecture. The pilot's commands are transmitted directly to the four Actuator Control Electronics (ACE) units, which are then routed through the ACE to the ARINC 629 data buses.

At the heart of the system there are three identical digital primary flight computers (PFC). Each PFC forms a channel so that the three separate PFCs provide three independent control channels in the primary flight control system.

The PFC architecture is shown inset and comprises three independent dissimilar processors being generated by

independent groups to the same requirement specification. The system normally operates with one processor in each PFC in command and with the other two processors acting as monitors. The PFCS is able to absorb multiple random component failures or a combination of a software generic error and random failures. Figure 4.38 illustrates the failure absorption capability.

In the very unlikely event of the PFCS becoming totally inoperable, a reversionary analogue command path is directly available through the ACE to provide aircraft control. Each ACE also contains a solid state pitch rate gyro for use when normal sources of body motion are not available to enable a pitch auto-stabilisation function to be maintained. In addition to this backup mode, an independent mechanical link is also provided via the stabiliser trim system and a pair of flight spoilers.

Fig. 4.38 Fault tolerance – lane failure

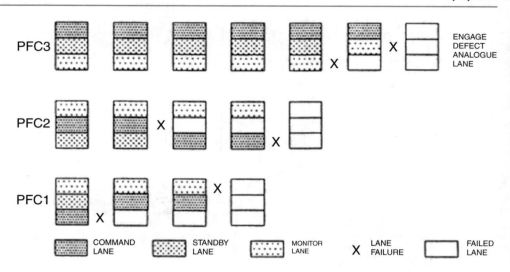

4.5 Digital Implementation

4.5.1 Advantages of Digital Implementation

Modern FBW flight control systems are implemented using digital technology, signals being transmitted as serial digital data using time division multiplexed data bus networks. The signal data are subsequently processed by digital microprocessors in the flight control computers, which carry out the following tasks:

(a) Voting, monitoring and consolidation
(b) Control law implementation
(c) Reconfiguration in the event of a failure
(d) Built-in test and monitoring

Primary FBW flight control systems using analogue computing technology are still in service but nowadays analogue technology is mainly used to provide a dissimilarly redundant backup system for protection from common mode failures, as just explained.

The advantages of modern digital implementation compared with analogue are overwhelming with the technology now available. For instance:

Hardware economy: One computer can control all three axes of control, whereas an analogue system requires dedicated hardware for each axis of control. The reduction in hardware weight and volume is of the order of 5:1 for a system of even modest complexity. The more complex systems could not be implemented economically using analogue technology.

Flexibility: Control laws and gearings (or gains) can be changed by software changes as opposed to hardware modifications giving greater flexibility during the design

and development phases. The introduction of modifications in service is also easier. The cost of software changes is not a trivial one, to say the least, but there is no doubt that it is still less than the cost of hardware modifications.

Reduced nuisance disconnects: Digital computation allows sophisticated voting and consolidation algorithms to be used, which minimise nuisance trip outs, or disconnects.

Smaller failure transients: Sophisticated consolidation algorithms can be implemented, which minimise the transient experienced on disconnecting a failed channel.

Built-in test capability: Very comprehensive self-test capabilities can be incorporated into the system for pre-flight checkout and maintenance.

Digital data buses: Very large reductions in the weight of cabling are achieved by the use of multiplexed data transmission and data bus networks. High-integrity data transmission can be achieved with very comprehensive self-checking and data validation capabilities. The use of three, or more, totally independent data bus networks enables the failure survival requirements to be met. Figure 4.2 illustrates a flight control data bus configuration.

4.5.2 Digital Data Problems

The use of digital data introduces particular problems, which need to be taken into account in a closed-loop digital control system. These are essentially due to the need to sample data and the frequency of sampling.

4.5.2.1 Aliasing

Figure 4.39 illustrates the effect of sampling a time-varying signal at too low a sampling frequency. The sampled output contains a much lower-frequency signal, which is not present in the real signal. The effect of sampling is to 'fold back' the

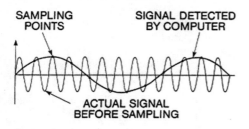

Fig. 4.39 Signal aliasing due to sampling

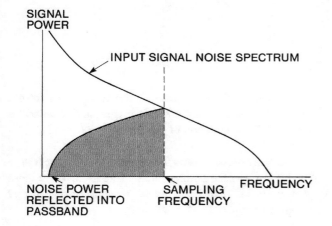

Fig. 4.40 Aliasing

high-frequency content of the real signal above the sampling frequency so that it appears in the frequency range below the sampling frequency (see Fig. 4.40). This effect is known as 'aliasing'.

It is thus necessary to filter the signal to attenuate any noise with a frequency content above the sampling frequency before sampling – such filters are known as anti-aliasing filters (see Fig. 4.41). Clearly, the higher the sampling frequency the better, so that the lag introduced by the anti-aliasing filter is very small over the range of control frequencies.

4.5.2.2 Data Staleness

In comparing the corresponding outputs of say four independent parallel computing channels, or lanes, it is possible for the output of one channel to be delayed by one iteration period before its output can be compared with the others. Its information is thus 'stale' by one iteration period. Hence if the input quantity is changing during the iteration period, the output of this 'stale' channel will differ from the others although no fault has occurred. The situation is illustrated in Fig. 4.42.

To overcome this problem, it is necessary to have some form of time synchronisation of the computer iteration periods. This iteration period synchronisation is generally implemented through software.

The alternative is to allow longer iteration periods so that all computers are processing the same data. This however introduces a latency or time delay in the output, which results in a phase lag and is de-stabilising.

4.5.2.3 Latency

This has been covered earlier in Sect. 4.3.2. Figure 4.22 illustrates the phase lag introduced by a time delay. This results in a phase lag in the output, which increases as the frequency of the input increases.

The amplitude, however, is not attenuated (as with a simple lag filter) and so the de-stabilising effect is increased.

4.5.3 Software

4.5.3.1 Introduction

Software generation is one of the most challenging tasks in the design of a high-integrity digital FBW flight control system and can account for 60–70% of the total engineering development costs of the complete FBW system. This is because of the size of the software which is required to carry out all the flight control functions and the problems of establishing the safety of the software.

It is not possible in this book to do more than introduce the reader to the problems and the methods being used to overcome them. The subject of safety critical software is an evolving discipline and is generating a large and increasing bibliography of books and papers. A list of some appropriate references is set out at the end of the chapter.

4.5.3.2 The Flight Control Software Functions

The main functions that are carried out by the software for the flight control computers can be divided into three basic areas which have a degree of interaction and comprise the follwing:

1. Redundancy management
2. Control laws
3. Built-in test

Redundancy management can account for 60–70% of the total software (and computer throughput) in the case of a modern civil FBW aircraft. The proportion is generally not quite so high in a military FBW aircraft – around 50% for a modern agile fighter.

Implementation of the flight control laws generally accounts for 25–30% of the total software in a modern civil FBW aircraft. The proportion in a modern agile fighter, however, can amount to around 40%.

Built-in test software for in-flight monitoring accounts for a smaller percentage of the total software in either case – around about 10%. However, during the pre-flight checking

Fig. 4.41 Anti-aliasing filter

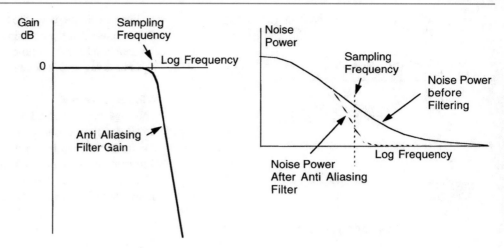

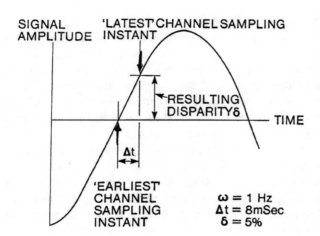

Fig. 4.42 Differential data staleness

stage on the ground the proportion of the operating software taken up by built-in test is much higher and can be as high as 40%.

Some appreciation of the functions carried out by the redundancy management software can be gained by considering the tasks involved in failure detection and isolation and reconfiguration in the event of a failure in the case of a quadruplex system with four independent 'lanes' of sensors and computers.

This includes tasks such as the following:

- *Sensor data validation*: Checking that the digital data from each individual sensor is correctly encoded and has not been corrupted within the data transmission system.
- *Sensor failure detection*: Detection of a failed sensor by cross-comparison of sensor data and majority voting.
- *Sensor failure isolation and system reconfiguration*: Isolation of a failed sensor and system reconfiguration to survive the failure and also to minimise the resulting transient.

- *Sensor consolidation*: Consolidation of the data from a group of sensors to use a single representative value for subsequent computation of the control laws.

A first-order list of sensors for a typical system is set out in Table 4.2 to show the numbers of sensors involved in these respective tasks. Most of these sensors would be at quadruplex or triplex level redundancy.

- *Cross-lane data transfer*: Data transfer between the individual computing lanes to enable cross-comparison of outputs.
- *Computer output voting and consolidation*: Cross-comparison of the individual computer outputs to detect and isolate a computer failure. Consolidation of the 'good' computer outputs to provide the control surface demand signals to transmit to the control surface actuator servos.
- *Computer iteration period synchronisation*: The iteration periods of the individual computers need to be loosely synchronised to avoid data staleness problems.
- *Recording of fault data* to enable the management of the maintenance function.
- *System status* indication to the crew.
- *Control* of the overall system.

Considering the large number of sensors involved plus all the other tasks described enables some appreciation to be obtained of the redundancy management software task. The safety critical nature of this software can be appreciated.

The processes involved in carrying out just one of the sensor failure detection and output consolidation tasks are shown in the flow diagram in Fig. 4.43.

The control law software is, of course, of an equally critical nature, together with the built-in test software. The control laws now being implemented in the new generation of flight control systems are of much greater complexity than in existing conventional aircraft.

Table 4.2 Pitch channel sensors

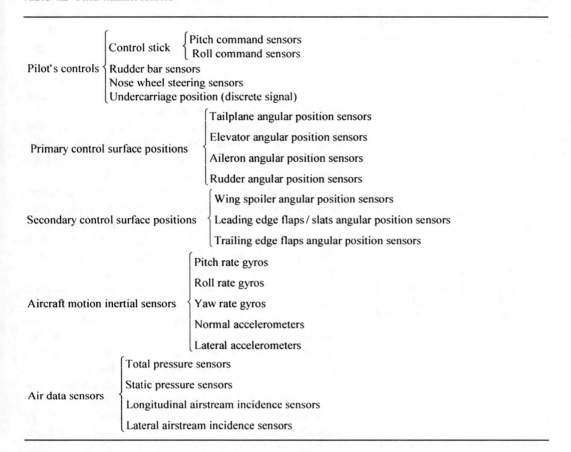

Pilot's controls	Control stick — Pitch command sensors / Roll command sensors
	Rudder bar sensors
	Nose wheel steering sensors
	Undercarriage position (discrete signal)
Primary control surface positions	Tailplane angular position sensors
	Elevator angular position sensors
	Aileron angular position sensors
	Rudder angular position sensors
Secondary control surface positions	Wing spoiler angular position sensors
	Leading edge flaps / slats angular position sensors
	Trailing edge flaps angular position sensors
Aircraft motion inertial sensors	Pitch rate gyros
	Roll rate gyros
	Yaw rate gyros
	Normal accelerometers
	Lateral accelerometers
Air data sensors	Total pressure sensors
	Static pressure sensors
	Longitudinal airstream incidence sensors
	Lateral airstream incidence sensors

The high-speed microprocessors now available enable very sophisticated control laws to be implemented and the aircraft to have a higher performance and safer handling characteristics as a result. The control law software is thus of increased complexity. However, the cost of generating this software, although significant, can be amortised over the production run of what is a more competitive aircraft and hence can be cost effective. This is because the hardware costs are substantially unchanged from earlier systems because of the continuing progress in processor power and microcircuit design and manufacturing technology, such as the extensive use of ASICs – Application Specific Integrated Circuits.

The control law software tasks can comprise the following:

- Complex gain scheduling of a large number of control terms as a function of height, airspeed, incidence and possibly attitude, and aircraft configuration. This can be particularly important in a military aircraft, which may be required to operate to the very edges of its flight envelope in combat when increased cross-coupling effects, etc., require changes in the control terms to be made. The

control transfer functions involve proportional, derivative and integral terms, which must be derived from the various motion sensors using suitable algorithms.
- Automatic limiting of the pilot's input commands to achieve 'carefree' manoeuvring to restrict the pilot from attempting to manoeuvre the aircraft into an unacceptable attitude or approach too near the stall or exceed the structural limits of the aircraft.
- Changing the control laws to maintain effective control in the event of sensor failures, etc.
- Control of the secondary control surfaces – leading edge flaps/slats, trailing edge flaps, spoilers.
- Control of engine thrust and possibly thrust vectoring.

4.5.3.3 The Software Development Process

The management of the software complexity is clearly one of, if not the, major tasks, and a phased development process as shown in the 'V' diagram in Fig. 4.44 is followed. The operating principle is a progressive breakdown of the functional requirements during the design phase followed by a progressive build-up of the testing of these requirements during the proving phase. The left side of the V follows design decomposition, the bottom of the V is the

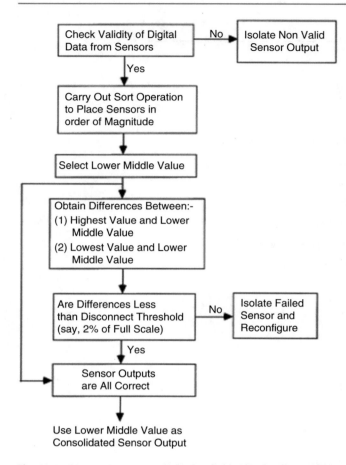

Fig. 4.43 Flow diagram of sensor voting and consolidation process

implementation of the software code and the right hand side the progressive build-up of testing through to customer acceptance. The process starts with the extraction of the specific flight control system requirements from the customer's Source Control Documentation (SCD) to derive the Top Level System Requirement Document (TLSRD). The system is then designed at architectural level to produce the Software Requirements Document (SRD). Further detailed software design then leads to the Software Design Documents (SWDD) for each module allowing each module to be coded. Software testing commences at module level in accordance with Software Module Test Plans (SMTP). Testing is continued in an incremental manner as the various functional areas are integrated in accordance with the Functional Area Test Plan (FATP). Functional system integration testing is conducted in a closed-loop configuration with all the flight control computer functions exercised under normal and failure conditions using a representative aircraft model. All design activities are monitored by an independent validation and verification team that also specifies and conducts all tests. After successful completion of their Functional Tests, units are then shipped to the aircraft manufacturer for further intensive testing in the Systems Integration Laboratory.

The correct specification of the flight control system requirements in a clear unambiguous form is the essential starting point and various methods and software tools are used in conjunction with work stations to assist in this process. They include the use of system description languages and formal methods. References to further reading on system description languages will be found at the end of the chapter, as it is beyond the scope of this book.

Formal methods will be discussed briefly in terms of their role in the development of safety critical software. It should also be pointed out that the use of formal methods in flight control system software design is still evolving.

Design techniques such as object oriented design (OOD) are used in both the top-level design and detailed software design phases. OOD is an iterative design technique which is repeated until it results in both a description in English of the software and its task and a Booch diagram which is equivalent to a block diagram of the software packages and their relationship with the hardware. The technique is powerful in the early design stages to bridge the gap between the customer and the software engineer with a representation understandable (and workable) by both. The OOD method involves considering the design in terms of 'objects' and associated 'actions' on the objects and this can ease the translation into the high-order language (HOL) Ada because 'objects' within the design become Ada packages.

A number of computer assisted software engineering (CASE) tools are available for running OOD on suitable workstations. Further reading on OOD is given at the end of the chapter.

The question of software languages is a crucial one in flight control systems. The software for digital flight control systems designed around the mid-1970s to the mid-1980s was generally written in Assembler language, usually for a specifically designed processor. (Available microprocessors at that time lacked the speed and throughput.) Assembler languages are essentially an assembly of specific instructions translated directly into machine code and have two major advantages – speed of execution and visibility of some of the computing processes within the computer from the point of view of failure modes and effects analysis. However, they lack flexibility, ease of use for handling complex problems, error checking capabilities and ease of transfer to run on other machines, which a high-order language (HOL) can provide.

The US DoD standard high-order language is Ada and its use is mandatory in US military avionic systems. It is also now widely used in the UK and is gaining increasing acceptance for civil as well as military avionic systems. For example, the software for the Boeing 777 primary flight computers is written in Ada and is converted into executable form for three different microprocessors using three independent Ada compilers. A compiler translates an HOL to machine executable instructions or code.

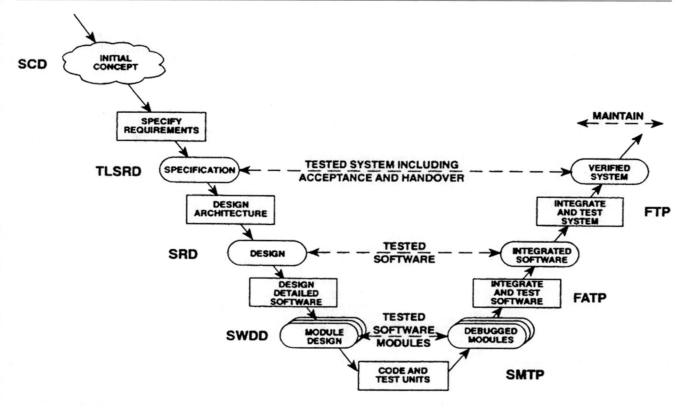

Fig. 4.44 Software life cycle

4.5.3.4 Software Validation and Verification

The safety critical nature of flight control software makes the validation of its integrity essential. Clearly, the software documentation and configuration control for all stages of the software life cycle must be fully compliant with the safety and airworthiness requirements of the regulatory authorities (e.g. CAA, FAA and MoD). Relevant standards are DO-178C, DOD-STD-2167 and DEF STAN 00–55 (see 'Further Reading').

The stages of reaching this stage follow a carefully structured design approach and may include the use of formal methods. Formal methods of software development involve the use of mathematical set theory and logic to specify the functionality of software. Specifications in such mathematical notations have some major benefits over English or diagrammatic notations in that they are unambiguous, they can be checked for completeness and their properties can be examined by mathematical methods. This leads to specifications that are more detailed but contain fewer errors, being more consistent and less likely to be misinterpreted. The net effect is that certain aspects of the system are fully defined thereby reducing the development risks. The mathematics used in formal methods to describe information and systems is known as discrete mathematics, rather than the 'continuous' mathematics used to describe physical processes. Logic and set theory are used to represent system states and operations. One such method is a mathematical language known as 'Z', which is used to specify functional requirements of systems. This has been developed by the Oxford University Research Group of IBM. Z is based on Zermelo–Fraenkel set theory, from which it derives its name and is a flexible mathematical notation with constructs for creating abstract or concrete specifications in a well-structured way. It should be pointed out, however, that the Z notation can take a little time getting used to by less mathematically orientated engineers. The formal verification of the design process is shown in Fig. 4.45 and appropriate references to formal methods are given at the end of the chapter.

It should be stressed again that the whole subject of safety critical software is a very large and complex one and is still evolving and maturing. The objective of this section is to try to introduce the reader to the subject and to indicate where appropriate references can be obtained.

4.5.3.5 Dissimilar or Multi-Version Software

The extreme difficulty, if not impossibility, of proving the integrity of a system using common software in its parallel redundant channels (or lanes) to the safety levels required by the civil regulatory authorities has led to the need for dissimilar redundancy as explained earlier. Two (or more) totally independent failure survival flight control computing systems

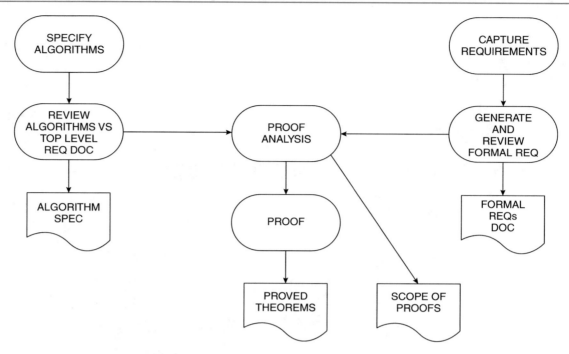

Fig. 4.45 Formal verification of design

are installed. Each system uses a different type of microprocessor and the software is written in different software languages by totally independent software teams using the most stringent methods and procedures. The process is shown below:

However, the degree of independence in writing dissimilar, or 'multi-version', software to meet a common system requirement is not 100%.

Some work carried out in the USA has shown that multiversion software written to meet a common stated requirement can contain errors which are not independent, that is, common errors can occur.

Part of the problem resides in the understanding of the stated system requirements, which may involve unforeseen problems and possible ambiguities in interpretation. The rigorousness and degree of control of the software development process is a further factor and the better this is, the fewer the errors anyway.

A conservative estimate is that multi-version software should achieve a 95% improvement in reducing the software errors present in a single software program – that is, only 1 in 20 of the errors present would be shared. An improvement of 20:1 by using multi-version software is clearly well worth having. It is also considered that the use of formal methods to define the system requirements and ensure these are soundly stated should further improve the confidence level in reducing the number of shared errors.

4.5.4   Failure Modes and Effects Analysis

An exhaustive 'failure modes and effects analysis' of the flight control system is required by the regulatory authorities before a Certificate of Airworthiness is granted. This embraces every part of the overall system including both hardware and software and power supplies. Any part of the system can fail from any cause and it must be shown that the overall system can survive this failure and maintain safe flight. The importance of dissimilar redundancy in this context is apparent.

The MTBF of each element in the system has to be established using well-validated statistical data where possible, and it must be shown that the system configuration will meet the overall safety and integrity requirements (as specified in Sect. 4.4.1).

4.6 Helicopter FBW Flight Control Systems

The advantages of FBW flight control have been referred to at the end of Chap. 3. There are no major differences in terms of the hardware, digital implementation and system architectures used in a helicopter FBW flight control system from those used in an FBW fixed wing aircraft. The differences lie in the following:

1. The means of generating the aerodynamic forces and moments by controlling the individual rotor blade angles, cyclically and collectively, through the swash-plate actuators – as opposed to moving the control surfaces in the case of a fixed wing aircraft.
2. Generating forward thrust, and hence forward speed, by controlling the direction and magnitude of the rotor thrust vector, as opposed to direct control of the forward speed of a fixed wing aircraft by controlling the engine thrust.
3. The helicopter dynamics and hence response to the control inputs differ from that of a fixed wing aircraft.
4. The speed and height envelope of a helicopter is, of course, very different, ranging from the hover to a maximum speed of around 20 knots sideways or backwards, and a forward maximum speed of currently around 200 knots. The maximum ceiling of a helicopter is typically around 20,000 ft. compared with up to around 65,000 ft. for a fixed wing aircraft.
5. In an FBW helicopter, control of the flight is exercised by the swash-plate control system and this must be of extremely high integrity for safe flight. In contrast, the typical large civil aircraft FBW system invariably has multiple control surfaces, so each individual control surface has a lower fault tolerance requirement than that of the flight control system as a whole. The certification of a civil helicopter FBW system thus needs particular attention to be paid to the swash-plate actuation and its control system in the areas of mechanical failure survival and generic software errors. Detailed treatment of these topics goes beyond the scope of this book.

The safety and integrity requirements for a military helicopter FBW system are more comparable with a military aircraft FBW system where the tailplane provides the means of control in pitch and there may be no possibility of providing redundant control in pitch via other control surfaces.

To return to the advantage of helicopter FBW control, the FBW system is a full time, full authority, failure survival system of very high integrity and disengagement transients do not need to be considered, unlike the auto-stabilisation systems covered in the previous chapter.

The preferred control system in the pitch axis is a pitch rate command system with attitude retention when the stick is released. This is more similar to conventional fixed wing FBW flight control systems and needs to incorporate proportional plus integral control terms in the pitch cyclic input to the rotor. The roll axis control is similarly a roll rate command system with roll attitude retention.

A block diagram of a helicopter FBW flight control system is shown in Fig. 4.46. The redundancy in terms of sensors, computers, actuators and data buses to enable the system to survive failures has been omitted for clarity.

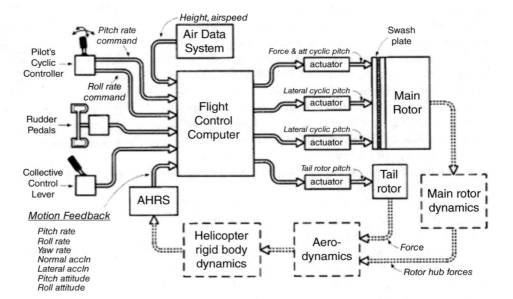

Fig. 4.46 Helicopter FBW flight control system

As explained earlier in Chap. 3, the basic means of control of the flight path and speed of a helicopter is to control the direction and magnitude of the rotor thrust vector. The direction of the rotor thrust vector is determined by the pitch attitude and roll attitude of the helicopter, which result from the application of the pitch and roll cyclic controls. The magnitude of the rotor thrust is determined by the collective pitch angle of the rotor blades.

Pitch rate and roll rate command control with attitude retention gives the pilot a very easy control to operate. The rate command loop makes the helicopter behave like a simple first-order system as far as the pilot is concerned when setting up the helicopter attitude – the FBW system suppressing all the 'nasty' cross-coupling effects which have to be corrected in a manually controlled helicopter.

The helicopter pitch attitude, in turn, controls the aircraft forward speed, as explained in Chap. 3. The collective pitch control provides a fast effective control of the magnitude of the rotor thrust – raise the collective lever and the helicopter climbs; lower the collective lever and the helicopter descends.

With FBW control, there is the option of developing more novel control algorithms. In the case of the yaw axis control with an FBW system, the tail rotor control can vary with airspeed, such that around the hover the foot pedals provide a yaw rate command. At high airspeed, however, the tail rotor is controlled automatically to minimise lateral acceleration or sideslip, while the effects of pedal deflection is to deliberately introduce additional yawing moments so as to enable flat turns for example, or to aid weapon aiming tasks.

It should be noted that, as mentioned in the previous chapter, the helicopter is basically asymmetric and this manifests itself into the need for cross-coupling in various areas. Collective to yaw coupling is necessary because as collective control is applied, the resultant torque reaction needs to be compensated via the tail rotor. In a manually flying control system, this compensation is normally implemented within the mechanical controls, at least to a first order. In an FBW helicopter, such cross-feeds can be more sophisticated than the simple proportional mechanical cross-feeds, hence minimising pilot workloads.

Collective to pitch control may also be necessary, to minimise the tendency of the tail end of the helicopter to droop during vertical climbs.

The rotor behaviour is also inherently asymmetric; at high airspeed the advancing blade(s) has a higher speed through the air than does the retreating blade(s). The airspeed of the helicopter needs to be limited such that the advancing blade tips do not become supersonic, while the retreating blades do not stall. Another effect of rotor asymmetry relates to high speed turning flight because of the tendency to over-torque or over-speed the rotor depending on the direction of the turn.

Such manifestations of rotor asymmetry can be dealt with in an FBW design to minimise pilot workload.

4.7 Active FBW Inceptors

Figure 4.47 shows the basic concepts of an active FBW inceptor. The spring and damper elements of a passive inceptor have been replaced by computer-controlled force feedback actuators on each axis. Algorithms within the controller provide the required force-displacement characteristics by means of algorithms based on measured force and displacement in each axis.

The active inceptor can produce the characteristics shown in Fig. 4.48 with such features as variable force gradients with intermediate 'soft' stops. Soft stops are intermediate force thresholds to bound the zone of 'carefree handling' while the variable gradient provides the desired 'feel' characteristics over the flight envelope.

Provision of such characteristics in a passive inceptor would be impractical because of the mechanical complexity involved.

With a digitally implemented controller it is a relatively simple task to modify the force-displacement characteristics of the inceptor during flight test without any mechanical changes to the unit.

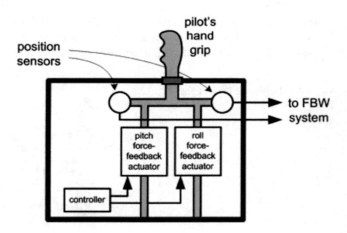

Fig. 4.47 Basic concept diagram of an active FBW inceptor

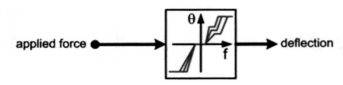

Fig. 4.48 Example of active force-displacement characteristic

In addition to the ability to vary the feel characteristics of the inceptor, it is also a simple task to provide such features as follows:

- Variable null position, to provide parallel trim and autopilot back drive.
- Facilities such as 'stick pusher' or 'stick shaker' to provide cues to the pilot relating to buffet onset or imminent stall or departure.
- The ability to couple together the displacements of a pair of inceptors.

It is this last facility which is seen as the major benefit of active inceptors compared with their passive counterparts. It is particularly relevant to trainer aircraft, where the student pilot is able to detect on his inceptor small forces and movements from his instructor's inceptor. Simple logic with the respective controllers allows the instructor to override the student should it prove necessary.

In transport aircraft, both the pilot and co-pilot are fully aware of each other's actions, much in the same way as with mechanically coupled traditional control wheel and pedestals, without the complexity entailed with mechanical cross-feeds.

Finally, the operating concept of active inceptors does not just apply to the primary pitch-roll control system in an aircraft, or the so-called pitch-roll cyclic in a helicopter. It can also be applied to rudder pedals, throttle controls and helicopter collective pitch controls (collective lever).

Figure 4.49 shows the basic concept of a force coupled pair of inceptors for a dual control system.

Active inceptors can be made relatively small and compact; however, this is not always the case. Both the Boeing 777 and 787 have an FBW flight control system but retain the traditional control yokes, which are coupled to active inceptors. The FBW control laws and the active force feel system give these aircraft very similar handling characteristics to other Boeing airliners with conventional mechanically signalled controls. This enables pilots to make a relatively easy transition between the Boeing airliners in an airline operator's fleet.

There are both active force/displacement inceptors and active force inceptors, but space constraints limit coverage to the more common force/displacement type. Figure 4.50 is a schematic diagram of an active displacement type of FBW inceptor.

Geared, brushless DC servo motors are shown as providing the actuation forces – the gear drive reducing the power requirements. Brushless DC servo motors are chosen because of their excellent torque characteristics and reliability. Very effective damping can be provided by the inherent electro-magnetic damping in the DC servo motor amplified by feedback within the servo amplifier.

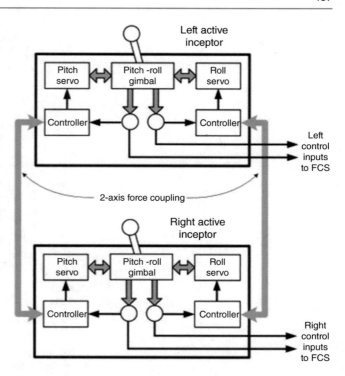

Fig. 4.49 Schematic diagram of a force coupled pair of inceptors for a dual control system

Consider what happens when the pilot exerts a gentle pressure on the stick to displace it from the normal central position. This force is sensed by the force sensor and the stick force/displacement controller generates zero stick-displacement demand at this point, as it is below the 'break-out' level.

The servo motor force balances the force the pilot is applying and the stick is displaced by a very, very small amount from the central position to provide the necessary motor current from the servo amplifier. (The servo loop has a relatively high gain, or 'stiffness'.) Further increasing the stick force will result in the break-out force being reached and continuing to increase the stick force is matched by the appropriate motor force. The Controller generates the appropriate stick-displacement demand corresponding to this force and the stick is driven by the servo motor so that the displacement coincides with the demanded value. (The controller is resident in the FBW Computers.)

The outputs of the stick-displacement sensors are majority voted and provide the pilot's command signals, for example pitch rate and roll rate, to the FBW computers.

The integrity aspects need to be considered and the following elements examined:

1. Mechanical integrity of the gear drives and gimbal bearings. The probability of a mechanical jam can be made very low, and in the unlikely event of a jam, the aircraft can be controlled by the force sensor outputs (typically duplex).

Fig. 4.50 Schematic diagram of a force/displacement FBW inceptor

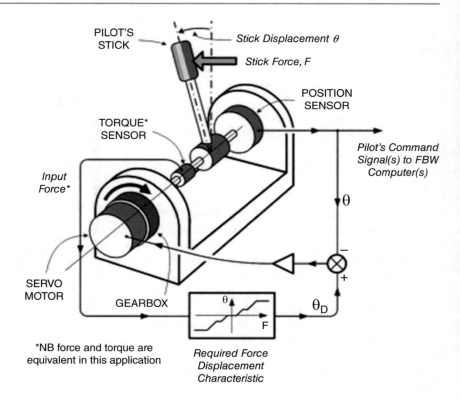

PILOT'S STICK

Stick Displacement θ

Stick Force, F

POSITION SENSOR

TORQUE* SENSOR

Pilot's Command Signal(s) to FBW Computer(s)

Input Force*

θ

SERVO MOTOR

GEARBOX

θ

F

θ$_D$

−

+

*NB force and torque are equivalent in this application

Required Force Displacement Characteristic

2. Failures in the servo amplifiers and gimbal servo motors. The MTBF of a servo amplifier and motor combination is of the order of 4000 hours, so that a duplex configuration meets the integrity requirements.

The servo motors can have dual windings, with one winding providing the means of control and the other winding providing standby control through a standby servo amplifier, so that only one servo motor per axis is required.

In the unlikely event of loss of the servo drives, the pilot can still move the stick and exercise a degree of control of the aircraft.

4.8   Fly-By-Light Flight Control

4.8.1   Introduction

Mention has already been made of the common mode failures that can be caused by severe electro-magnetic interference (EMI) if the equipment is not effectively screened from EMI.

Electro-magnetic interference can arise from the following:

Lightning strikes: Very large electro-magnetic pulses (EMPs) with electrical field strengths of hundreds of volts per metre can be produced and a very wide spectrum of electro-magnetic radiation frequencies generated.

Overflying high-power radio/radar transmitters: Several cases have occurred of aircraft experiencing severe transients in the automatic flight control system placing the aircraft temporarily out of control while overflying high-power radar/radio transmitters. This has been due to the susceptibility of the flight control system analogue electronics to EMI because of inadequate electro-magnetic screening. Current digital flight control systems are designed to much higher electro-magnetic compatibility (EMC) specifications (that is, the ability to withstand high levels of EMI), and should be able to withstand such environments.

Failures in the electro-magnetic screening system so that the avionic equipment becomes susceptible to internal sources of electro-magnetic radiation such as radio and radar equipment, electrical generators, etc., or external sources of EMI. Failure of the electro-magnetic screening can result from breaks or high resistance in the Earth connections to the aircraft structure and the electrical bonding of the structure due to corrosion, poor quality installation, etc. Such failures can be of a dormant nature and might only become apparent when an abnormal EMI source is encountered.

Vicinity to a nuclear explosion and consequent emission of gamma radiation and an EMP of very high electric field strength. Military avionic equipment is specially designed to survive up to a certain (classified) level of these effects. This attribute is referred to as 'nuclear hardness' and is achieved by the design of the circuits, use of nuclear hard components, circumvention techniques and very efficient electro-magnetic screening. The subject is beyond the scope of this book and much of the information on the subject is of a highly classified nature. However, a military aircraft with a nuclear hard flight control system must have a very low susceptibility to EMI in order to meet the nuclear hardness specifications.

Electronic units and their enclosures, or 'boxes', can be designed to have very efficient electro-magnetic screening. However, the 'Achilles heel' of the system can be the interconnecting cables which link together all the elements of the system. These interconnecting cables must have very efficient electro-magnetic screening, which incurs a significant weight penalty. The incoming wires from the cables to the units (or boxes) must also be terminated with EM filters, which present a low impedance to EMI-induced transient voltages in the cables. The transient voltages are thus attenuated before they can affect the circuits inside the box. (It should be noted that the wires from the cable connectors and the EM filters are housed within a screened enclosure around the connector known as an 'EMC vault', so that there is no aperture for EMI radiation to enter the box.)

The use of digital data transmission enables error check encoding to be incorporated in the digital data word format so that any corruption of the digital data (e.g. by EMI) can be detected and the data ignored. However, the data are lost and the system 'frozen' for the period of the interference so that only a short transient loss could be tolerated. The integrity of the screening of the interconnecting cables must thus be very high indeed to meet the flight safety requirements.

The use of optical data transmission whereby digital (or analogue) data are transmitted as a modulated light intensity signal along a fibre optic cable overcomes these problems. The optical transmission media is unaffected by EMI and very much higher data rates can be achieved, for example, GHz bandwidth (109 Hz). Current avionic data bus systems using fibre optic data transmission operate in the 20–100 Mbits/s bracket. (The principles of fibre optic data transmission and optical data buses are covered in Chap. 9.)

The fibre optic cable is also much lighter than the equivalent screened cables it replaces. The weight saving can be significant and, as mentioned earlier, weight saved on equipment installation is geared up by a factor of about 10 on the overall aircraft weight. A further driver to the adoption of optical data transmission is the extensive use of composite materials in the aircraft structure and the consequent loss of the screening provided by an aluminium skin, which forms a 'Faraday cage' as far as external electric fields are concerned.

The use of fibre optic links to interconnect the basic elements of the flight control system – pilot's stick, sensors, flight control computers, actuator control electronics, etc. – has become known as a 'fly-by-light' flight control system and is covered in more detail in Sect. 4.8.2.

It should be noted, however, that while a number of prototype aircraft have flown with flight control systems using fibre optic links, there are no production aircraft in service with a fly-by-light flight control system at the time of writing this book. This is because electrical bus systems have proved entirely adequate and interconnection with the bus is relatively simple. Fibre optic connectors and passive couplers are not readily available and are relatively expensive.

4.8.2 Fly-By-Light Flight Control Systems

A fly-by-light flight control system configuration is shown schematically in Fig. 4.51. The accompanying redundancy is omitted for clarity. The fibre optic links interconnecting the units of the flight control system eliminate the possibility of propagating electrical faults between units, as the optical fibre is an insulator.

Fibre optic links can be bi-directional and can also be used to convey the system status to the pilot's control/display

Fig. 4.51 Fly-by-light flight control system (redundancy omitted for clarity)

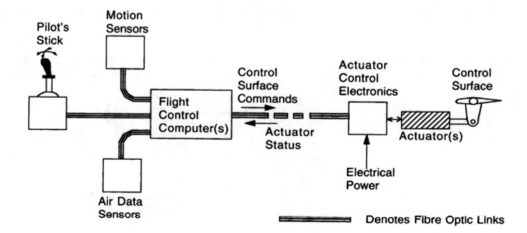

panel. For instance, 'wrap round' tests can be carried out to check that the data have reached their destination and the status or 'health' of the control surface actuation system can be checked by monitoring the servo error signals and actuator hydraulic pressures.

The actuator control electronics and actuator are located in fairly close proximity to each other and the only connections required are the fibre optic cables and electric power supplies, the large wiring harness required in earlier analogue systems being eliminated. Such systems are sometimes referred to as 'smart actuators'.

A further advantage of fibre optic data transmission is the ability to use 'wavelength division' multiplexing whereby a single fibre can be used to transmit several channels of information as coded light pulses of different wavelengths (or colours) simultaneously.

The individual data channels are then recovered from the optically mixed data by passing the light signal through wavelength selective passive optical filters, which are tuned to the respective wavelengths. WDM has a very high integrity as the multiplexed channels are effectively optically isolated.

It is interesting to note that one of the first applications of fly-by-light technology has been on the earliest flying vehicle, namely the airship. Non-rigid airships may be acquiring a new lease of life as long-duration platforms for airborne radar surveillance systems. The airship is able to carry a large high-power radar antenna within its envelope and is able to remain airborne for several days (with occasional in-flight refuelling while hovering over a supply ship). It can also carry the relatively large crew required for 24-hour operation in considerable comfort.

Fully powered controls are required for the airship's large control surfaces because of the need for autopilot operation on long-duration missions and auto-stabilisation to counter an inherent pitch instability and provide yaw damping. The long control cables required from the gondola to the control surfaces (100–200 metres on the projected large airships) and the probability of experiencing lightning strikes make an optically signalled flight control system an attractive solution.

It is interesting to note that 'proof of concept' Skyship 600 airship first flew with an optically signalled flight control system in October 1988, nearly 40 years ago.

4.8.3 Optical Sensors

There are a large number of sensors involved in the flight control system for a representative civil and military aircraft – many of these sensors being at quadruplex level. The approximate location of the sensors is indicated and it can be seen that many are located near the extremities of the aircraft and possibly less well screened than fuselage mounted equipment. The extensive use of composites can further degrade electro-magnetic screening in the vicinity of such sensors.

The use of what are referred to as passive optical sensors offers particular advantages in such applications as these sensors are not affected by EMI. Passive optical sensors are defined as sensors which do not require electrical supplies or any electronic processing at the sensor, the output of the sensor being an optical signal which is modulated by the quantity being measured. The processing of this optical signal is carried out by a separate electronic unit which is fully screened from EMI, the only connection between the sensor and the electronic unit being an optical fibre cable. The basic concept of such sensors is shown in Fig. 4.52.

There are no optical flight control sensors installed in production aircraft at the time of writing, but it is clearly a possible further development.

4.9 Automatic Flight Control of Vectored Thrust Aircraft

The coverage of automatic flight control and fly-by-wire manoeuvre command control has up to now been confined to conventional aircraft configurations where the engine thrust acts in a fixed direction along the forward axis of the aircraft. Although fly-by-wire flight control has enabled aerodynamically unstable aircraft to achieve a fast well-damped

Fig. 4.52 Passive optical sensor concept

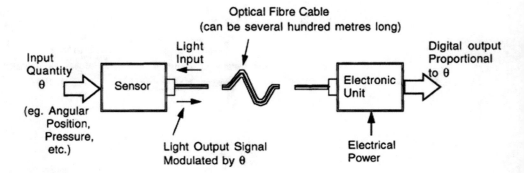

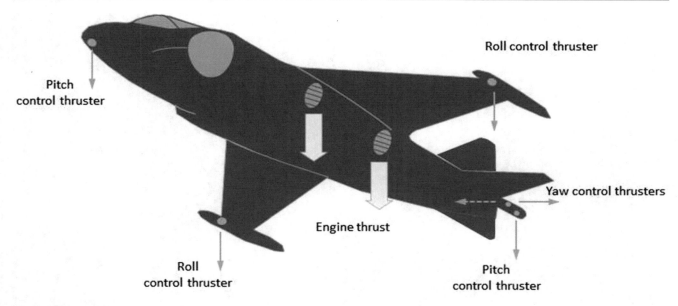

Fig. 4.53 Harrier jump jet showing forward and rear nozzles

response to a pitch or roll rate command, the pilot still needs to coordinate the control exerted about the pitch, roll and yaw axes, and also control the engine thrust in order to execute a level coordinated term with zero sideslip.

The advent of aircraft exploiting vectored thrust that is changing the direction of the engine thrust and its magnitude to achieve short take-off and vertical landing has resulted in a major increase in pilot workload in operating the additional controls required.

4.9.1 The BAE Systems Harrier

The first successful aircraft with short take-off and vertical landing capability was the BAE Systems Harrier – a brilliant design and a prime example for what can be achieved by close collaboration between top class engine and airframe designers and their teams.

Figure 4.53 illustrates the Harrier configuration. The propulsion system is provided by a single high thrust, bypass fan jet engine, the Rolls Royce Pegasus. The Pegasus is a high bypass ratio fan jet engine; nearly 50% of the engine thrust is generated by the bypass fan. The bypass fan air is expelled through a rearward facing nozzle on each side of the fuselage at the front of the engine. The exhaust jet pipe is bifurcated and a nozzle on each side of the fuselage expels the hot exhaust gases. The four nozzles are controlled by air-driven servo motors and the pilot is able to move all four nozzles together to point downwards through an angle of 90 degrees plus or minus 8 degrees. The thrust can thus be given a forward or rearward component. A key feature is the thrust lines of all four nozzles go through the centre of gravity of the aircraft and do not exert any out of balance forces on the aircraft.

The aerodynamic controls become ineffective at low airspeeds and control moments are excited by reaction jets, which are situated at the nose, tail and wing tips of the aircraft. The reaction jet control valves are indicated by small black circles in Fig. 4.53.

The reaction jet control valves are mechanically interlinked with the flight controls and supplied with compressed air from the bypass fan. The compressed air supply for the reaction jets is shut off during normal aerodynamic flight and only comes into operation as the aerodynamic controls begin to lose their effectiveness.

The engineering elegance of the Harrier control system can be appreciated; only one additional control is needed to achieve vertical flight, namely, the nozzle angle control lever. The workload in flying the Harrier during the vertical landing phase is, however, very high. The pilot has to alternately move the nozzles with his left hand while stabilising the aircraft in pitch with right hand movements and also operate the throttle when necessary to increase the thrust. Harrier pilots are selected on their ability to control the Harrier under all circumstances and are rightly proud of their ability.

4.9.2 The Lockheed Martin F-35B Lightning2 Joint Strike Fighter

A design competition was held in the 1990s for the next-generation Strike Fighter where stealth was a prime requirement. A common design that would meet the requirements of the US Air Force, The US Navy and the US Marine Corps was required and the project was given the title of Joint Strike Fighter. The winner of the competition was Lockheed Martin that have developed three basic variants, the F-35A for the US Air Force, the F-35C for the US Navy and the F-35B for the

Fig. 4.54 First sea trials of F-35 aboard HMS Queen Elizabeth. (Courtesy of http://savethe royalnavy.org/wp)

US Marine Corps. The F-35A and the F-35C are very similar. The F-35C is strengthened to enable catapult launch from the carrier and an arrestor hook is fitted.

The F-35B for the US Marine Corps however is required to have a short take-off and vertical landing capability. This is achieved by using a high-thrust turbo jet engine and extending the turbine shaft forward to drive a two-stage compressor fan just behind the pilot and with the fan jet directed vertically downwards.

Figure 4.54 shows the first sea trials of F-35 aboard HMS Queen Elizabeth. The fan jet provides approximately 40% of the required lift force to support the weight of the aircraft. The remaining 60% is provided by turning the articulated jet pipe downwards so that the jet provides a lift force behind the centre of gravity, balancing the lift force from the fan jet in front of the CG.

The compressor fan is coupled to the turbine through a clutch, which is engaged at low engine speeds and then when engaged it can be mechanically locked so that the full shaft power from the turbine of 25,000 shaft horsepower can be transmitted to drive the compressor fan. The fan is a two-stage fan driven through a bevel gear drive in contra-rotating directions. Compressor thrust from the fan is controlled by an inlet guide vane where the opening can be varied to control the compressor thrust.

Increasing the guide vane opening increases the compressor thrust. The guide vane opening is controlled by a servo actuator controlled by a command signal from the engine control computer. The response to the guide vane opening is fast and the lag is small.

The jet pipe thrust has a varying area nozzle, which is controlled by an actuator; increasing the nozzle area increases the thrust, and, conversely, reducing the nozzle area reduces the thrust. The response to the nozzle area change is fast and the lag is small.

As the air speed decreases, the aerodynamic forces and moments decrease rapidly and alternative means of control about the aircraft pitch and yaw axes are required.

A pitching moment can be created by a 'seesaw' type action, by increasing or reducing the inlet guide vane opening and at the same time reducing or increasing the jet pipe nozzle area. The response of the aircraft to both control actions is fast and the lag is small. Large pitching moments can be exerted. It should be noted that the total thrust exerted by the jet engine stays the same and the height does not change. Increasing the engine thrust causes the aircraft to ascend and conversely reducing the engine thrust will cause the aircraft to descend. The aircraft is stabilised and controlled about the roll axis by reaction jets under the outer wing. These reaction jets are powered by the engine compressor.

The aircraft is stabilised and controlled about the yaw axis by reaction jets in a similar manner to the roll axis.

The workload in manoeuvring the aircraft for a vertical landing involves the aerodynamic controls, the vector thrust, and the reaction jets and the engine thrust controls. The manoeuvring task has to be automated as the pilot is unable to cope with the workload of all these controls. During the vertical landing phase, the aerodynamic controls start to lose their effectiveness as the airspeed decreases. The

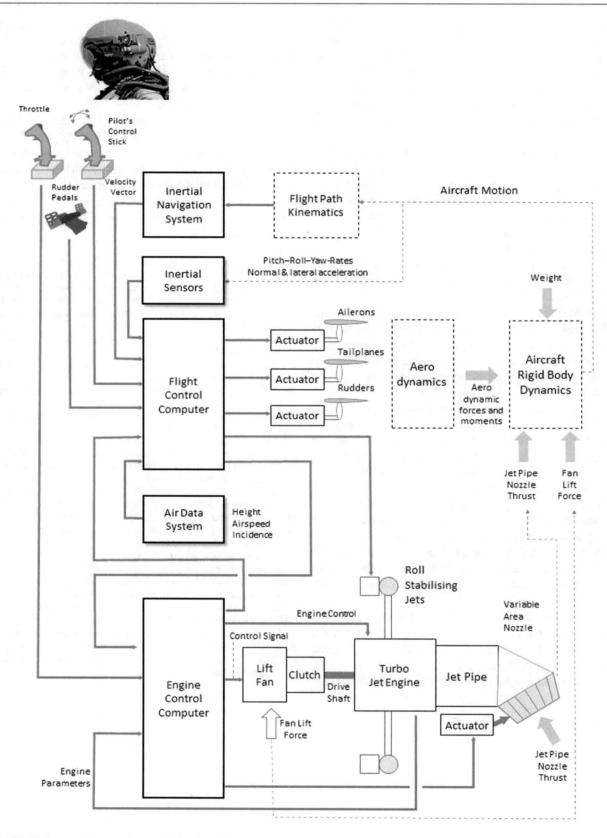

Fig. 4.55 Flight control system for vertical landing of vector thrust aircraft

aerodynamic controls and vector thrust controls also can cross-couple and this cross-coupling needs to be minimised by appropriate control signals. The manoeuvring task has thus been automated by the flight control system and the pilot commands the aircraft flight path velocity vector and the automatic flight control system operates all the controls so that the aircraft velocity vector follows that commanded by the pilot.

The aircraft velocity vector is measured by the Inertial Navigation System and is displayed on the helmet mounted display. The pilot steers the aircraft by the control stick – moving the stick backwards to raise the flight path marker, or forward to lower it. Sideways movement of the stick will steer the aircraft to the left or to the right. The aircraft speed is controlled by the throttle.

The pilot's control stick and throttle are computer-controlled force sticks that provide pilot 'feel' as described earlier in the chapter, in Sect. 4.7.

The system provides the pilot with excellent control and handling characteristics throughout the vertical landing. Figure 4.55 is a possible configuration for the automated flight control system based on published or deduced material. The essential redundancy required to meet the failure survival requirements has been omitted for clarity.

Figure 4.55 shows the cross-feed data links between the flight control computer and the engine control computer. These enable the feed forward or feedback of control signals between the two computers to minimise the cross-coupling between all the control systems. It would seem to be very likely that the F-35B exploits the 'H Infinity Matrix' to derive the optimum control signals for the feed forward and feedback control signals to produce the excellent control and handling characteristics which have been achieved.

It should be noted that the UK is an active participant in the F-35 programme and Lockheed Martin are fully aware of the results and techniques adopted in the VAAC Harrier Programme.

In normal flight the lift fan on the F-35B is declutched and all three variants are very similar. The major difference is the performance penalty in terms of payload due to the additional weight of the lift fan. Details of the fly-by-wire flight control systems for the three variants are not available. It would seem reasonable, however, to assume that a pitch rate command and roll rate command system as discussed earlier in the chapter are installed, and the rudder is automatically controlled to suppress any yaw or sideslip. It would seem likely that the F-35B flight control system is reconfigured to that used in the F-35 A and B.

(The flight control system configuration shown in Fig. 4.55 is only used in the vertical landing phase.)

The pilot's control stick and throttle units have active force inceptors described earlier in the chapter in Sect. 4.4. and provide the pilot with the appropriate 'feel'.

Further Reading

Barnes, J.G.P.: Programming in Ada. Addison Wesley

Brire, D., Favre, C., Traverse, P.: A Family of Fault-Tolerant Systems: Electrical Flight Controls, from Airbus A320/330/340 to Future Military Transport Aircraft, ERA Avionics Conference (1993)

Coad, P., Yourden, E.: Object Oriented Analysis. Yourdon Press/Prentice Hall (1991)

Davis, A.M.: Software Requirements – Analysis and Specification. Prentice Hall (1990)

D'Azzo, J.J., Houpus, C.H.: Linear Control System Analysis and Design. McGraw-Hill

DeMarco, T.: Structured Analysis and Design. Yourdon Press Computing Press (1978)

Hatley, D.J., Pirbhai, I.A.: Strategies for Real-Time System Specification. Dorset House (1988)

IEE Control Engineering Theory 57: Flight Control Systems – Practical Issues in Design and Implementation, Pratt R.W. (ed.) (2000)

Ince, D.C.: An Introduction to Discrete Mathematics and Formal System Specification. Oxford University Press (1988)

Interim Defence Standard DEF STAN OO-55: The Procurement of Safety Critical Software in Defence Equipment, 5 April 1991

McLean, D.: Automatic Flight Control Systems. Prentice Hall

Military Standard DOD-STD-2167A: Defense System Software Development, 29 February 1988

Mirza, N.A.: Primary Flight Computers for the Boeing 777, ERA Avionics Conference 1992 ERA Report 92–0809

Ramage, J.K.: AFTI/F16 automated maneuvering attack system configuration and integration. In: Proceedings of the IEEE 1986 National Aerospace and Electronics Conference. Dayton, Ohio

RTCA-EUROCAE DO178B/ED-12B: Software Considerations in Airborne Systems and Equipment Certification, December 1992

Rushby, J.: Formal Methods and the Certification of Critical Systems, Technical report CSL-93-7 December 1993 Computer Science Laboratory SRI International

5.1 Introduction

Gyroscopes (hereafter abbreviated to gyros) and accelerometers are known as inertial sensors. This is because they exploit the property of inertia, namely, the resistance to a change in momentum, to sense angular motion in the case of the gyro and changes in linear motion in the case of the accelerometer. They are fundamental to the control and guidance of an aircraft. For example, in an FBW aircraft the rate gyros and accelerometers provide the aircraft motion feedback which enables a manoeuvre command control to be achieved and an aerodynamically unstable aircraft to be stabilised by the flight control system (as explained in Chap. 4).

Gyros and accelerometers are also the essential elements of the spatial reference system or attitude and heading reference system (AHRS) and the inertial navigation system (INS). They largely determine the performance and accuracy of these systems and account for a major part of the system cost.

The AHRS and INS share common technology and operating principles. This chapter covers gyros and accelerometers and attitude derivation from strap-down gyros and accelerometers. It provides the basic background to inertial navigation and AHRS systems, which are covered in Chap. 6.

5.2 Gyros and Accelerometers

5.2.1 Introduction

The accuracy requirements for gyros and accelerometers can differ by several orders of magnitude depending on the application. Table 5.1 shows the accuracy requirements for a typical FBW flight control system and a 1 NM/hour strap-down INS as representing the lower and upper ends of the performance spectrum.

Not surprisingly, the costs can also differ by an order of magnitude or more.

The quest for and attainment of the accuracies required for inertial navigation have involved many billions of dollars expenditure worldwide and the continual exploitation of state-of-the-art technology. For example, the world's first laser was demonstrated in 1960, and the first experimental ring laser gyro (RLG) was demonstrated in 1963 with an accuracy of a few degrees per hour. The first production RLG-based inertial navigation systems (which require 0.01°/hour accuracy) went into large-scale civil airline service in 1981. Strap-down RLG-based IN systems now dominate the market, a technology revolution.

The gyros used in older designs of control and guidance systems are what is termed 'spinning rotor gyros' as they exploit the angular momentum of a spinning rotor to sense angular motion. However, their mechanical complexity and inherent failure modes constrain their cost and reliability and hence cost of ownership. Power consumption and run-up time are further limiting parameters of this technology.

High accuracy IN systems require the angular momentum gyros to be isolated from airframe angular motion because of their limited dynamic range. This is achieved by mounting them on a gimbal suspended stable platform. The reliability and cost of ownership of a stable platform INS is thus inevitably limited by its complexity.

Gyros operating on 'solid state' principles and which can be body mounted (i.e. strapped-down) have thus been developed because of their intrinsically higher reliability and lower cost of ownership. Optical gyros such as the ring laser gyro, and more recently, the fibre optic gyro, have displaced the spinning rotor gyro.

For these reasons and space constraints, angular momentum gyros have been left out of this edition. (References given in 'Further reading' at the end of this chapter.)

Table 5.1 Accuracy requirements

		Flight control system	Strap-down INS
Gyro	Scale factor	0.5%	0.001% (10 ppm)
	Zero offset/rate uncertainty	1°/min	0.01° /hour
Acctr	Scale factor	0.5%	0.01% (100 ppm)
	Zero offset/bias stability	5×10^{-3} g	5×10^{-5} g (50 µg)

At the lower end of the accuracy spectrum, micro-machined vibrating mass rate gyros exploiting semiconductor manufacturing technology have become well established. Avionic applications include flight control and solid state standby artificial horizons. They are extremely reliable and rugged devices with low power consumption and offer the major advantage of relatively low cost.

It is interesting to note that the name gyroscope is derived from the Greek words *gyros*, meaning rotation and *skopein*, meaning to view, and is the name given by the French physicist, Jean Foucault, to the gimbal mounted spinning flywheel he used to measure the Earth's rotation in 1852; the world's first gyro. To view rotation is thus a good description of the function of all types of gyro.

5.2.2   Micro Electro-Mechanical Systems (MEMS) Technology Rate Gyros

These gyros exploit the effects of the Coriolis forces which are experienced when a vibrating mass is subjected to a rate of rotation about an axis in the plane of vibration. There are two basic configurations which are being exploited: a tuning fork configuration and a vibrating cylinder configuration.

Both configurations can exploit micro-machining technology and integrated circuit manufacturing methods enabling miniature, extremely robust sensors to be constructed which have no wearing parts in their mechanism and with MTBFs in excess of 100,000 hours, and, very importantly, they can be of relatively low cost.

This fabrication technology has been given the acronym 'MEMS', that is, Micro Electro-Mechanical Systems in the USA, and these sensors are 'MEMS sensors'.

They are being exploited not only in avionic applications such as flight control, standby attitude and heading reference systems, and missile mid-course guidance, but also in the automobile industry in car stability enhancement systems. Other applications include robotics, factory automation, instrumentation, GPS augmentation for – vehicle location systems – navigation systems – precision farming – antenna stabilisation – camera stabilisation – autonomous vehicle control. In short, there is a very large market for these rate gyros, apart from aerospace applications, and this drives the costs down. These developments in gyro fabrication technology have come into large-scale exploitation over the last decade and have had a revolutionary impact. For the first time, miniature solid state rate gyros with MTBFs in excess of 100,000 hours became available at relatively low costs. Costs, of course, are dependent on the accuracy requirements and the size of the order, but for a mass market application such as the automobile industry, the costs must be of the order of 50–100 US dollars, or less, to be competitive in a very highly competitive market.

It is interesting to note that nature has evolved such sensors over many million years in the common house fly which uses vibrating *halteres* to sense rate of rotation. The fly has been described as a 'control configured flying vehicle' in its unsurpassed manoeuvrability, as anyone who has tried to swat a fly can confirm. The halteres are effectively small flexible rods with a mass at the end and have evolved from a vestigial second pair of wings. Amputation of the halteres results in unstable flight – just like the loss of the rate gyros in an aerodynamically unstable FBW aircraft.

Because of space constraints, attention has been concentrated on the quartz tuning fork type of micro-miniature rate gyro. This appears to be the most widely used type at the time of writing but no disparagement is implied in any way of the vibrating cylinder type.

The basic principle of the tuning fork rate gyro is briefly explained as follows. Refer to Fig. 5.1a, which shows a single vibrating tine. If the base is rotated about the X axis while the mass is oscillating in the Y direction, the Coriolis forces induced will cause oscillations to be built up along the Z axis. Providing the elastic support is exactly symmetrical; such oscillations will be in phase with the original Y motion. The plane of oscillation tends to stay fixed in space. Such a simple instrument suffers from the unacceptable characteristic that the smallest linear motions applied to its base cause unacceptably large errors. To overcome the effect of base motion, it is necessary to use balanced oscillations in which the oscillations of one mass are counter-balanced by equal and opposite motion of a second equal mass as shown in Fig. 5.1b. The amplitude of the torsional oscillation about the X axis, or input axis, is directly proportional to the input rate. To obtain maximum sensitivity, the size and shape of the tuning fork stem is such that the torsional vibration frequency of the tuning fork about its axis of symmetry is identical to the flexural frequency of the tuning fork. This amplitude is measured by suitable sensors to give an output signal proportional to input rate. Figure 5.1c outlines the basic theory. Note: Corolios accelerations are discussed further in Chap. 6, Sect. 6.2.3.2.

Fig. 5.1 Tuning fork rate gyro

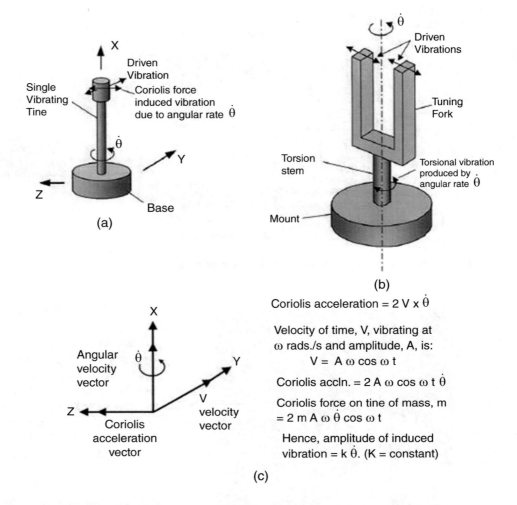

Coriolis acceleration = 2 V x $\dot{\theta}$

Velocity of time, V, vibrating at
ω rads./s and amplitude, A, is:
$$V = A\,\omega\,\cos\,\omega\,t$$

Coriolis accln. = 2 A ω cos ω t $\dot{\theta}$

Coriolis force on tine of mass, m
= 2 m A ω $\dot{\theta}$ cos ω t

Hence, amplitude of induced
vibration = k $\dot{\theta}$. (K = constant)

(c)

Figures 5.2 and 5.3 show the 'Gyro Chip' vibrating quartz tuning fork rate gyro developed by the BEI Systron Donner Inertial Division; the author is indebted to the company for permission to publish this information.

The basic configuration is shown schematically in Fig. 5.2 and comprises a vibrating quartz tuning fork to sense angular rate which is coupled to a similar fork as a pick-up to produce the rate output signal.

The piezo-electric drive tines are driven by an oscillator to vibrate at a precise amplitude. An applied rotation rate about an axis parallel to the vibrating tines causes a sine wave of torque to be produced resulting from the Coriolis acceleration as explained earlier. This oscillatory torque in turn causes the tines of the Pick-up Fork to move up and down in and out of the plane of the fork assembly. This causes an electrical output signal to be produced by the Pick-up Amplifier which is amplified and phase sensitive rectified to provide a DC signal which is directly proportional to the input rate. (The sign of the DC output signal changes sign when the input rate reverses due to the action of the phase sensitive rectifier.)

The pair of tuning forks and their support flexures and frames are batch fabricated from thin wafers of single crystal piezo-electric quartz and are micro-machined using photo-lithographic processes similar to those used to produce millions of digital quartz wristwatches each year. Figure 5.8 shows a fabrication wafer.

Standard rate ranges are from ±50°/s up to 1000°/s. Threshold resolution is ≤0.004°/s (less than 15°/hour, Earth's rate of rotation) and the bandwidth (90° lag) is greater than 60 Hz. Start-up time is less than 1 s and power consumption is less than 0.7 Watt. Non-linearity ≤0.05% of full scale, and MTBF is over 100,000 hours.

To achieve the highest accuracy requires accurate modelling in the system computer of the temperature dependent errors.

Several manufacturers use quartz tuning fork rate gyros for deriving the aircraft attitude in their solid state standby display instruments – a good testimony to their performance and attributes of small size, low power consumption, very high reliability and competitive cost.

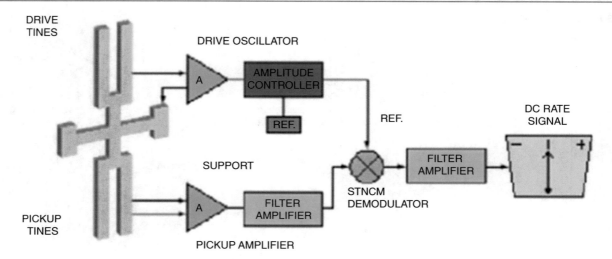

DRIVE
TINES

DRIVE OSCILLATOR

AMPLITUDE
CONTROLLER

REF.

REF.

DC RATE
SIGNAL

SUPPORT

STNCM
DEMODULATOR

FILTER
AMPLIFIER

PICKUP
TINES

FILTER
AMPLIFIER

PICKUP AMPLIFIER

Fig. 5.2 Quartz rate sensor. (Courtesy of Systron Donner Inertial Division)

Fig. 5.3 Quartz rate sensor fabrication wafer with device overlaid. (Courtesy of Systron Donner Inertial Division)

It should be noted that the tuning fork gyro is not a new concept. F W Meredith took out a patent for such a device in 1942 when he was working at the RAE (Royal Aircraft Establishment), Farnborough. Further development was carried out at the RAE in the late 1950s by Messrs. G.H. Hunt and A.E.W. Hobbs who demonstrated a performance of less than 1°/hour drift.

It is micro-machining technology and semi-conductor fabrication methods which have enabled the advances to be achieved.

5.2.3 Optical Gyroscopes

5.2.3.1 Introduction

Optical gyroscopes such as the ring laser gyro and the fibre optic gyro measure angular rate of rotation by sensing the resulting difference in the transit times for laser light waves travelling around a closed path in opposite directions. This time difference is proportional to the input rotation rate and the effect is known as the 'Sagnac effect' after the French physicist G. Sagnac who, in fact, demonstrated that rotation rate could be sensed optically with the Sagnac interferometer as long ago as 1913. Figure 5.4 illustrates schematically the basic configurations of the Sagnac interferometer.

The Sagnac effect time difference, ΔT, between the clockwise (cw) and anti-clockwise (acw) paths is given by

$$\Delta T = \frac{4A}{c^2}\dot{\theta} \qquad (5.1)$$

where A is the area enclosed by the closed path, c the velocity of light and $\dot{\theta}$ the angular rate of rotation about an axis normal to the plane of the closed path.

The difference in optical path length ΔL, where L is the perimeter of the path, is given by.

$$\Delta L = c\Delta T$$
$$\Delta L = \frac{4A}{c}\dot{\theta} \qquad (5.2)$$

A rigorous derivation of the above formulae requires the use of the general theory of relativity. A simpler kinematic

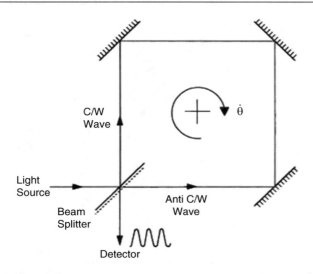

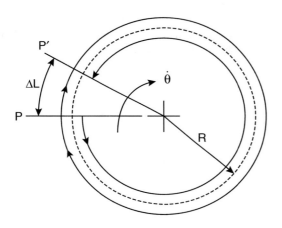

Fig. 5.4 Sagnac interferometer. The clockwise and anti-clockwise waves interfere to produce a fringe pattern which shifts when the interferometer is subjected to input rate, $\dot{\theta}$

Fig. 5.5 Sagnac effect

explanation, however, is given below for the case of a circular path in vacuo.

Referring to Fig. 5.5, consider a photon of light starting from P and travelling round the perimeter in a cw direction and a photon starting from P travelling in the acw direction.

In the absence of an input rate, the transit times of the two photons will be identical and equal to $2\pi R/c = T$. Now consider the path rotating at a rate $\dot{\theta}$. In time T, P has moved to P^1 and the path length for the cw photon is equal to $(2\pi R + R\dot{\theta}T)$ and the path length for the acw photon is equal to $(2\pi R - R\dot{\theta}T)$. The difference in transit time,

$$\Delta T = \frac{\left(2\pi R + R\dot{\theta}T\right)}{c} - \frac{\left(2\pi R - R\dot{\theta}T\right)}{c}$$

substituting $T = 2\pi R/c$ yields

$$\Delta T = \frac{4\pi R^2}{c^2}\dot{\theta} \tag{5.3}$$

and

$$\Delta T = \frac{4\pi R^2}{c}\dot{\theta} \tag{5.4}$$

$A = \pi R^2$ hence formulae (5.3) and (5.4) are identical with (5.1) and (5.2).

It can be shown that these formulae are unchanged when the optical path comprises a medium of refractive index n such as when an optical fibre provides the closed path.

The Sagnac effect is very small for low rates of rotation. Michelson and Gale in 1925 used a Sagnac interferometer with a rectangular cavity 600 m × 330 m to measure the Earth's rotation rate and measured a path difference of only 1/4 fringe.

The invention of the laser in 1960 radically changed the situation and enabled optical gyroscopes based on the Sagnac effect to be developed by exploiting the laser's ability to provide a source of highly coherent light with very stable wave length characteristics.

The first experimental ring laser gyro was demonstrated in the USA by Macek and Davis in 1963 and has since been developed by a number of companies and establishments worldwide to large-scale production status. Many tens of thousands of RLGs are operating in strap-down INS systems and have established high accuracy with better than 0.01°/hour bias uncertainty and MTBFs in excess of 60,000 hours. The RLG is described in the next section and is basically an active resonant system with the laser cavity forming the closed optical path. Input rotation rates are measured by the difference in the resonant frequencies of the clockwise and anti-clockwise paths resulting from the difference in the path lengths produced by the rotation.

The development of low loss single mode optical fibre in the early 1970s for the telecommunications industry enabled Sagnac effect fibre optic gyros to be developed. These use an external laser diode source together with suitable beam splitting optics to launch the laser light so that it travels in cw and acw directions through a cylindrical coil comprising many turns of optical fibre. The effective area of the closed optical path is thus multiplied by the number of turns in the coil. Path lengths of hundreds of metres are achievable.

The first FOG was demonstrated in the USA by Vali and Shorthill in 1976. Development and production of the passive interferometer type of FOG, or IFOG, has since taken place in many companies and establishments worldwide.

The IFOG operates basically as a Sagnac interferometer and measures the phase shift in the fringe pattern produced by the input rotation rate. The accurate measurement of very small Sagnac phase shifts has required the development of special techniques which will be explained. The IFOG was initially developed to an AHRS level of accuracy and the first large production orders for IFOG-based AHRS were placed in 1991. Development to inertial accuracy has since taken place in many organisations and IFOG-based IN systems are now available.

The resonator type of fibre optic gyro, or RFOG, operates in a similar manner to the RLG but with the resonant ring of optical fibre driven by an external laser diode. Two acousto-optical modulators are used to shift the laser frequency injected into the cw and acw paths, respectively, with the modulation frequencies being controlled by servo loops to maintain the resonance peak condition. The difference in the resonant frequencies of the two paths is then directly proportional to the input rotation rate as with the RLG. Development of the RFOG appears to be on the 'back-burner' at the present time, as the IFOG type is giving an excellent all-round performance.

Attention has thus been concentrated on the RLG and the IFOG.

5.2.3.2 The Ring Laser Gyro

The basic elements of the RLG are shown schematically in Fig. 5.6. The two counter rotating laser beams are generated from the lasing action of a helium–neon gas discharge within the optical cavity, the triangular closed path being formed by reflecting mirrors at each corner of the triangle. This closed path forms the resonant cavity and the longitudinal mode frequency, f, is determined by the cavity optical path length, L, being given by $f = nc/L$ where n is an integer and c the velocity of light. At zero input rotation rate, the cw and acw path lengths are equal and there is zero difference between the frequencies of the cw and acw waves. When the RLG is rotated about an axis normal to the plane of the closed path there is a difference in the path length of the cw and acw travelling waves, as shown earlier, which causes a frequency difference between the two waves. This frequency difference is measured by allowing a small percentage of the two laser beams to be transmitted through one of the mirrors. A corner prism is generally used to reflect one of the beams so that it can be combined with the other to generate a fringe pattern at the read-out detector photo-diodes. An input rotation rate causes the fringe pattern to move relative to the read-out photo-diodes at a rate and in a direction proportional to the frequency difference (positive or negative). A sinusoidal output signal is generated by each fringe as it passes by the photo-diodes. These are spaced so that there is a 90° phase difference between their outputs so that the direction of rotation can be determined from which photo-diode output is leading. The photo-diode outputs are then converted into positive (or negative) pulses by suitable pulse triggering and direction logic circuits.

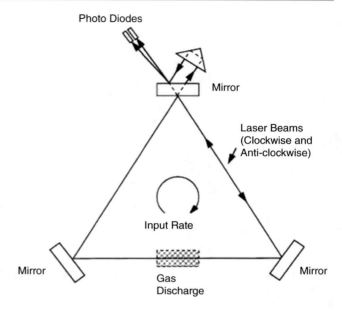

Fig. 5.6 Laser gyro schematic

The frequency difference, Δf, resulting from a difference in optical path length, ΔL, is given by

$$\frac{\Delta T}{L} = \frac{\Delta f}{f}$$

From Eq. (5.2)

$$\Delta T = \frac{4A}{c}\dot{\theta}$$

whence

$$\Delta f = \frac{4Af}{cL}\dot{\theta} \qquad (5.5)$$

The wavelength of the laser transition $\lambda = c/f$, hence

$$\Delta f = \frac{4A}{\lambda L}\dot{\theta} \qquad (5.6)$$

That is,

$$\Delta f = K_0\dot{\theta} \qquad (5.7)$$

where K_0 is the gyro scale factor $= 4A/\lambda L$.

The gyro thus behaves as an *integrating rate gyro*

$$\int_0^T \Delta f dt = K_0 \int_0^\theta d\theta \qquad (5.8)$$

The angle turned through about the gyro input axis in the time period, T, is equal to the net number of positive (or negative) pulses counted in that period. The RLG thus provides a direct digital output of the input angular rotation with a process of inherent integration carried out in the optical domain.

The high sensitivity of the RLG can be seen from the following example. A typical RLG with an equilateral triangular cavity of perimeter, $L = 0.2$ m (20 cm) has an area, $A = 1.9245 \times 10^{-3}$ m^2 (19.245 cm^2) and laser transition wavelength, $\lambda = 0.633$ microns (0.633 $\times 10^{-6}$ m).

The RLG scale factor is

$$\frac{4 \times 1.9245 \times 10^{-3}}{0.633 \times 10^{-6} \times 0.2} = 60825 \, \text{Hz/rad per s}$$

and hence 1 pulse count is 1/60825 rad = 3.39 arc seconds.

This resolution can be increased by a factor of 4 by triggering on the positive and negative going zero crossings of both detectors.

An alternative and equally valid explanation of the RLG which some readers may prefer is as follows. The two counter propagating waves set up by the lasing action will beat together and set up a standing wave pattern in the cavity. This standing wave pattern remains fixed in space irrespective of the angular rotation of the cavity. An observer rotating with the cavity, in this case the read out photodiodes, will thus see a succession of light and dark fringes as the read-out moves past the spatially fixed standing wave. Figure 5.7 illustrates the principle.

The RLG behaves in an analogous way to an incremental optical encoder with the encoder shaft angle fixed in space and so measures the angle through which the gyro has been rotated. The fact that the standing wave pattern remains fixed in space as the cavity rotates can be interpreted as a manifestation of the inertia which electro-magnetic energy possesses in view of its mass equivalence.

Figure 5.8 illustrates the construction of a typical RLG. The equilateral triangular cavity is machined from a solid block of a vitro-ceramic material known as 'CERVIT' which has a very low coefficient of thermal expansion. The three mirrors are mounted on the block by optical contact for stability and ruggedness. The mirrors have multi-dielectric coatings and have a reflectivity of more than 99.9%. The mirror system, being a critical area of RLG technology, absorption, and above all scattering, must be minimised.

The optical path length of the cavity is adjusted by means of a piezo-electric transducer attached to one of the cavity mirrors. The position of the mirror being controlled by a servo loop so that the laser oscillates at its peak average power.

High voltages of the order of 2000 volts dc are applied between the two separate anodes and the common cathode to ionise the He–Ne gas mixture and provide the required lasing action. The system uses a common cathode and two separate

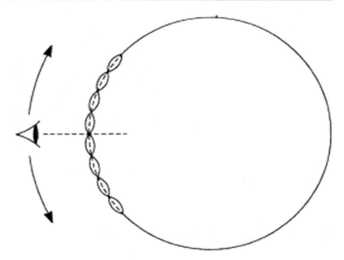

Fig. 5.7 Laser gyro principle

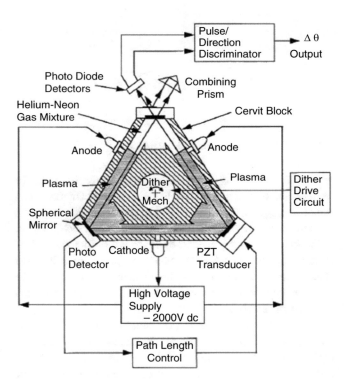

Fig. 5.8 Ring laser gyro

anodes so that Langmuir flow effects in each plasma arm are cancelled out by control currents which balance the currents in the two discharges. These flow effects can cause bias shifts due to differential changes in the refractive index of the light travelling along the cw and acw paths.

A major problem which has had to be overcome with the RLG is the phenomenon known as 'lock-in'. This arises because of imperfections in the lasing cavity, mainly in the mirrors, which produce back scattering of one beam into the

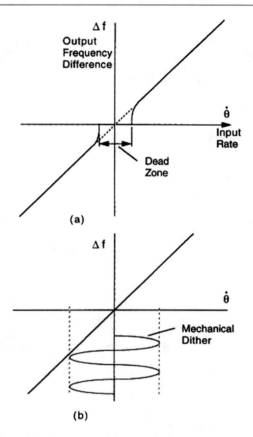

Fig. 5.9 Lock-in effect of mechanical dither

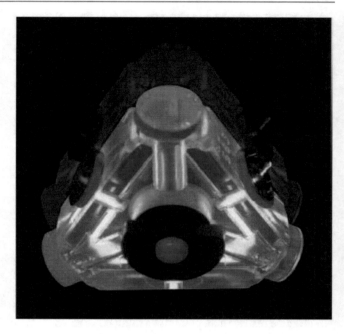

Fig. 5.10 Ring laser gyro. (Courtesy of Honeywell)

other. The resulting coupling action tends to pull the frequencies of the two beams together at low rotation rates producing a scale factor error. For input rates below a threshold known as the 'lock-in rate', the two beams lock together at the same frequency so that there is zero output and a dead zone results. Figure 5.9a illustrates the effect of lock-in under steady input rate conditions. This lock-in dead zone is of the order of 0.01 to 0.1°/s compared with 0.01°/hour accuracy required for an INS.

A very effective method of overcoming this problem is to mechanically dither the laser block about the input axis at a typical frequency of about 100 Hz with a peak velocity of about 100°/s (corresponding to an amplitude of 1.5 arc seconds approximately). The amplitude of the dither rate and acceleration are chosen so that the dwell time in the lock-in zone is so short that lock-in cannot occur. Figure 5.9b illustrates the effect of dither on removing the dead zone. The dither signal can be removed from the output by mounting the read-out reflector prism on the gyro case and the read-out photo-diodes on the block so as to produce an optical cancellation of the dither signal. Alternatively the read-out prism and read-out photo-diodes can both be mounted on the block and the unwanted dither signal can be removed by a digital filter. Some reduction in gyro

bandwidth is incurred, however, because of the digital filter response.

Alternative 'solid state' techniques are also employed to overcome lock-in. The mechanical dither technique, however, is by far the most widely used technique at the current time. Although a criticism could be made in terms of 'technical elegance' of a solid state system which depends on mechanical dither, the facts are that zero failures in the dither system in over several hundred million hours have been experienced.

Figure 5.10 is a photograph of a current, widely used, inertial quality ring laser gyro and shows the glowing laser beams in the optical cavity.

The photograph, by courtesy of Honeywell, is of the Honeywell GG 1320 AN Digital Laser Gyro. This gyro is used in the Honeywell Air Data Inertial Reference Systems, ADIRS, installed in many civil airliners. It is also used in the INS systems Honeywell produce for military aircraft. A slightly lower accuracy version of the GG1320 AN Digital Laser Gyro is also used in what is termed a 'Super AHRS'.

RLG performance characteristics. The general performance characteristics of a ring laser gyro are summarised below.

High accuracy – The RLG meets the dynamic range for a pure IN system of being able to measure angular rates from 0.01°/hour to 400°/s to the required accuracy – a dynamic range of 10^8:1.

Insensitivity to acceleration – The RLG has no acceleration sensitive bias errors, as it is based on optical effects rather than inertial effects.

Very high rate range – This is limited only by the noise/bandwidth characteristics of the read out electronics: ±1000°/s is no problem.

Very high scale factor accuracy – Errors are in the 5–10 ppm bracket.

Negligible warm up time – Full gyro operation from the instant of turn-on.

Excellent turn-on to turn-on performance – Performance capabilities can be maintained over several years without calibration.

Random noise uncertainty – This is measured in 'degrees per √hour', and is one of the RLG's most important error characteristics. The error is significantly higher than experienced with angular momentum gyros. It affects the system heading determination in the gyro compassing phase during the initial alignment process. This is because it extends the time required to filter the Earth's rate signal from the gyro noise in order to determine the initial heading.

Very high reliability – MTBFs in excess of 60,000 hours are being demonstrated in large scale service. 100,000 hours MTBF is being approached.

Life – The laser beams ultimately destroy the mirrors but this is over a very long period. A life equivalent to over 100 years usage is being claimed.

Volume – This is determined by the path length, typically 20 cm for inertial performance.

5.2.3.3 The Interferometric Fibre Optic Gyro

The implementation of the interferometric type of FOG is best explained in a series of stages starting with the simple basic system shown in Fig. 5.11.

Light from the laser diode source is passed through a first beam splitter and a single optical mode is selected. The light passes through a second beam splitter and propagates in both directions around the fibre coil. In the absence of rotation, as already explained, the transit times are identical so that when the light arrives back at the second beam splitter, perfect constructive interference occurs with accompanying fringe pattern. The gyro output signal is obtained by directing the returning light via the first beam splitter to a photo-detector. As explained, an input rotation rate about an axis normal to the plane of the coil results in a difference in the transit times between the clockwise and anti-clockwise beams as given by Eq. (5.1), namely,

$$\Delta T = \frac{4A}{c^2}\dot{\theta}$$

Fig. 5.11 Basic fibre optic gyro

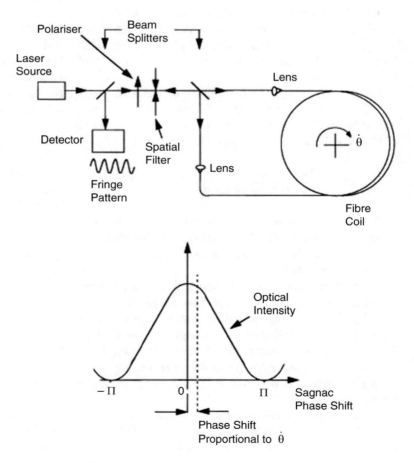

If the fibre coil has N turns, $A = \pi R^2 N$ where R is the mean radius of the coil, and N can be expressed in terms of the length of the coil, L and the mean radius R, that is,

$$N = \frac{L}{2\pi R}$$

hence

$$\Delta T = \frac{LD}{c^2}\dot{\theta} \qquad (5.9)$$

where D is the coil diameter $= 2R$.

This transit time difference results in the Sagnac phase shift, given by

$$\Phi_s = \omega\Delta T$$

where ω is the angular frequency of the laser source $= 2\pi c/\lambda$, and λ is the laser wavelength.

Hence

$$\Phi_s = \frac{2\pi LD}{\lambda c}\dot{\theta} \qquad (5.10)$$

This phase shift between the clockwise and anti-clockwise travelling light waves results in a reduction in the intensity of the light at the detector. This change in intensity is very small for useful rotation rates and has required the development of special techniques.

For example, a typical FOG with an optical path length $L = 200$ m, coil diameter, $D = 70$ mm and source laser wavelength $= 1.3$ microns, when subjected to an input rate of $10°/\text{sec}$ will have a Sagnac phase shift of

$$\Phi_s = \frac{2\pi \times 200 \times 70 \times 10^{-2}}{1.3 \times 10^{-6} \times 3 \times 10^{-8}} \times 10 = 2.26°$$

Resolution of Earth's rate ($15°/\text{hour}$) requires the measurement of a phase shift of 3.38 arc seconds. Resolution of $0.01°/\text{hour}$ seems barely credible at first sight but is achievable as will be shown.

The next problem to be overcome with the simple interferometer shown in Fig. 5.11 is that the relationship of the intensity of the fringe pattern to the Sagnac phase shift (which is proportional to the input rate) is a 'one plus cosine' law (see Fig. 5.11). This results in a minimum sensitivity at low input rates and it is thus necessary to introduce a phase shift of $90°$ so that the sensor is operated in the region of maximum sensitivity.

The technique generally used is an alternating phase bias modulation method whereby a 'phase modulator' is placed at the end of the fibre loop, as shown in Fig. 5.12. The two counter-propagating waves receive the same phase

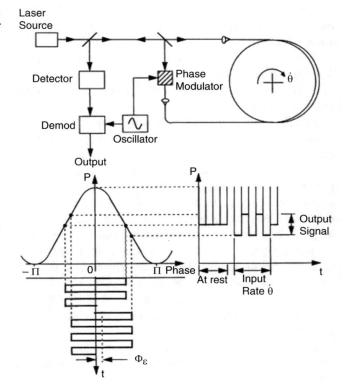

Fig. 5.12 Phase modulation – open-loop IFOG

modulation but at different times (clockwise wave at the end of the transit round the coil, anti-clockwise wave at the beginning of the transit). The phase modulator is driven at a frequency such that half a cycle corresponds to the time delay for the light to travel round the coil. (For a 200 m FOG this is a modulation frequency of about 500 kHz.)

A non-reciprocal modulation is thus produced – a modulation which affects the clockwise direction of propagation differently from the anti-clockwise direction of propagation. Application of an input rotation results in a modulated output signal whose amplitude is proportional to the input rate (see Fig. 5.12). This improved system as it stands is an open-loop system with attendant disadvantages such as the following:

(a) Output stability and scale factor is dependent on the stability of the gains of the various amplifiers preceding the demodulator, the demodulator stability and the stability of the intensity of the laser light source.

(b) The non-linear output characteristic with the phase shift resulting from large input rates (output characteristic is sinusoidal).

The inherent limitations on scale factor stability are overcome by closed-loop control, shown in concept in Fig. 5.13a. The output of the demodulator is passed through a servo amplifier which then drives a non-reciprocal phase transducer placed within the fibre interferometer. In this way the sensor is always operated at null by generating a suitable

Fig. 5.13 Closed-loop
interferometric FOG

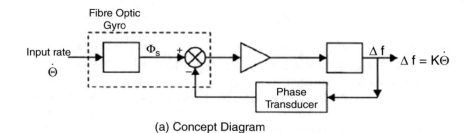

(a) Concept Diagram

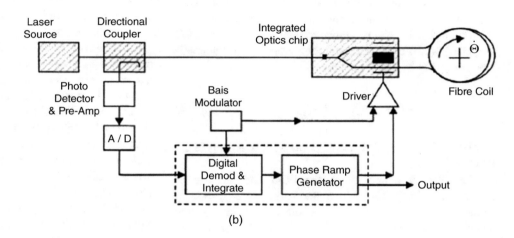

(b)

non-reciprocal phase shift which is made equal and opposite to that generated by the input rate by the action of the closed loop. The output of the system is then the output of the non-reciprocal phase transducer.

Figure 5.13b shows a widely used basic architecture for a typical closed-loop IFOG. This uses a multi-function integrated optics chip for polarisation, beam splitting the light into clockwise and anti-clockwise waves and phase modulation. The light intensity output signal from the photo-detector is converted to a digital value by an A to D converter followed by digital demodulation and integration. The loop is closed by driving the integrated optics phase modulator with a voltage ramp whose slope is proportional to the input rate. This ramp produces a non-reciprocal phase shift between the light waves so as to restore and maintain the sensor at the zero rotation condition.

The advantages of the closed-loop system are as follows:

(a) Output is independent of light source intensity variations as the system is always operated at null.
(b) Output is independent of the gains of individual components in the measurement system as long as a very high open-loop gain is maintained.
(c) Output linearity and stability depend only on the phase transducer.

The voltage ramp driving the phase modulator generates a frequency difference to null the phase difference produced by the input rate.

Frequency difference and phase shift are directly related. The phase shift, Φ, that light experiences in propagating along a single mode optical fibre of length, L, and refractive index, n, is given by

$$\Phi = 2\pi f \frac{L}{c/n}$$

(f is the frequency of the light, velocity of propagation in the fibre = c/n).

Hence

$$d\Phi = \frac{2\pi nL}{c} df$$

Sagnac phase shift,

$$\Phi_s = \frac{2\pi LD}{\lambda c} \dot{\theta}$$

hence at null

$$\frac{2\pi nL}{c} \Delta f = \frac{2\pi LD}{\lambda c} \dot{\theta}$$

from which

$$\Delta f = \frac{D}{n\lambda} \dot{\theta} \qquad (5.11)$$

That is,

$$\Delta f = K_1 \dot{\theta}$$

where $K_1 = D/n\lambda$.

The output characteristic is thus the same as the RLG and the IFOG behaves as an integrating rate gyro.

The scale factor stability is dependent on the stability of D, n and λ.

For example, FOG already used $D = 70$ mm, $\lambda = 1.3$ mm, $n = 1.5$. Substituting these values, one output pulse corresponds to an angular increment of 5.75 arc seconds.

Equation (5.11) can be shown to be the same as the RLG output Eq. (5.6) by multiplying the numerator and denominator of (5.11) by $\pi D/4$, namely,

$$\Delta f = \frac{D}{n\lambda} \frac{\pi D/4}{\pi D/4} \dot{\theta} = \frac{4A}{\lambda(n\pi D)} \dot{\theta} \qquad (5.12)$$

The equivalent optical perimeter ($n\pi D$) corresponds to the RLG path length, L.

The inherent simplicity of the FOG and the use of *Integrated Optic Chips* where the various electro-optical elements are integrated on to a substrate leads to low manufacturing costs and the IFOG is a highly competitive gyro in terms of cost, reliability, ruggedness and performance. Integrated three axis versions are being produced with the three FOGs sharing the same laser diode source.

Figure 5.14 is a photograph of the prototype of the inertial measuring unit (IMU) of the fibre optic gyro based Attitude Heading Reference System developed by the US subsidiary of the former Smiths Industries in the UK. The mechanical simplicity of an IFOG can be appreciated from the photo; basically a few hundred metres of single mode optical fibre wound on a spool. The sophistication resides in the electro-optical devices and the processing chips.

Fig. 5.14 Fibre optic gyro inertial measuring unit. (Courtesy of Smiths Industries)

The IMU incorporates three interferometric fibre optic gyros and three accelerometers; the three FOGs share the same laser source. The FOG triad system is configured to share many of the optical and electrical components between the three FOG axes by time sequencing the processing of the optical signals for each axis using optical switches. By time sharing components, the cost, weight and volume are significantly reduced while reliability is increased by reducing the number of parts. The optical switches also replace the conventional beam splitters and increase the signal to noise ratio in the detected signal.

The fibre coils of the gyro are comprised of 400 metres of non-polarisation pre-serving single mode optical fibre wound on a 3.5 cm diameter spool. The quoted gyro performance is 0.8°/hour (1 sigma), scale factor accuracy 60 ppm (1 sigma) and random walk 0.04°/hour.

The IMU weighs 420 gm and is 245 cm^3 in volume; power consumption is 7.5 Watts.

The major improvement in all aspects compared with the previous gimballed AHRS units using angular momentum gyros can be appreciated.

To achieve inertial quality performance has required particular attention to minimising non-reciprocal optical error sources such as polarisation. Detailed discussion of these error sources and methods of minimising their effects is beyond the scope of this book, and the reader is asked to accept their resolution as a fact.

The effect of environmental noise sources must also be minimised in a FOG design in order to achieve inertial quality performance. This is because these non-reciprocal effects introduced would otherwise swamp the small Sagnac phase shifts being measured. Environmental noise sources include the following:

- Temperature variations – Introducing time-dependent temperature gradients along the optical fibre. Operating environments can range from −40 to +70 °C with ramp rates of 0.5 to 1 °C/minute.
- Acoustic noise – Optical fibre is acoustically sensitive (sensitive to sound pressure) and in fact fibre optic sensors are widely used in sonar arrays because of this sensitivity.
- Mechanical vibration – Introducing mechanical stresses in the fibre with accompanying non-reciprocal effects. Environments of 6 to 9 g can be experienced.
- Magnetic fields – A magnetic field acting parallel to the optical fibre results in a Faraday rotation of the state of polarisation in the fibre. This is a non-reciprocal effect which results in an unwanted phase shift. Fluctuating magnetic fields of up to 10 Gauss can be experienced about any axis. Mu-metal magnetic screening will almost certainly be necessary.

Very high inertial accuracy performance has been achieved by several manufacturers of closed-loop IFOGs using polarisation maintaining (PM) optical fibre in the sensing coil and production status achieved.

In a paper published in 2000, Honeywell reported very encouraging progress using depolarised fibre in conjunction with polarisation randomisation in the loop. Depolarised fibre is widely used in the telecommunications industry and is of much lower cost than PM optical fibre.

A Honeywell brochure produced in the early 2000s for an IFOG-based inertial reference system for space and strategic applications, quotes a bias stability <0.0003°/hour; average random walk <0.0001°/hour. Scale factor stability, < 1 ppm (short term) and a quantisation of 0.0004 arc sec/LSB.

It is worth discussing briefly the relative merits of inertial IFOGs and the RLG.

The RLG is very reliable, very well proven (around 30 years operating experience), meets the accuracy requirements and has over 90%, or more, of the aviation INS market, at the present time. There is no incentive to change such a well-established and highly reliable system at the moment. Its manufacture, however, involves very precise machining operations and exceptionally high quality mirrors.

The IFOG does not require the precision machining processes involved with the RLG and it exploits relatively low cost integrated chips.

Time alone will tell which type of gyro will achieve the largest share of the aviation INS market in the future.

It should be pointed out, however, that there are no low cost inertial accuracy gyros, irrespective of the technology. This is because of the extreme accuracy required from them. Some cost less than others, but they still do not come cheaply.

5.2.4 Accelerometers

5.2.4.1 Introduction: Specific Force Measurement

The acceleration of a vehicle can be determined by measuring the force required to constrain a suspended mass so that it has the same acceleration as the vehicle on which it is suspended, using Newton's law: force = mass × acceleration.

The measurement is complicated by the fundamental fact that it is impossible to distinguish between the force acting on the suspended mass due to the Earth's gravitational attraction and the force required to overcome the inertia and accelerate the mass so that it has the same acceleration as the vehicle. The vehicle acceleration, $\mathbf{a}$, being produced by the vector sum of the external forces acting on the vehicle, namely, the propulsive thrust, $\mathbf{T}$, lift, $\mathbf{L}$, drag, $\mathbf{D}$ and the gravitational force, $m\mathbf{g}$, acting on the aircraft mass, m.

(The bold print denotes the quantities are vector quantities.)

$$\mathbf{T} + \mathbf{L} + \mathbf{D} + m\mathbf{g} = m\mathbf{a}$$
$$\mathbf{a} = \frac{(\mathbf{T} + \mathbf{L} + \mathbf{D})}{m} + \mathbf{g} \qquad (5.13)$$

The vector sum of the external forces excluding the gravitational force divided by the aircraft mass, that is $(\mathbf{T} + \mathbf{L} + \mathbf{D})/m$, is known as the 'specific force'. The force, $\mathbf{F_a}$, required to constrain the suspended mass, m_a, is thus given by

$$\mathbf{F_a} + m_a\mathbf{g} = m_a\mathbf{a}$$
$$\frac{\mathbf{F_a}}{m_a} + \mathbf{g} = \mathbf{a} = \frac{(\mathbf{T} + \mathbf{L} + \mathbf{D})}{m} + \mathbf{g}$$

Hence

$$\mathbf{F_a} + m_a\frac{(\mathbf{T} + \mathbf{L} + \mathbf{D})}{m} \qquad (5.14)$$

Therefore, the accelerometer will thus measure the specific force component along its input axis and *not* the vehicle acceleration component.

It is thus essential to know the magnitude and orientation of the gravitational vector with respect to the accelerometer input axes in order to compute the vehicle acceleration components from the accelerometer outputs.

Only if the accelerometer input axis is exactly orthogonal to the gravity vector (i.e. horizontal) so that there is zero gravitational force component will the accelerometer measure the vehicle acceleration component along its input axis.

5.2.4.2 Simple Spring Restrained Pendulous Accelerometer

A simple spring restrained pendulous accelerometer is shown schematically in Fig. 5.15. This comprises an unbalanced pendulous mass which is restrained by the spring hinge so that it can only move in one direction, that is, along the input axis. The spring hinge exerts a restoring torque which is proportional to the angular deflection from the null position. When the case is accelerated the pendulum deflects from the null position until the spring torque is equal to the moment required to accelerate the centre of mass of the pendulum at the same acceleration as the vehicle. This simple type of accelerometer is typically oil filled to provide viscous damping so that the transient response is adequately damped. An electrical position pick-off measures the deflection of the pendulum from the null position and provides the output signal.

Fig. 5.15 Simple spring restrained pendulous accelerometer

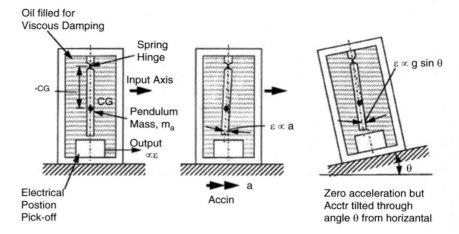

The transfer function of the simple accelerometer described is a simple quadratic lag filter of the type

$$\frac{\text{Output}}{\text{Input Accln.}} = \frac{K_0}{D^2 + 2\zeta\omega_0 D + \omega_0^2} \qquad (5.15)$$

K_0 is the accelerometer scale factor, and the undamped natural frequency $\omega_0 = \sqrt{K/I}$, where K is the torsional spring stiffness of hinge and I is the moment of inertia of pendulum about hinge axis.

The simple spring restrained accelerometer is used in applications where the accuracy requirements are not demanding, for example, 1 to 2% bracket.

The maximum angular deflection of the pendulum from the null position is limited to about $\pm 2°$ to minimise cross-coupling errors. This results in a maximum cross-axis coupling error of sin 2°, that is, approximately 3.3% of any acceleration acting along an axis normal to the input axis. The acceleration input range thus determines the spring stiffness, or rate, and hence the bandwidth of the sensor. For example, consider an accelerometer with a maximum input range of say ± 10 g and an undamped natural frequency of 25 Hz. The same basic accelerometer designed for an input range of ± 3 g would have a bandwidth of only 14 Hz approximately, being reduced by a factor $\sqrt{3/10}$.

5.2.4.3 Closed-Loop Torque Balance Accelerometer

The above shortcomings in terms of accuracy and bandwidth can be overcome by closed-loop torque balance operation of the accelerometer. The deflection of the pendulum from its null position under an input acceleration is sensed by the position pick off and the pick off signal after amplification and suitable dynamic compensation is used to control a precision torque motor to maintain the pendulum at or very near to its null position. Measurement of the torque exerted by the torque motor to balance the inertia torque resulting from accelerating the pendulous mass so that it has the same

acceleration as the vehicle then enables the acceleration to be determined. (Strictly speaking, the specific force is measured.) Pulse torque operation of the capture amplifier is generally used giving a pulse rate output which is directly proportional to the input acceleration. Each pulse represents a velocity increment, for example 0.05 m/s (0.1 knots approx.), providing an inherent integration.

The construction of a typical modern torque balance pendulous accelerometer (also referred to as a 'servo accelerometer') is shown in Fig. 5.16. The accelerometer consists basically of a beam fabricated from fused quartz which is suspended within the case by a very low stiffness flexural hinge, as shown in the illustration. Quartz exhibits zero hysteresis and consequent increased bias stability compared with a metal spring hinge. A capacitive position pick-off is used to measure the displacement of the pendulous mass (also referred to as the 'proof mass', or 'seismic mass') from its null position. (Older generation inertial accelerometers generally use inductive pick-offs for this function.) Torques are applied by means of a moving coil/permanent magnet torquer (similar to a loud-speaker driver) with the coils fixed to the beam. Thin flexible conducting ligaments enable electrical connections to be made to the torquer coils and the capacitive pick-off plates.

A mechanical damper is generally incorporated to supplement the dynamic compensation in the capture amplifier in providing a well damped response, typically around 0.5 critically damped. Bandwidth is typically around 500 Hz A temperature sensor is generally incorporated in high accuracy accelerometers to enable temperature dependent scale factor errors to be corrected. Typical size is around 2.54 cm (1 in) diameter and 2.54 cm (1 in) length.

Accelerometers used in very severe environments with very high vibration and shock levels are generally oil filled to increase their robustness.

'Solid state' accelerometers fabricated from silicon using semi-conductor manufacturing technology are now being widely used for lower accuracy applications. The technology

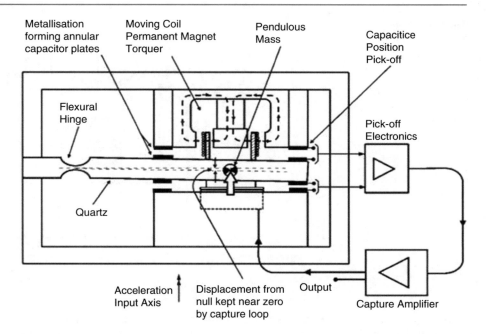

Fig. 5.16 Torque balance pendulous accelerometer schematic

Metallisation forming annular capacitor plates

Moving Coil Permanent Magnet Torquer

Pendulous Mass

Capacitice Position Pick-off

Flexural Hinge

Pick-off Electronics

Quartz

Acceleration Input Axis

Displacement from null kept near zero by capture loop

Output

Capture Amplifier

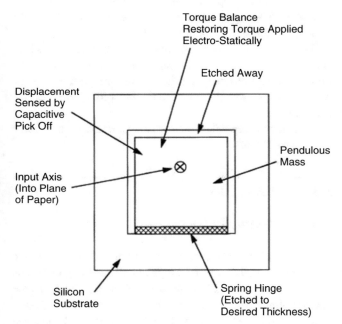

Torque Balance Restoring Torque Applied Electro-Statically

Etched Away

Displacement Sensed by Capacitive Pick Off

Pendulous Mass

Input Axis (Into Plane of Paper)

Silicon Substrate

Spring Hinge (Etched to Desired Thickness)

Fig. 5.17 'Solid state' dry accelerometer construction

offers very small size and low manufacturing costs. Figure 5.17 illustrates the construction of a 'solid state' silicon accelerometer. The capacitive pick-off senses the deflection of the pendulum under acceleration and electrostatic forces are exerted to maintain the pendulum near the null position by applying DC voltages to the case mounted electrodes by means of the caging loop.

The torque balance loop of an electro-magnetic torque balance type of pendulous accelerometer is illustrated in

block diagram form in Fig. 5.18, for readers who wish to derive the accelerometer transfer function.

The accelerometer loop gain must be sufficiently high to keep the pendulum excursions from the null position very small under the vibration environment. Significant rectification errors can otherwise be introduced as the accelerometer will sense a component of the vibration at right angles to its input axis which is proportional to the pendulum displacement from null. A simplified explanation of this is set out below.

Assume for simplicity (and worst case) that the vibration is at 45° to the accelerometer input axis so that the along and cross-axis vibration components are equal and are given by

$$a = a_{max} \sin \omega t$$

where a_{max} is the peak value of the vibration acceleration, a, and ω is the angular frequency.

The displacement of the pendulum from null is given by

$$\varepsilon = \frac{a_{max}}{K} \sin(\omega t + \Phi)$$

where K is dependent on the loop gain at frequency ω and Φ is the phase shift between ε and a.

Sensed cross-axis acceleration is $\varepsilon \cdot a_{max} \sin \omega t$

$$= \frac{a_{max}^2}{K} \sin \omega t \cdot \sin(\omega t + \phi)$$

The product term $[\sin \omega t \cdot \sin(\omega t + \varphi)]$ has an average value of $\cos \Phi / 2$ so that a proportion of the input vibration is rectified. This error source contributes to the 'g^2 error

Fig. 5.18 Torque balance
accelerometer loop

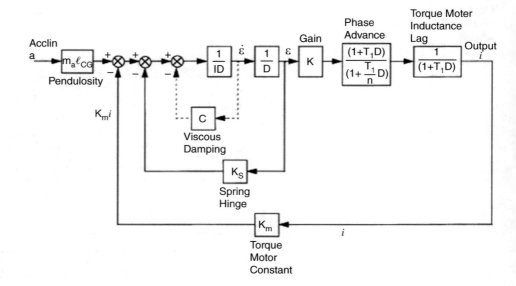

sensitivity' of the accelerometer as it is proportional to the square of the vibration acceleration.

5.2.5 Skewed Axes Sensor Configurations

One of the attractive features of a strap-down system is that economical failure absorption configurations are available. One such configuration is an arrangement of six single axis rate gyros and six accelerometers, with their input axes skewed with respect to the principal axes of the vehicle so that they form equal solid angles, see Fig. 5.19. The configuration is sometimes referred to as a 'dodecahedron' configuration as the sensor input axes are normal to the faces of a solid dodecahedron. Motion about any principal axis is sensed by four sensors so that it is possible to detect and isolate up to two failures without loss of capability. The sensor supplies and electronics are entirely independent, so that common failures are precluded. Each gyro in the dodeca-hedron configuration senses two body rates and each body rate is sensed by four of the six gyros.

The gyro outputs denoted by letters A to F are functions of the body rates.

$$A = p \sin \alpha + r \cos \alpha$$
$$B = -p \sin \alpha + r \cos \alpha$$
$$C = p \cos \alpha + q \sin \alpha$$

$$D = p \cos \alpha - q \sin \alpha$$
$$E = q \cos \alpha + r \sin \alpha$$
$$F = q \cos \alpha - r \sin \alpha$$

where $\alpha = 31.716747°$, the angle between the gyro input axes and the body axes. There are thus six equations in three

unknowns which can give two independent measures of each body rate. Sophisticated algorithms can be used to isolate a failed gyro and re-combine the data from the remaining sensors in an optimum way.

There are other configurations of skewed sensors which are also used. For example, three orthogonal sensors with their input axes aligned with the aircraft principal axes and a fourth sensor with its input axis skewed at 45° to the principal axes.

Two such independent sensor sub-assemblies can tolerate three gyro failures.

5.3 Attitude Derivation

5.3.1 Introduction

The measurement of the aircraft's attitude with respect to the horizontal plane in terms of the pitch and bank angles and its heading, that is, the direction in which it is pointing in the horizontal plane with respect to North, is essential.

The spatial attitude of an aircraft is specified by the three Euler angles which are defined in Chap. 3, Sect. 3.4.2. Referring to Fig. 3.17, the aircraft is first rotated in the horizontal plane through the heading (or azimuth) angle, ψ, then rotated about the pitch axis through the pitch angle, θ, and then rotated about the roll axis through the bank angle, Φ. The rotations must be made in that order as they are non-commutative, that is, a different orientation would be obtained if the rotations were made in a different order.

There are two basic inertial mechanisations which are used to derive the Euler angles to the required accuracy. The first method, which is known as a *stable platform* system, has the gyros and accelerometers mounted on a platform which is suspended in a set of gimbals. The gyros then

Fig. 5.19 Skewed axes sensors – dodecahedron configuration

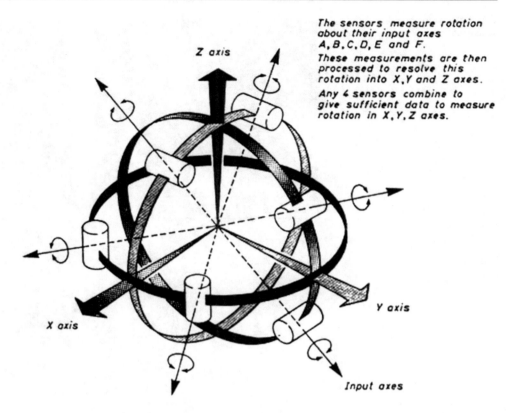

The sensors measure rotation about their input axes A, B, C, D, E and F.

These measurements are then processed to resolve this rotation into X, Y and Z axes.

Any 4 sensors combine to give sufficient data to measure rotation in X, Y, Z axes.

Z axis

Y axis

X axis

Input axes

control the gimbal servos so that the platform maintains a stable orientation in space irrespective of the aircraft manoeuvres. Angular position pick-offs on the gimbals then provide a direct read-out of the Euler angles. A typical stable platform is illustrated in Fig. 5.20.

The second method is known as a *strap-down* system as it has the gyros and accelerometers mounted on a rigid frame or block which is directly fixed, that is, 'strapped-down', to the airframe. The gyros and accelerometers thus measure the angular and linear motion of the aircraft with respect to the aircraft's body axes. The Euler angles are then computed from the body rate information by the system computer.

It should be noted that a stable platform system and a strap-down system are mathematically equivalent systems. The function of the mechanical gimbals is carried out by the strap-down system computer which in effect contains a mathematical model of the gimbals.

The stable platform mechanisation of an IN system is also easier to visualise and so is used to explain basic IN principles, such as Schuler tuning, in the next chapter.

5.3.2 Strap-Down Systems

Figure 5.21 illustrates the strap-down AHRS or INS concept. The derivation of the Euler angles from the body rates of rotation is explained below.

5.3.2.1 Attitude Algorithms

The Euler angles, Φ, θ, ψ are defined in Chap. 3, Sect. 3.4.2 and the equations relating the body angular rates p, q, r to the Euler angle rates are derived and comprise Eqs. (3.16), (3.17), (3.18). Equations for $\dot{\Phi}$, $\dot{\theta}$, $\dot{\psi}$ can be derived from these equations by suitable algebraic manipulation and are set out as follows:

$$\dot{\Phi} = p + q \sin \Phi \tan \theta + r \cos \Phi \tan \theta \qquad (5.16)$$

$$\dot{\theta} = q \cos \Phi - r \sin \Phi \qquad (5.17)$$

$$\dot{\psi} = q \sin \Phi \sec \theta + r \cos \Phi \sec \theta \qquad (5.18)$$

These equations can be expressed more compactly in matrix form

$$\begin{bmatrix} \dot{\Phi} \\ \dot{\theta} \\ \dot{\psi} \end{bmatrix} = \begin{bmatrix} 1 & \sin \Phi \tan \theta & \cos \Phi \tan \theta \\ 0 & \cos \Phi & -\sin \Phi \\ 0 & \sin \Phi \sec \theta & \cos \Phi \sec \theta \end{bmatrix} \begin{bmatrix} p \\ q \\ r \end{bmatrix} \qquad (5.19)$$

The Euler angles can then be derived from the Euler angle rates by a process of integration using as initial conditions a known attitude at a given point in time. This process becomes meaningless, however, at $\theta = 90°$ when $\tan \theta$ and $\sec \theta$ become infinite. (This is the mathematical equivalent of

Fig. 5.20 Stable platform schematic

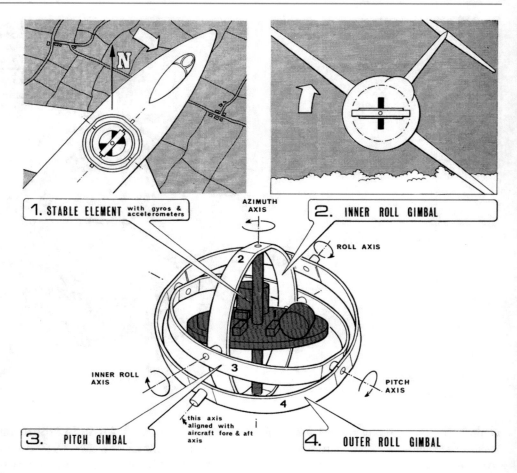

1. STABLE ELEMENT with gyros & accelerometers

2. INNER ROLL GIMBAL

3. PITCH GIMBAL

4. OUTER ROLL GIMBAL

AZIMUTH AXIS

ROLL AXIS

INNER ROLL AXIS

PITCH AXIS

this axis aligned with aircraft fore & aft axis

'gimbal lock'.) The use of the three-parameter Euler algorithms is therefore generally limited to pitch angles between $\pm30°$ as the error equations are unbounded, and to avoid mathematical singularities.

A fully manoeuvrable system is therefore required where there are no restrictions on the pitch angles. The limitations of the three-parameter Euler system are overcome by the use of what is known as the *Euler four symmetrical parameters* to define the vehicle attitude.

It can be shown that an axis set may be moved to any required orientation by a single rotation about a suitably positioned axis. Let this axis make angles $\cos^{-1}\alpha$, $\cos^{-1}\beta$ and $\cos^{-1}\gamma$ with the inertial axes OX_0, OY_0 and OZ_0, respectively. Let a single rotation, μ, about this axis bring a moving axis set from OX_0, OY_0, OZ_0 into coincidence with OX, OY, OZ, the set whose orientation it is desired to specify.

The Euler four symmetrical parameters are given by

$$\left.\begin{aligned} e_0 &= \cos \mu/2 \\ e_1 &= \alpha \sin \mu/2 \\ e_2 &= \beta \sin \mu/2 \\ e_3 &= \gamma \sin \mu/2 \end{aligned}\right\} \tag{5.20}$$

Then e_0, e_1, e_2, e_3 can be used to specify the attitude of the vehicle with respect to OX_0, OY_0, OZ_0.

The following relationships with the Euler angles can be derived:

$$\sin \theta = 2(e_0 e_2 - e_3 e_1) \tag{5.21}$$

$$\tan \psi = \frac{2(e_0 e_3 + e_1 e_2)}{e_0^2 + e_1^2 - e_2^2 - e_3^2} \tag{5.22}$$

$$\tan \Phi = \frac{2(e_0 e_1 + e_1 e_3)}{e_0^2 - e_1^2 - e_2^2 + e_3^2} \tag{5.23}$$

It can be shown that

$$\begin{bmatrix} \dot{e}_0 \\ \dot{e}_2 \\ \dot{e}_2 \\ \dot{e}_3 \end{bmatrix} = \frac{1}{2} \begin{bmatrix} -e_1 & -e_2 & -e_3 \\ e_0 & -e_3 & e_2 \\ e_3 & e_0 & -e_1 \\ -e_2 & e_1 & e_0 \end{bmatrix} \begin{bmatrix} p \\ q \\ r \end{bmatrix} \tag{5.24}$$

Because four parameters are being used to describe the orientation when only three are necessary, a constraint equation exists of the form

Fig. 5.21 Strap-down system schematic

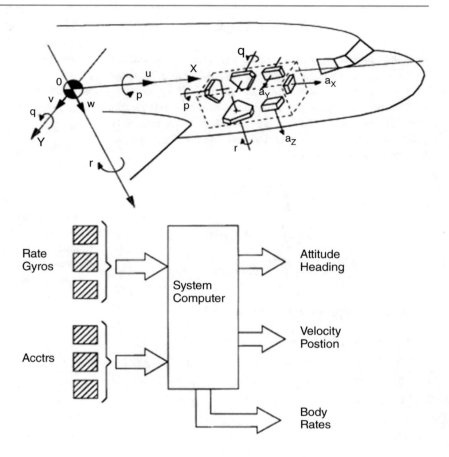

$$e_0^2 + e_1^2 + e_2^2 + e_3^2 = 1 \qquad (5.25)$$

These two equations have great advantages over the equivalent Euler angle equations:

1. They apply to all attitudes.
2. The error equations are bounded by the constraint equation.
3. The numerical value of each parameter always lies in the range -1 to $+1$, so easing the scaling problems in the computing mechanisation

Equation (5.24) can be re-arranged into the form

$$\begin{bmatrix} \dot{e}_0 \\ \dot{e}_1 \\ \dot{e}_2 \\ \dot{e}_3 \end{bmatrix} = \frac{1}{2} \begin{bmatrix} 0 & -p & -q & -r \\ p & 0 & r & -q \\ q & -r & 0 & p \\ r & q & -p & 0 \end{bmatrix} \begin{bmatrix} e_0 \\ e_1 \\ e_2 \\ e_3 \end{bmatrix} \qquad (5.26)$$

This can be written more compactly as

$$\dot{\mathbf{X}} = \mathbf{A}\mathbf{X}$$

where

$$\mathbf{X} = \begin{bmatrix} e_0 \\ e_1 \\ e_2 \\ e_3 \end{bmatrix} \text{ and } \mathbf{A} = \frac{1}{2} \begin{bmatrix} 0 & -p & -q & -r \\ p & 0 & r & -q \\ q & -r & 0 & p \\ r & q & -p & 0 \end{bmatrix}$$

(The bold capital letters denote that $\mathbf{X}$ and $\mathbf{A}$ are matrices.)

This equation can be solved by approximate integration techniques by assuming p, q, r are constant over a short period of time Δt from time t_n to time t_{n+1}.

The predicted value of $\mathbf{X}$ at time t_{n+1}, $\mathbf{X}_{n+1}$ is given by

$$\mathbf{X}_{n+1} = \mathbf{X}_n + \dot{\mathbf{X}}_n \Delta t = \mathbf{X}_n + \mathbf{A}\mathbf{X}_n \Delta t$$

That is,

$$\mathbf{X}_{n+1} = (\mathbf{1} + \mathbf{A}\Delta t)\mathbf{X}_n$$

($\mathbf{1}$ is the unity matrix). $(\mathbf{1} + \mathbf{A}\Delta t)$ is in fact an approximation to the *transition matrix* which relates the value of the state vector at time t_{n+1} to the value at time t_n.

The incremental angular rotations measured about the roll, pitch and yaw axes are denoted by ΔP, ΔQ, ΔR, respectively. (The pulse torqued DTG, the RLG and the IFOG all function as integrating rate gyros.)

$$\Delta P = \int_{t_n}^{t_{n+1}} p\, dt = p\Delta t$$

$$\Delta Q = \int_{t_n}^{t_{n+1}} q\, dt = q\Delta t$$

$$\Delta R = \int_{t_n}^{t_{n+1}} r\, dt = r\Delta t$$

Hence the approximate transition matrix is

$$\begin{bmatrix} 1 & 0 & 0 & 0 \\ 0 & 1 & 0 & 0 \\ 0 & 0 & 1 & 0 \\ 0 & 0 & 0 & 1 \end{bmatrix} + \frac{1}{2}\begin{bmatrix} 0 & -\Delta P & -\Delta Q & -\Delta R \\ \Delta P & 0 & \Delta R & -\Delta Q \\ \Delta Q & -\Delta R & 0 & \Delta P \\ \Delta R & \Delta Q & -\Delta P & 0 \end{bmatrix}$$

Hence

$$\begin{bmatrix} \dot{e}_0 \\ \dot{e}_1 \\ \dot{e}_2 \\ \dot{e}_3 \end{bmatrix}_{t_{n+1}} = \begin{bmatrix} 1 & -\Delta P/2 & -\Delta Q/2 & -\Delta R/2 \\ \Delta P/2 & 1 & \Delta R/2 & -\Delta Q/2 \\ \Delta Q/2 & -\Delta R/2 & 1 & \Delta P/2 \\ \Delta R/2 & \Delta Q/2 & -\Delta P/2 & 1 \end{bmatrix}\begin{bmatrix} e_0 \\ e_1 \\ e_2 \\ e_3 \end{bmatrix}_{t_n}$$

$$(5.27)$$

The performance of the numerical integration algorithm can be further improved by using second-order Runge–Kutta algorithms or fourth-order Runge–Kutta algorithms where very high accuracy is required, and by decreasing the integration time increment, Δt with a more powerful computer.

The constraint equation $e_0^2 + e_1^2 + e_2^2 + e_3^2 = 1$ is used to correct the transition matrix for the accumulated computational errors in the integration process to maintain orthogonality of the computed axes.

The four symmetrical Euler parameters are also mathematical *quaternions* as it can be shown that they are made up of the sum of a scalar quantity and a vector with orthogonal components.

The values of the Euler angles ψ, θ and Φ are then calculated from Eqs. (5.22), (5.21) and (5.23).

The corresponding initial values of the Euler parameters when the initial values of the Euler angles are ψ, θ, Φ, are given by

$$\left.\begin{aligned} e_0 &= \cos \psi/2 \cos \theta/2 \cos \Phi/2 + \sin \psi/2 \sin \theta/2 \sin \Phi/2 \\ e_1 &= \cos \psi/2 \cos \theta/2 \cos \Phi/2 - \sin \psi/2 \sin \theta/2 \cos \Phi/2 \\ e_2 &= \cos \psi/2 \sin \theta/2 \cos \Phi/2 + \sin \psi/2 \cos \theta/2 \sin \Phi/2 \\ e_3 &= \sin \psi/2 \cos \theta/2 \cos \Phi/2 - \cos \psi/2 \sin \theta/2 \sin \Phi/2 \end{aligned}\right\}$$

$$(5.28)$$

Worked example on attitude algorithm. A simple worked example is set out below to show how the attitude is derived from the Euler four parameter transition matrix to give a practical appreciation of the maths involved in evaluating the matrix for those readers less familiar with matrices.

Assume at time t_n the values of the four Euler parameters are

$$\begin{bmatrix} e_0 \\ e_1 \\ e_2 \\ e_3 \end{bmatrix} = \begin{bmatrix} 0.9 \\ 0.3 \\ 0.2 \\ 0.3 \end{bmatrix}$$

(Check $e_0^2 + e_1^2 + e_2^2 + e_3^2 = 0.81 + 0.09 + 0.01 + 0.09 = 1$.)

The values of the body angular rates are as follows:

$$p = 1\,\text{rad/s} \quad (57.3°/\text{s})$$
$$q = 0.04\,\text{rad/s} \quad (2.3°/\text{s approx.})$$
$$r = 0.05\,\text{rad/s} \quad (2.9°/\text{s approx.})$$
$$\text{iteration period } \Delta t = 0.01\,s$$
$$\text{whence} \qquad \Delta P = 0.01\,\text{rad}$$
$$\Delta Q = 0.0004\,\text{rad}$$
$$\Delta R = 0.0005\,\text{rad}$$

The Euler angles at time t_n are calculated using Eqs. (5.21), (5.22) and (5.23) and substituting the appropriate values of e_0, e_1, e_2, e_3, from which

$$\begin{cases} \psi = 36.869898° \\ \theta = 0° \\ \Phi = 36.869898° \end{cases}$$

The transition matrix is

Fig. 5.22 Strap-down
'equivalent stable platform'

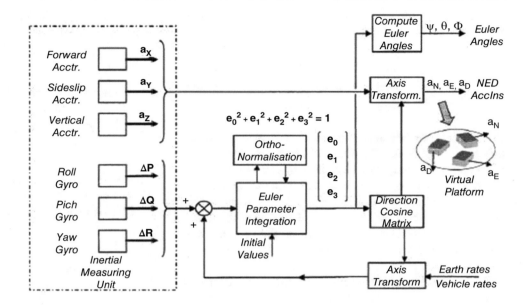

$$\begin{bmatrix} 1 & -\Delta P/2 & -\Delta Q/2 & -\Delta R/2 \\ \Delta P/2 & 1 & \Delta R/2 & -\Delta Q/2 \\ \Delta Q/2 & -\Delta R/2 & 1 & \Delta P/2 \\ \Delta R/2 & \Delta Q/2 & -\Delta P/2 & 1 \end{bmatrix}$$

$$= \begin{bmatrix} 1 & -0.005 & -0.0002 & -0.00025 \\ 0.005 & 1 & 0.00025 & -0.0002 \\ 0.002 & -0.00025 & 1 & 0.005 \\ 0.00025 & 0.0002 & -0.005 & 1 \end{bmatrix}$$

Hence at time t_{n+1}

$$e_0 = 1(0.9) - 0.005(3) - 0.0002(0.1) - 0.00025(0.3$$

That is, $e_0 = 0.898405$.

Evaluating the updating matrix yields

$$\begin{bmatrix} e_0 \\ e_1 \\ e_2 \\ e_3 \end{bmatrix}_{t_{n+1}} = \begin{bmatrix} 0.898405 \\ 0.304465 \\ 0.101605 \\ 0.299785 \end{bmatrix}$$

Check

$$e_0^2 + e_1^2 + e_2^2 + e_3^2 = 1.000025$$

The small error in the constraint equation is due to using a first-order approximate solution.

The Euler angles at time t_{n+1} are

$$\psi = 36.918525\,°$$
$$\theta = 0.000963\,°$$
$$\Phi = 37.442838\,°$$

The process is repeated at each iteration period.

5.3.2.2 Generation of Strap-Down 'Equivalent Stable Platform'

Figure 5.22 illustrates the computational processes involved in generating an 'equivalent stable platform' within the strap-down system computer.

The integration of the Euler parameter rates to yield the Euler parameters, as explained, enables the Direction Cosine Matrix (DCM) to be computed for the aircraft axes to local North, East, Down (NED) axes conversion, and vice versa.

This enables the strap-down accelerometer measurements, a_x, a_y, a_z, of the aircraft acceleration components along the aircraft body axes to be converted into the aircraft acceleration components along the local NED axes, a_N, a_E, a_D. The computing system thus provides the equivalent of accelerometers mounted on a stable platform with their input axes aligned with the local NED axes.

The Direction Cosine Matrix for converting to local NED axes from aircraft body axes is

$$[\text{DCM}]_{\text{la}}$$

$$= \begin{bmatrix} e_0^2 + e_1^2 - e_2^2 - e_3^2 & 2(e_1 e_2 - e_0 e_3) & 2(e_0 e_2 + e_1 e_3) \\ 2(e_0 e_3 + e_1 e_2) & e_0^2 - e_1^2 + e_2^2 - e_3^2 & 2(e_2 e_3 - e_0 e_1) \\ 2(e_1 e_3 - e_0 e_2) & 2(e_0 e_1 + e_2 e_3) & e_0^2 - e_1^2 - e_2^2 + e_3^2 \end{bmatrix}$$

$$(5.29)$$

Thus the aircraft acceleration components with respect to local NED axes are given by

$$\begin{bmatrix} a_N \\ a_E \\ a_D \end{bmatrix} = [\text{DCM}]_{la} \begin{bmatrix} a_x \\ a_y \\ a_z \end{bmatrix} \quad (5.30)$$

Gyros measure angular motion with respect to an inertial axis frame, that is, an axis frame which is fixed in space. It is therefore necessary to rotate the gyro-derived reference frame at the appropriate rates so that it stays aligned with the local NED axes because of the Earth's rotation and the vehicle's motion over the spherical surface of the Earth. The latter corrections are known as the vehicle rate corrections. Earth's rate and vehicle rate corrections are covered in Sect. 5.3.4.

In the case of a stable platform, these corrections are made by processing the vertical and azimuth gyros at the appropriate rates, as explained earlier.

In the case of a strap-down system, the Earth's rate and vehicle rate corrections are converted by the appropriate DCM to the coordinate frame rates with respect to aircraft axes. These coordinate frame rates are then summed with the aircraft body rates, p, q, r, so that the subsequent integration will yield the aircraft attitude with respect to local NED axes (knowing the initial conditions).

The DCM for converting the co-ordinate frame rates for the Earth's rate and vehicle rate from local NED axes to aircraft body axes is given by

$$[\text{DCM}]_{al}$$
$$= \begin{bmatrix} e_0^2 + e_1^2 - e_2^2 - e_3^2 & 2(e_0 e_3 + e_1 e_2) & 2(e_1 e_3 - e_0 e_2) \\ 2(e_1 e_2 - e_0 e_3) & e_0^2 - e_1^2 + e_2^2 - e_3^2 & 2(e_0 e_1 + e_2 e_3) \\ 2(e_0 e_2 + e_1 e_3) & 2(e_2 e_3 - e_0 e_1) & e_0^2 - e_1^2 - e_2^2 + e_3^2 \end{bmatrix}$$
$$(5.31)$$

It is shown in Sect. 5.3.4.2 that the Earth's rate corrections about the North, East, Down axes are $\Omega \cos \lambda$, 0, $\Omega \sin \lambda$, respectively, where Ω is the Earth's rate of rotation and λ is the latitude.

The vehicle rate corrections are V_E/R, $-V_N/R$, $-V_E/R \tan \lambda$ about the North, East, Down axes where V_N and V_E are the components of the aircraft's velocity along the North and East axes and R is the radius of the Earth.

Hence, the vehicle rate and Earth's rate corrections about the aircraft X, Y, Z axes are given by

$$[\text{DCM}]_{al} \begin{bmatrix} (V_E/R + \Omega \cos \lambda) \\ -V_N/R \\ (-V_E/R \tan \lambda + \Omega \sin \lambda) \end{bmatrix}$$

The initial value of the direction cosine matrix between the aircraft and the local NED axes is given by

$$[\text{DCM}]_{la}$$
$$= \begin{bmatrix} c\theta c\psi & s\Phi s\theta c\psi - c\Phi s\psi & c\Phi s\theta c\psi + s\Phi s\psi \\ c\theta c\psi & s\Phi s\theta c\psi + c\Phi s\psi & c\Phi s\theta c\psi - s\Phi c\psi \\ -s\theta & s\Phi c\theta & c\Phi c\theta \end{bmatrix}$$
$$(5.32)$$

where s is sine and c is cosine.

The initial value of the direction cosine matrix between the local NED axes and the aircraft axes is given by

$$[\text{DCM}]_{al}$$
$$= \begin{bmatrix} c\theta c\psi & c\theta s\psi & -s\theta \\ s\Phi s\theta c\psi - c\Phi s\psi & s\Phi s\theta s\psi + c\Phi c\psi & s\Phi c\theta \\ c\Phi s\theta c\psi + s\Phi s\psi & c\Phi s\theta s\psi - s\Phi c\psi & c\Phi c\theta \end{bmatrix}$$
$$(5.33)$$

5.3.2.3 Digital Processing of Attitude Algorithms

The strap-down system accuracy is critically dependent on the accuracy of the digital processing involved in the numerical solution of the attitude algorithms. Computer performance was in fact a major limitation in the implementation of strap-down systems until the advent of high-speed microprocessors.

The major error sources in the digital processing are briefly discussed below, together with the methods used to minimise these errors.

1. *Commutation.* Errors arise in digital processing because of the non-commutativity of angular motion. A different attitude is attained if a series of rotations about the body axes are made in a different order. Commutation errors are introduced in the numerical processing because for each solution update of the transformation matrix, the processor operates sequentially on the input angular rotation increments in a fixed order. This may not correspond to the actual time ordered sequence in which the angles were accumulated in the vehicle. Commutation errors can thus arise in deriving the vehicle attitude particularly when subjected to cyclical or repetitive inputs such as conical motion, which can result in a rectification drift of the transformation matrix. These commutation errors can be minimised by decreasing the angle increment size involved in each computation. Coning motion is briefly covered in the next section.

2. *Integration.* Errors arise in the integration because of using discrete approximate solutions of a continuous process and equations which have no analytic solution. These

errors are minimised by using sophisticated integration algorithms (e.g. Runge–Kutta fourth order) and by increasing the update rate.

3. *Round off.* These errors are due to the finite resolution of data and rounding off the computations to the value of the least significant bit. This causes a random walk error build up in multiple computations. Double precision working (32 bit) for critical operations is used to minimise these errors.

4. *Quantisation.* These errors result from the process of converting the analogue output of the sensors into discrete increments which can be input as a series of pulses into the digital computer. This quantisation process causes the input to the computer to always lag the outputs of the sensors. The average value of this quantisation (or sensor storage error) is one half of the quantisation angle. This information is not lost but received in the next update cycle and results in a quantisation noise in the output that needs to be modelled for the best performance. These errors are minimised by decreasing the quantisation angle and mathematical modelling.

5.3.3 Coning Motion

A brief explanation of coning motion is set out below in view of its importance as a possible source of error. Coning motion results from the vehicle being subjected to angular oscillations of the same frequency about two orthogonal axes with a 90° phase difference between the two motions. The motion of the third orthogonal axis of the vehicle describes the surface of a cone – hence the term conical motion (see Fig. 5.23).

Suppose the motion measured by the gyros is as follows:

$$p = 0$$
$$q = A\omega \cos \omega t$$
$$r = A\omega \sin \omega t$$

referring to Eq. (5.17)

$$\dot{\theta} = q \cos \Phi - r \sin \Phi$$

assuming the steady-state Euler angles are all zero and θ and Φ are small angles

$$\dot{\theta} = q = A\omega \cos \omega t$$

integrating yields

$$\theta = A \sin \omega t$$

referring to Eq. (5.16)

$$\dot{\Phi} = p + q \sin \Phi \tan \theta + r \cos \Phi \tan \theta$$

For small angles $\sin \Phi = \Phi$, $\tan \theta = \theta$, $\cos \Phi = 1$ approximately and neglecting second-order terms

$$\dot{\Phi} = r\theta = A\omega \sin \omega t \cdot A \sin \omega t$$

That is,

$$\dot{\Phi} = \frac{A^2 \omega}{2}(1 - \cos 2\omega t) \tag{5.34}$$

hence, average value of rectified rotation rate,

$$\dot{\Phi} = \frac{A^2 \omega}{2} \tag{5.35}$$

The errors introduced by coning motion in strap-down systems arise essentially from errors in computing the motion profile due to the non-commutativity of angular rotations.

For example, a circular coning motion resulting from angular motions of 1° amplitude and 1 Hz frequency measured by gyros with a scale factor error of 0.01% results in a rectified drift rate equal to

$$\frac{0.01}{100} \times \frac{1}{2} \times \left(\frac{1}{57.3}\right)^2 \times 2\pi \times 1 \times 57.3 \times 60 \times 60$$

degrees/hour, that is, 0.02°/hour, which is significant in an INS.

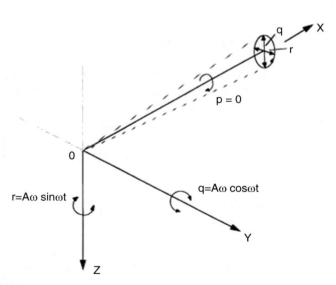

Fig. 5.23 Conical notion

5.3.4 Attitude with Respect to Local North, East, Down Axes

5.3.4.1 Introduction

The aircraft *Pitch* and *Bank* angles are required with respect to the local level plane, that is, a plane which is normal to the local vertical, defined as a line through the aircraft to the Earth's centre.

The aircraft *heading* angle is generally required with respect to true North, that is, the direction of the local meridian pointing towards the North pole. (The local meridian being a circle round the Earth which passes through the North and South Poles and the aircraft's present position.)

It is therefore necessary to convert the gyro derived data, which is with respect to inertial axes, to an Earth referenced axis frame.

There are three basic directional references which are used to align an AHRS or INS. These comprise the following:

1. Earth's gravitational acceleration vector – This is sensed by the accelerometers and enables the local vertical to be determined.
2. Earth's angular velocity vector – This can be measured by the gyros, provided they are of sufficient accuracy, and enables the aircraft heading to be determined by a process known as 'gyro compassing'. It will be shown that a heading accuracy of 0.1°, however, requires the gyro bias uncertainty to be less than 0.01°/hour. The technique is therefore not suitable for AHRS using lower accuracy gyros.
3. Earth's magnetic field – This enables a heading reference to generally within 0.7° accuracy to be established at latitudes below 60° North or South of the equator.

Figure 5.24 illustrates the basic Earth reference axis frame, generally referred to as a 'local level, North slaved axis frame' or 'North, East, Down axes – NED'.

5.3.4.2 Angular Rate Corrections for the Earth's Rotation

Figure 5.24 and accompanying vector diagram show the Earth's angular velocity vector, Ω, can be resolved into two components.

(a) Component about the North pointing axis is $\Omega \cos \lambda$, where λ is the latitude angle at the aircraft's position.
This correction must be applied to the computer-derived axes of the virtual stable platform to maintain a level state. As an example, at a latitude of 51°30′N (London area), the Earth's rate component about the North axis is 15 cos 51°30′degrees per hour = 9.3377°/hour ($\Omega = 15$°/hour).

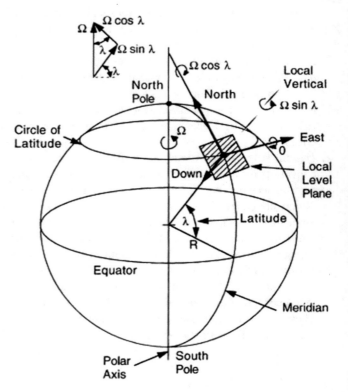

Fig. 5.24 Earth's rotation rate components

(b) Component about the local vertical axis is $\Omega \sin \lambda$.
This correction must be applied to the virtual stable platform, if it is desired to maintain it pointing North.

It should be noted that there is zero component of the Earth's angular velocity about the East pointing axis and this forms the basis of the gyro compassing alignment technique which is explained later.

5.3.5 Vehicle Rate Corrections

Figure 5.25 illustrates how the local vertical rotates in space as the aircraft flies over the surface of the Earth because the Earth is spherical. The angular rates of rotation of the local vertical with respect to inertial axes are equal to:

- V_N/R about the East pointing axis, where V_N is the northerly component of the aircraft's velocity and R is the radius of the Earth.
- V_E/R about the North pointing axis, where V_E is the easterly component of the aircraft's velocity.

These angular rates are referred to as the vehicle rates.

It is therefore necessary to rotate the virtual platform about the East pointing and North pointing axes at angular rates of

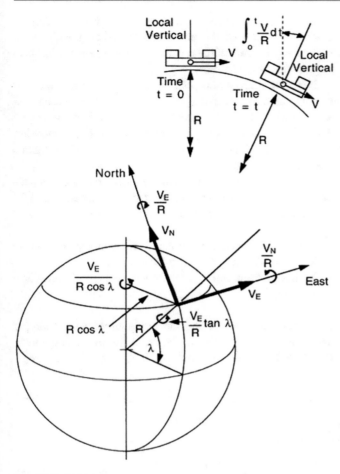

Fig. 5.25 Vehicle rates

The use of very high accuracy gyros and accelerometers enables the velocity components to be derived directly from the accelerometers and fed back to correct the vertical reference so as to form a Schuler tuned system, and is explained in the next chapter.

There is also a vehicle rate correction, $(V_E/R) \tan \lambda$, about the local vertical (down) axis which must be applied to the virtual platform in order to maintain the north reference. Figure 5.25 shows the three vehicle rate correction terms V_N/R, V_E/R, $(V_E/R) \tan \lambda$. The $(V_E/R) \tan \lambda$ term is derived by resolving the vehicle rate component about the Earth's polar axis $V_E/(R \cos \lambda)$, through the latitude angle, namely,

$$\frac{V_E}{R} \cos \lambda \cdot \sin \lambda = \frac{V_E}{R} \tan \lambda$$

It can be seen that the $(V_E/R) \tan \lambda$ term increases rapidly at high latitudes and becomes infinite at the poles when $\lambda = 90°$. Alternative co-ordinate systems are used to overcome this limitation at high latitudes ($>75°$). This is explained in the next chapter. It should be noted that this term is generally referred to as the 'meridian convergence term'.

5.3.5.1 Vertical Monitoring

As mentioned in the introduction, the basic vertical reference is the Earth's gravitational vector. The platform (real or virtual) can thus be initially aligned locally level using the accelerometers to measure the tilt errors from the local vertical. The accelerometers act in a sense as spirit levels. This assumes the process is carried out when the aircraft is stationary on the ground, or in straight and level flight at constant speed and the aircraft is not accelerating. However, any aircraft accelerations would result in the accelerometers defining a deflected or 'dynamic' vertical because the accelerometers are unable to distinguish between the gravity component and an acceleration component.

An ideal vertical reference system using gyros with negligible drift or bias uncertainties once initially aligned with the local vertical would continue to measure the aircraft's attitude with respect to the local vertical using the gyros alone, assuming the Earth's rate and vehicle rate corrections were applied without error. In practice, however, gyro drift and errors in applying the necessary correction terms would cause a progressive build up in the vertical error unless some means for monitoring the vertical error and applying correcting feedback were incorporated. In fact, a compromise between accelerometer measurement and gyro measurement of the vertical is always used.

The next chapter explains how Schuler tuning enables an INS to maintain a very accurate vertical reference irrespective of the aircraft's acceleration (refer Sect. 6.2.1).

V_N/R and V_E/R, respectively, in order to maintain the virtual platform locally level as the aircraft moves over the spherical surface of the Earth.

The magnitude of these vehicle rate correction terms is more easily appreciated by expressing the aircraft's velocity in knots (nautical miles/hour) bearing in mind 1 nautical mile is equal to 1 minute of arc at the Earth's surface.

For example, an aircraft with a ground speed of 600 knots and a track angle of 030° has a northerly velocity component $V_N = 600 \cos 30°$, that is, 519.615 knots and an easterly velocity component $V_E = 600 \sin 30°$, that is, 300 knots.

A correction rate equal to 519.615/60, that is, 8.66°/hour, about the East axis and 300/60, that is, 5°/hour, about the North axis, is required in order to maintain the platform (real, or virtual in the case of a strap-down system) locally level.

An external source of velocity such as a Doppler radar, or less accurately air data, can be used to derive the northing and easting velocity components for the vehicle rate correction terms. The Doppler or air data derived velocity components would generally be combined with the inertially derived velocity components from the accelerometers as explained later.

AHRS systems use an independent source of the aircraft's velocity such as a Doppler, or (less accurately) the Air Data System to monitor the vertical reference. The aircraft's velocity components are derived from the AHRS accelerometers in the same way as an INS and the velocity differences are then used to correct the vertical reference. The method is known as Doppler (or Air Data)/Inertial velocity mixing and is covered in the next chapter (Sect. 6.4.2).

5.3.5.2 Azimuth Monitoring

The very high accuracy gyros and accelerometers, which are essential for Schuler tuning, enable gyro compassing to be used to align the heading reference to within 0.1° of true North. The very low gyro drift rate enables an accurate heading reference to be maintained for several hours without monitoring. Gyro compassing is covered in the next chapter (Sect. 6.2.3).

The gyros used in an AHRS have too high a drift rate for gyro compassing to be effective for initial alignment. Once aligned, the heading accuracy will also degrade with time because of the gyro drift rate which can range from 5°/hour to 0.1°/hour, depending on the AHRS accuracy grade. A magnetic heading reference is used for initial alignment and to provide a long-term monitoring of the heading by combining the gyro heading with the magnetic heading. This is explained in the next chapter (Sect. 6.4.2).

5.3.6 Introduction to Complementary Filtering

Inertial velocity mixing exploits the technique of *complementary filtering* for combining the data from two independent sources of information. Other examples include the following:

- Combining a magnetic heading reference with a gyro heading reference.
- Combining barometrically measured altitude or radar altimeter measured altitude, with inertially derived altitude.

Sophisticated mixing systems employing Kalman filters to control the gains of the mixing loops are frequently used, particularly in view of the computing power now available to implement such systems.

It is appropriate at this point to discuss the basic principles of complementary filtering. A simple first-order mixing system is used as an example to keep the maths relatively simple. The build up to the more powerful second-order, third-order, fourth-order and above systems with variable mixing gains can then be appreciated.

In this simple example, the outputs from the two independent systems measuring a quantity θ are designated θ_1 and θ_2, respectively. Their characteristics are as follows:

- θ_1 has a low noise content and a fast response to changes in θ. It is also subject to drift with time.
- θ_2 has a high noise content but has good long-term accuracy.

A method of combining the two sources with a simple first-order mixing system is shown in the block diagram in Fig. 5.26. It can be seen from Fig. 5.26 that

$$\theta_0 = \theta_1 + K \int (\theta_2 - \theta_0) dt \qquad (5.36)$$

from which

$$\theta_0 = \frac{TD}{1 + TD}\theta_1 + \frac{1}{1 + TD}\theta_2 \qquad (5.37)$$

where $T = 1/K$. That is

$$\theta_0 = F_1(D)\theta_1 + F_2(D)\theta_2 \qquad (5.38)$$

where $F_1(D) = TD/1 + TD$, which is the transfer function of a first-order high pass filter (or 'wash-out' filter) and $F_2(D) = 1/1 + TD$ which is the transfer function of a first-order low pass filter.

Fig. 5.26 Simple first-order mixing

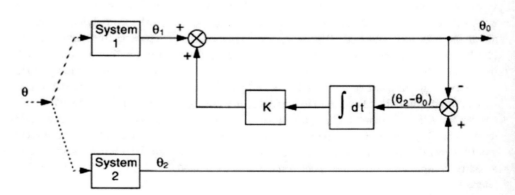

Fig. 5.27 Complementary filtering

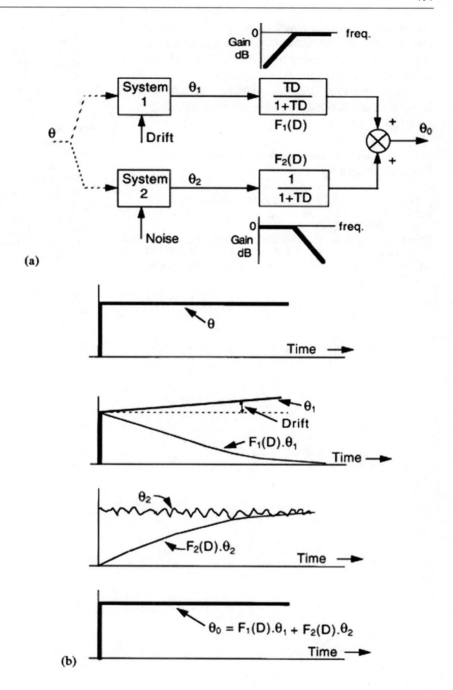

(a)

(b)

Figures 5.27a, b show the basic concept and the response of the various elements of the system to a step input in θ.

The output, θ_0, is thus the sum of two components:

(i) The quantity θ_1 is coupled through a simple high pass filter with a transfer function $F_1(D) = T D/1 + T D$ which passes dynamic changes in θ without lag but has zero output at dc. A constant drift is, therefore dc blocked and has no effect on θ_0. A change in drift will, however, cause a transient change in θ_0.

(ii) The quantity θ_2 is coupled through a simple low pass filter with a transfer function $F_2(D) = 1/1 + T D$ which filters the noise present in θ_2 but preserves the dc accuracy.

The output θ_0 thus reproduces the input θ from the sum of the two complementary filtered outputs of θ_1 and θ_2 without lag and with a low noise content.

A constant drift in θ_1 is dc blocked but will produce a steady-state offset in θ_0 to back-off this drift. To eliminate this offset requires a parallel integral term leading to a second-order system. Other errors such as bias errors may require a third integral term to be added to the system leading to a third-order mixing system and so on.

It can be seen with this simple example that the complementary filtered output is superior to either source on its own.

Further Reading

AGARD Lecture Series No. 95: Strap-Down Inertial Systems

Bergh, R.A.: Dual-ramp closed-loop fiber-optic gyroscope. Proc. SPIE. **1169**, 429–439 (1989)

Ebberg, A., Schiffner, G.: Closed-loop fiber-optic gyroscope with a sawtooth phase modulated feedback. Opt. Lett. **10**, 300–302 (1985)

Lefevre, H: The Fiber Optic Gyroscope. Artech House, Boston/London (2014)

Lefevre, H., Martin, P., Morisse, J., Simonpietri, P., Vvivenot, P., Arditty, H.J.: High dynamic range fiber gyro with all-digital processing. SPIE. **1367**, 70–80 (1990)

Matthews, A.: Fiber optic based inertial measuring unit. **SPIE, 1173**, Fiber Optic Systems for Mobile Platforms III (1989)

Page, J.L.: Multiplexed approach to the fiber optic gyro inertial measuring unit. In: Proceedings OE/FIBERS 1990 SPIE Conference, San Jose, California, September 1990

Ribes, M., Spahlinger, G., Kemmler, M.W.: 0.1°/hr DSP-controlled fiber optic gyroscope. SPIE. **2837**, 199–207 (1996)

Sanders, G.A., Szafraniec, B., Strandjord, L., Bergh, R., Kalisjek, A., Dankwort, R., Lange, C., Kimmel, D.: Progress in high performance fiber optic gyroscopes. In: 116/OWB1-1: Selected Topics in Advanced Solid State a Fiber Optic Sensors: Year 2000

Siouris, G.M.: Aerospace Avionic Systems. Academic Press, Cambridge, Massachusetts, USA (1993)

Wrigley, W., Hollister, W.M., Denhard, W.G.: Gyroscopic Theory, Design and Instrumentation. MIT Press Cambridge, MA, USA (1969). MIT

6.1 Introduction and Basic Principles

6.1.1 Introduction

The dictionary definition of navigation is a good one.

Navigation – The act, science or art of directing the movement of a ship or aircraft. Thus, navigation involves both control of the aircraft's flight path and the guidance for its mission.

The measurement of the aircraft's attitude with respect to the horizontal plane in terms of the pitch and bank angles and its heading, that is, the direction in which it is pointing in the horizontal plane with respect to North, is essential for both control and guidance.

This information is vital for the pilot in order to fly the aircraft safely in all weather conditions, including those when the normal visibility of the horizon and landmarks is poor or not available, for example in haze or fog conditions, flying in cloud and night flying. Attitude and heading information is also essential for the key avionic systems which enable the crew to carry out the aircraft's mission. These systems include the autopilot system (e.g. Attitude and Heading Hold modes and Autol) navigation system and the weapon aiming system. The information is also required for pointing radar beams and infrared sensors.

Accurate knowledge of the aircraft's position in terms of its latitude/longitude co-ordinates, ground speed and track angle, height and vertical velocity is also equally essential for the navigation of the aircraft.

The need for accurate and high integrity navigation is briefly summarised below. For civil aircraft, the density of air traffic on major air routes requires the aircraft to fly in a specified corridor or 'tube in the sky', these air routes being defined by the Air Traffic Control authorities. Not only must the aircraft follow the defined three-dimensional flight path with high accuracy, but there is also a fourth dimension, namely, that of time, as the aircraft's arrival time must correspond to a specified time slot.

High accuracy navigation systems are thus essential and form a key part of the flight management system. This is covered in Chap. 8.

For military operations, very accurate navigation systems are essential to enable the aircraft to fly low and take advantage of terrain screening from enemy radars, to avoid known defences and in particular to enable the target to be acquired in time. The aircraft flies fast and very low so that the pilot cannot see the target until the aircraft is very near to it. There may be then only about 6–10 seconds in which to acquire the target, aim and launch the weapons. It is thus necessary to know the aircraft's position near the target area to within 100 m accuracy. This enables the target sight line to be continually computed (knowing the target co-ordinates, including target height, and the aircraft's height) and a target marker symbol to be displayed on the HUD. This should be near the target and the pilot then slews the marker symbol to exactly overlay the target. This corrects the errors and initialises the weapon aiming process.

The use of stand-off weapons which are released several kilometres away from the target also requires an accurate knowledge of the aircraft's position in order to initialise the mid-course inertial guidance system of the missile (the terminal homing phase is achieved with a suitable infrared or microwave radar seeker system).

Clearly the integrity of the navigation system must be very high in both civil and military aircraft as large navigation errors could jeopardise the safety of the aircraft.

There are two basic methods of navigation, namely, dead reckoning (DR) navigation and position fixing navigation systems. Both systems are used to achieve the necessary integrity.

As briefly explained in Chap. 1, DR navigation is the process of continually computing a vehicle's position as the journey proceeds from a knowledge of the starting point position and the vehicle's speed and direction of motion and the time elapsed. It is essentially an incremental process

© The Author(s), under exclusive license to Springer Nature Switzerland AG 2023

R. P. G. Collinson, *Introduction to Avionics Systems*, https://doi.org/10.1007/978-3-031-29215-6_6

of continually estimating the changes that have occurred and updating the present position estimate accordingly.

The main types of airborne DR navigation systems are categorised below on the basis of the means used to derive the velocity components of the aircraft.

In order of increasing accuracy these are as follows:

1. *Air data-based DR navigation.* The basic information used comprises the true airspeed (from the air data computer) with wind speed and direction (forecast or estimated) and the aircraft heading from the Attitude Heading Reference System (AHRS).
2. *Doppler/heading reference systems.* These use a Doppler radar velocity sensor system to measure the aircraft's ground speed and drift angle. The aircraft heading is provided by the AHRS.
3. *Inertial navigation systems.* These derive the aircraft's velocity components by integrating the horizontal components of the aircraft's acceleration with respect to time. These components are computed from the outputs of very high accuracy gyroscopes and accelerometers which measure the aircraft's angular and linear motion.
4. *Doppler inertial navigation systems.* These combine the Doppler and INS outputs, generally by means of a Kalman filter, to achieve increased DR navigation accuracy.

The primary DR navigation system which is also the primary source of very accurate attitude and heading information is the Inertial Navigation System (INS). The term Inertial Reference System (IRS) is also used in civil aircraft. The IRS can have a lower inertial navigation accuracy of up to 4 NM/hour error compared with 1–2 NM/hour for a typical INS. The attitude and heading accuracy, however, is still very high. The terminology INS/IRS is used in this chapter to show they are essentially the same.

A fundamental feature of INS/IRS systems is that they are *Schuler* tuned systems. This is achieved by precise feedback of the components of the aircraft's velocity in the local level plane so that the system tracks the local vertical as the aircraft moves over the Earth's surface. *Schuler* tuning requires very accurate gyros and accelerometers and precise computing. The pay-off, however, is that it enables a very accurate vertical reference (fraction of a minute of arc errors) to be derived which is independent of the aircraft's acceleration. A very accurate heading reference is established in the pre take-off initial alignment phase using gyro compassing. The very low gyro drift rates (less than 0.01°/hour drift uncertainty) enables an accurate heading reference to be maintained for several hours.

The INS/IRS also derives the aircraft's velocity vector in conjunction with the Air Data System which provides barometric height information. This is of great assistance to the pilot when displayed on the HUD (Chap. 2). Accurate velocity vector information is also essential for the aiming of

unguided weapons (guns, bombs and rockets) in military aircraft. The strapdown configuration of the gyros and accelerometers, which is now used in modern systems, enables the INS/IRS to provide body angular rates and linear acceleration components for the flight control system.

The INS/IRS is thus a key aircraft state sensor for both military and civil aircraft. Large civil airliners operating on long haul, over water routes, have triple IRS installations to ensure availability and the ability to detect failures or degradation of performance by cross-comparison. They are also a key sensor for the automatic flight control system.

Smaller, short haul civil airliners are now generally equipped with a dual IRS installation with a 2.5 to 4 NM/hour navigation accuracy. Lower accuracy systems cost less, the flight time is shorter and they operate over routes which are well covered by ground-based radio navigation aids. They are also equipped with GPS which can be used to correct the IRS through a Kalman filter.

A number of manufacturers now provide a combined Air Data System and Inertial Reference System as a single unit known as an 'Air Data Inertial Reference System', ADIRS. This has a number of advantages such as lower cost, lower weight and occupies less space. Only one box is required and the power supply conditioning electronics, computer and most of the interfacing electronics can be shared by the two functions. Barometric height information is also essential for the vertical inertial channel.

Attitude Heading Reference Systems are of lower accuracy than the INS/IRS and generally provide a secondary source of attitude and heading information. They use less accurate but lower cost gyros and accelerometers which are insufficiently accurate for *Schuler* tuning to be effective. Long-term air data velocity mixing is used to constrain the vertical errors and long-term magnetic monitoring from a magnetic heading reference is used to constrain the heading errors. In general these systems are an order of magnitude (or more) less accurate than an INS/IRS. They are, however, of considerably lower cost than INS/IRS systems and the strapdown configuration adopted together with the use of solid state gyros and accelerometers ensures they have very high reliability.

They provide the primary source of attitude and heading in some aircraft such as helicopters and smaller civil aircraft.

The heading information from the AHRS is used for air data-based DR navigation (generally a reversionary mode) and in the Doppler-based DR navigation systems which are installed in many helicopters. It should be noted that although Doppler radar velocity sensors are widely installed in helicopters, they are no longer fitted in the new generation of military helicopters and strike fighters. This is because the radar energy emissions from the Doppler increase the risk of detection. 'Stealth' is the name of the game. Civil aircraft used to have Dopplers but these were superseded in the early 1970s with the introduction of civil IN systems.

A fundamental characteristic of DR navigation systems is that their accuracy is time dependent. For example, a good quality INS has an accuracy of 1 NM/hour so that the aircraft position uncertainty after 5 hours would be 5 NM.

A position fixing system is thus required to constrain the error growth of a DR navigation system and correct the DR position errors.

A position 'fix' can be obtained by recognising a prominent landmark or set of terrain features either visually, possibly using an infrared imaging sensor, or from a radar-generated map display. Alternatively, a suitable position fixing navigation system can be used. In general, several dissimilar systems may be used to provide the necessary confidence in the overall navigation accuracy.

Position fixing navigation systems depend on external references to derive the aircraft's position. For example, radio/radar transmitters on the ground, or in satellites whose orbital positions are precisely known. Unlike DR navigation systems, their errors are not time dependent. The errors are also independent of the aircraft's position in most position fixing systems.

The main position fixing navigation systems in current use are briefly summarised below.

1. Range and Bearing ('R/θ') Radio Navigation Aids

These comprise VOR/DME and TACAN.

VOR (VHF omni-directional range) is an internationally designated short-distance radio navigation aid and is an integral part of Air Traffic Control procedures.

DME (distance measuring equipment) is co-located with VOR and provides the distance from an aircraft to the DME transmitter.

TACAN is the primary tactical air navigation system for the military services of the USA and NATO countries. It is often co-located with civil VOR stations, such combined facilities being known as VORTAC stations. The systems are line of sight systems and provide the slant range of the aircraft from the ground station using a transmitter/transponder technique. A rotating antenna system at the TACAN ground station (or beacon) enables the aircraft to measure the bearing of the TACAN beacon to an accuracy of about 2°, using a principle analogous to a lighthouse.

Navigation position accuracy of VOR/DME and TACAN is of the order of one to two miles.

There used to be other radio navigation systems in operation which used a chain of ground based transmitting stations. The systems used pulse transmissions and operated by measuring the phase differences between the signals received from the different transmitter stations. They are referred to as *hyperbolic radio navigation systems*. These systems have been progressively discontinued over the years; OMEGA was discontinued around 2000 and the last remaining system, LORAN C, was very recently discontinued in February 2010.

The systems go back to the GEE system, developed by the UK in WW2, and which entered operational service with the RAF in 1942, effecting a dramatic improvement in navigation accuracy at a critical time. They are all now of historical interest only and will not be discussed further.

2. Satellite Navigation Systems – GPS (Global Position System)

GPS is the most important and accurate position fixing system developed to date. It is being used by every type of vehicle – aircraft – ships – land vehicles. Civilian use is now very widespread, for example, GPS receivers are fitted in very many cars, vans and lorries and they are readily affordable for ramblers.

The equipment required by the GPS user is entirely passive and requires a GPS receiver only. Electronic miniaturisation has enabled very compact and light weight GPS receivers to be produced. The full positional accuracy of 16 m (3D) and velocity accuracy of 0.1 m/s is now available to civil users (previously, only military users were able to achieve this accuracy). Precise time to within a few billionths of a second is also available.

The use of GPS in conjunction with a ground station system which transmits corrections to the user system, known as Differential GPS, has enabled a positional accuracy of 1 m to be achieved.

Construction of an independent, European system known as the Galileo satellite navigation system has begun and is scheduled to come into operation in 2016. The system is inter-operable with GPS and with 40 orbiting satellites will provide an accuracy worldwide of the order of 1 m.

3. Terrain Reference Navigation (TRN) Systems

Terrain reference navigation systems derive the vehicle's position by correlating the terrain measurements made by a sensor in the vehicle with the known terrain feature data in the vicinity of the DR estimated position. The terrain feature data are obtained from a stored digital map database.

Figure 6.1 shows the information flow from the inertial sensor systems, air data system(s) and the position fixing radio navigation systems to the user systems.

The accuracy of the INS/IRS can be greatly improved by combining the inertially derived position data with that from a position fixing navigational source, such as GPS. This is achieved by using a statistical predicting filter known as a Kalman filter which provides an optimum combination of the two sources of data. The resulting combination is superior to either source on its own and retains the best features of each system. An introduction to Kalman filters is provided in Sect. 6.3.

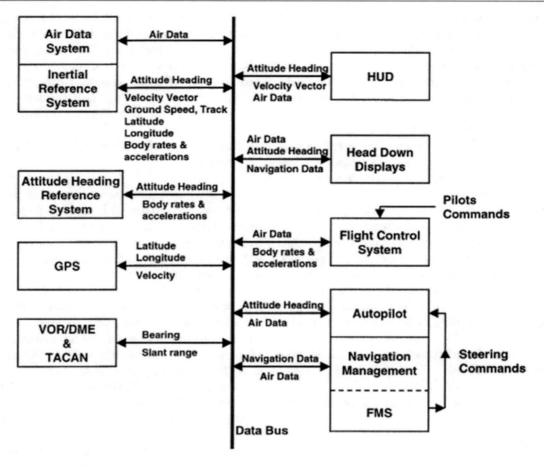

Fig. 6.1 Navigation system information flow to user systems

6.1.2 Basic Navigation Definitions

A very brief review of the terms and quantities used in navigation is set out below. Position on the Earth's surface is generally specified in terms of *latitude* and *longitude* co-ordinates which provide a circular grid over the surface of the Earth. The Earth is basically a sphere – the variation in the radius of the Earth is only about 40 NM in a radius of 3.438 NM at the equator, being slightly flattened at the poles (this variation is taken into account in the navigation computations).

Referring to Fig. 6.2, latitude and longitude are defined with respect to the polar axis, the equator and the prime meridian. A *meridian* is a circle round the Earth passing through the North and South poles. The *prime meridian* is the meridian passing through Greenwich which provides the datum for measuring the longitude. The *latitude* of a point on the Earth's surface is the angle subtended at the Earth's centre by the arc along the meridian passing through the point and measured from the equator to the point. The range of latitude angles is from 0° to 90° North and 0° to 90° South. The *longitude* of a point on the Earth's surface is the angle

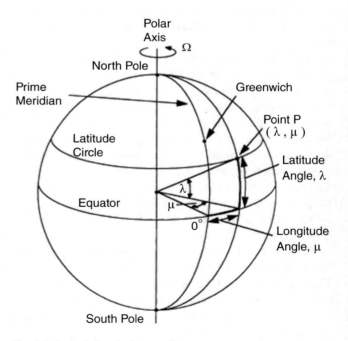

Fig. 6.2 Latitude/longitude co-ordinates

Fig. 6.3 Spherical triangles

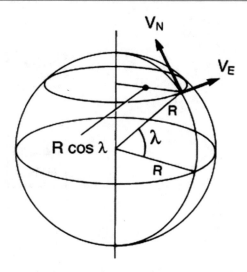

Fig. 6.4 Derivation of rates of change of latitude and longitude

subtended at the Earth's centre by the arc along the equator measured East or West of the prime meridian to the meridian passing through the point. The range of longitude angles to cover all points on the Earth's surface is thus 0° to 180° east of the prime meridian and 0° to 180° west of the prime meridian. Latitude and longitude are expressed in degrees, minutes of arc and seconds of arc.

Great circles are circles on the surface of a sphere with their centre at the centre of the sphere, that is, the plane of a great circle passes through the Earth's centre.

Meridians and the equator are thus great circles. Parallels of latitude which are circles round the Earth parallel to the equator, are, however, small circles.

The shortest distance between two points on the surface of a sphere is a great circle, hence navigation routes try to follow a great circle path. Navigation between points on the Earth's surface thus involves the solution of spherical triangles as shown in Fig. 6.3 (a spherical triangle being defined as a triangle on a sphere whose sides are part of great circles. Their solution is achieved using the well-established formulae of spherical trigonometry).

6.1.3 Basic DR Navigation Systems

The basic principles of deriving a DR navigation position estimate are explained below. The following quantities are required:

1. Initial position – latitude/longitude.
2. The northerly and easterly velocity components of the aircraft, V_N and V_E.

Referring to Fig. 6.4 it can be seen that the rate of change of latitude is

$$\dot{\lambda} = \frac{V_N}{R} \qquad (6.1)$$

the rate of change of longitude is

$$\dot{\mu} = \frac{V_E}{R \cos \lambda}$$

That is,

$$\dot{\mu} = \frac{V_E}{R} \sec \lambda \qquad (6.2)$$

The change in latitude over time, t, is thus equal to $1/R \int_0^t V_N dt \; dt$ and hence the present latitude at time t can be computed given the initial latitude λ_0. Similarly, the change in longitude is equal to $1/R \int_0^t V_E \sec \lambda dt$ and hence the present longitude can be computed given the initial longitude, μ_0, namely,

$$\lambda = \lambda_0 + \frac{1}{R} \int_0^t V_N dt \qquad (6.3)$$

$$\mu = \mu_0 + \frac{1}{R} \int_0^t V_E \sec \lambda dt \qquad (6.4)$$

It can be seen that a mathematical singularity is approached as λ approaches 90° and sec λ approaches infinity. This method of computing the latitude and longitude of the DR position is hence limited to latitudes below 80°. A different co-ordinate reference frame is used to deal with high latitudes as will be explained later.

Fig. 6.5 Doppler/heading
reference DR navigation system

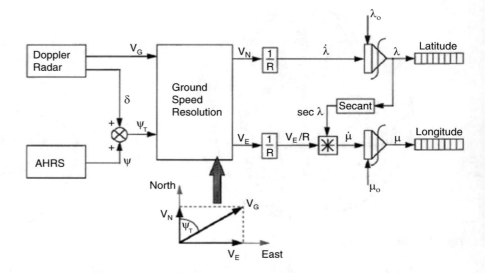

The basic computational processes in a DR navigation system using a Doppler/heading reference system are shown in Fig. 6.5. In the case of a Doppler/heading reference system, the ground speed V_G and drift angle δ are measured directly by the Doppler radar velocity sensor system.

The AHRS system provides an accurate measurement of the heading angle, ψ, and hence the track angle, ψ_T can be obtained from

$$\psi_T = \psi + \delta$$

The northerly velocity component of the aircraft, V_N, and the easterly velocity component, V_E, are then derived by resolution of the ground speed vector, V_G (see inset diagram on Fig. 6.5).

Hence

$$V_N = V_G \cos \psi_T$$

$$V_E = V_G \sin \psi_T$$

In the case of an air data-based DR navigation system the northerly and easterly velocity components of the aircraft can be derived as follows:

1. The horizontal velocity component V_H of the true airspeed V_T is obtained by resolving V_T through the aircraft pitch angle θ, namely,

$$V_H = V_T \cos \theta$$

2. The northerly and easterly velocity components of the airspeed are then derived by resolving V_H through the aircraft heading angle ψ, viz.namely,

$$\text{Northerly airspeed} = V_H \cos \psi$$

$$\text{Easterly airspeed} = V_H \sin \psi$$

3. The forecast (or estimated) wind speed V_W and direction ψ_W is resolved into its northerly and easterly components, viz.namely,

$$\text{Northerly wind component} = V_W \cos \psi_W$$

$$\text{Easterly wind component} = V_W \sin \psi_W$$

4. The northerly and easterly velocity components of the aircraft are then given by:

$$V_N = V_H \cos \psi + V_W \cos \psi_W \tag{6.5}$$

$$V_E = V_H \sin \psi + V_W \sin \psi_W \tag{6.6}$$

Such a system provides a reversionary DR navigation system in the absence of Doppler (or an INS) and would generally be used in conjunction with a radio navigation system.

6.2   Inertial Navigation

6.2.1   Introduction

It is instructive to briefly review the reasons for the development of inertial navigation and its importance as an aircraft state sensor.

The attributes of an ideal navigation and guidance system for military applications can be summarised as follows:

- High accuracy
- Self-contained
- Autonomous – does not depend on other systems
- Passive – does not radiate
- Unjammable
- Does not require reference to the ground or outside world

In the late 1940s, these attributes constituted a 'wish list' and indicated the development of inertial navigation as the only system which could be capable of meeting all these requirements. It was thus initially developed in the early 1950s for the navigation and guidance of ballistic missiles, strategic bombers, ships and submarines (Ships Inertial Navigation System, SINS).

Huge research and development programmes have been carried out worldwide involving many billions of dollars expenditure to achieve viable systems. For instance, as mentioned earlier over three orders of magnitude improvement in gyro performance from 15°/hour to 0.01°/hour drift uncertainty was required.

Precision accelerometers had to be developed with bias uncertainties of less than 50 μg. The major task of achieving the required computational accuracies had to be solved and in fact the first digital computers operating in real time were developed for IN systems. Once these problems had been solved, however, the INS can provide the following:

- Accurate position in whatever co-ordinates are required – latitude/longitude.
- Ground speed and track angle.
- Euler angles: heading, pitch and roll to very high accuracy.
- Aircraft velocity vector (in conjunction with the air data system).

Accurate velocity vector information together with an accurate vertical reference is essential for accurate weapon aiming and this has led to the INS being installed in military strike aircraft from the early 1960s onwards as a key element of the navigation/weapon aiming system.

The self-contained characteristics of an inertial navigation system plus the ability to provide a very accurate attitude and heading reference led to the installation of IN systems in long range civil transport aircraft from the late 1960s. They are now very widely used in all types of civil aircraft.

6.2.2   Basic Principles and Schuler Tuning

A good question to ask an interviewee who is interested in entering the avionics systems field is 'How can you measure the motion of a vehicle and derive the distance it has travelled without any reference to the outside world?'

The answer is that you can sense the vehicle's acceleration (and also the gravitational vector) with accelerometers. If the vehicle's acceleration components can then be derived along a precisely known set of axes, successive integration of the acceleration components with respect to time will yield the velocities and distances travelled along these axes. This is true provided the initial conditions are known, that is, the vehicle velocity and position at the start time.

Figure 6.6 illustrates the basic concepts of deriving the velocity and distance travelled of the vehicle from its acceleration components.

Put like this it seems simple. Any errors, however, in deriving the aircraft acceleration components from the accelerometer outputs will be integrated with time, producing velocity errors which in turn are integrated with time generating position errors.

For example, a constant accelerometer bias error, B (which can be equated to an initial tilt error), will result in a distance error which is equal to $\int \int B \, dt \, dt$, that is, $Bt^2/2$. The resulting distance error is thus proportional to the square of the elapsed time. An accelerometer bias error of 10^{-3} g will produce a distance error of 0.45 km after 5 minutes and 1.8 km after 10 minutes, for example.

Errors in deriving the orientation of the accelerometer input axes with respect to the local vertical, that is, tilt errors, introduce gravitational acceleration errors in the acceleration

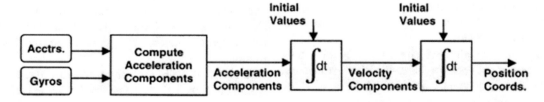

Fig. 6.6 Basic principles of inertial navigation

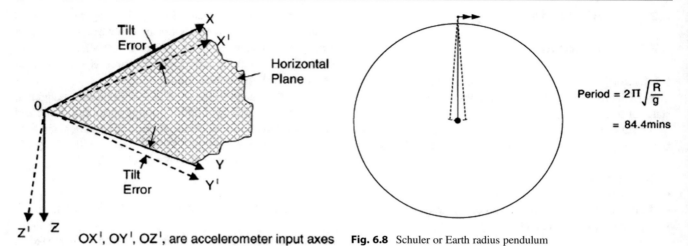

OX', OY', OZ', are accelerometer input axes

Fig. 6.7 Tilt errors

Fig. 6.8 Schuler or Earth radius pendulum

$$\text{Period} = 2\Pi \sqrt{\frac{R}{g}}$$

$$= 84.4 \text{mins}$$

measurements. Refer to Fig. 6.7. A tilt error, $\Delta\theta$, produces a gravitational error equal to $g\Delta\theta$ ($\Delta\theta$ is a small angle). A constant gyro drift rate, W, will cause the tilt error, $\Delta\theta$, to increase linearly with time ($\Delta\theta = Wt$) resulting in a gravitational acceleration error gWt. This results in a distance error $\int\int gWt\,dt\,dt$, that is, $gWt^3/6$. The distance error in this case is proportional to the cube of the time. For example, a gyro drift rate of 1°/hour results in a positional error of 0.2 km after 5 minutes and 1.6 km after 10 minutes. (Relatively large sensor errors are used in the above examples to show the importance of sensor accuracy.)

Clearly these types of error propagation are unacceptable except in applications where the time of flight is short (e.g. mid-course guidance of short range missiles) or where the system could be frequently corrected by another navigation system. It should be noted that these error propagations approximate closely to the initial error propagation with Schuler tuning over the first few minutes. This is because the correcting action exerted by Schuler tuning on the tilt errors takes place over an 84.4 minutes period and so is small over the first few minutes.

It will be shown that Schuler tuning provides an undamped closed-loop corrective action to constrain the system tilt errors so that they oscillate about a zero value with an 84.4 minutes period. The 'INS vertical' follows the local vertical irrespective of the vehicle accelerations so that the system behaves like a Schuler pendulum. Schuler tuning completely changes the error propagation characteristics from those just described and enables a viable unaided navigation system to be achieved. It is a vital and fundamental ingredient of any IN system.

The concept of a pendulum which would be unaffected by accelerations, and would always define the local vertical was originated by the Austrian physicist Max Schuler in 1924. This arose from studies he carried out on the errors in marine gyro compasses resulting from ship manoeuvres and the

effect of ship accelerations on the pendulous gyro compass system. Schuler conceived the idea of a pendulum which would be unaffected by acceleration by considering the behaviour of a simple pendulum whose length was equal to the radius of the Earth (see Fig. 6.8). The 'plumb bob' of such a pendulum would always be at the Earth's centre and would hence define the local vertical irrespective of the motion, or acceleration, of the point of suspension of the pendulum – any disturbance to the pendulum bob would cause it to oscillate about the local vertical with a period equal to $2\pi \sqrt{R/g}$, where R is the radius of the Earth, g is the gravitational acceleration.

Using the accepted values of R and g gives the period of the *Earth radius pendulum (or Schuler pendulum)* as 84.4 minutes.

A simple pendulum of length equal to the Earth's radius is obviously not feasible. If, however, a pendulous system could be produced with an exact period of 84.4 minutes then it would indicate the local vertical irrespective of the acceleration of the vehicle carrying it.

In 1924, the technology was not available to implement a practical system and in fact it was not until the early 1950s that the technology became available to implement a practical Schuler tuned IN system. The first successful airborne demonstration of a Schuler tuned inertial navigation system was achieved by Dr. Charles Stark Draper and his team at Massachusetts Institute of Technology (MIT) Boston, USA, around 1952.

The basic principles of any IN system are to derive the components of the aircraft's acceleration along locally level axes, generally the North and East axes, using an orthogonal set of accelerometers and gyros to measure the aircraft's motion.

Integration with respect to time of these acceleration components then gives the aircraft's North and East velocity components, knowing the initial conditions. The aircraft's position in terms of its latitude and longitude co-ordinates can then be derived (Sect. 6.1.3). As already explained, it is essential that the system is Schuler tuned to bound the tilt angle errors in deriving the local level plane.

The Schuler tuned stable platform was the only viable way of accurately deriving the North and East components of the aircraft's acceleration from the early 1950s to the late 1970s. This is because of limitations in angular momentum gyro technology.

Ring laser gyros of inertial accuracy achieved production status in the late 1970s and enabled a strap-down IN system to be implemented. As explained earlier, RLG-based strap-down IN systems have superseded the stable platform-based systems.

The implementation of a Schuler tuned strap-down INS and a stable platform INS are mathematically identical. The Schuler tuned strap-down INS can thus best be explained and visualised by considering the system to have a mathematical model of a stable platform, that is, a virtual stable platform, within the system computer (Sect. 5.3.2.2 and Fig. 5.22 in Chap. 5).

Figure 6.9 is a block diagram of a Schuler tuned strap-down IN system. The virtual accelerometer outputs are derived from the body mounted accelerometer outputs by the axis transformation processes explained in Sect. 5.3.2.2. Their outputs are the same as real accelerometers would have, when mounted on a real stable platform.

The North and East accelerometer outputs, however, are sensed along North, East, Down (N, E, D), axes which are rotating slowly in space with the Earth's rotation. These axis rotations introduce Coriolis acceleration components, as

explained later in the chapter. The aircraft also experiences small centrifugal acceleration components as it follows the Earth's curvature. The Coriolis and centrifugal acceleration components are small, < 0.05 m/s/s, but nevertheless need to be computed, and the real, or virtual, accelerometer outputs corrected before integrating with respect to time to obtain the aircraft's North and East velocity components.

The virtual platform is very accurately aligned with the local NED axes during the initial alignment and gyro compassing phase. Thereafter, it is rotated at the appropriate Earth's rotation rate components and the appropriate vehicle rate components so as to maintain alignment with the local NED axes.

The vehicle rate components are derived by integrating the corrected North and East accelerometer outputs with respect to time, and this inertially derived vehicle rate feedback provides the Schuler pendulum characteristic. The local value of the Earth's radius, R, is used together with the local value of the gravitational constant, g, to achieve accurate Schuler tuning. (The Schuler period is $2\pi \sqrt{(R/g)}$.)

Figure 6.10 is basically similar to Fig. 6.9 and is intended to aid visualisation of the process of tracking the local vertical as the aircraft moves over the Earth's surface and the Schuler pendulum characteristics.

Errors in the derivation of the vehicle rate terms will result in the platform tilting from the horizontal. These errors result from the following:

1. Accelerometer bias errors which result in bias errors, B_N and B_E, in the North and East acceleration components derived from the strap-down accelerometers.

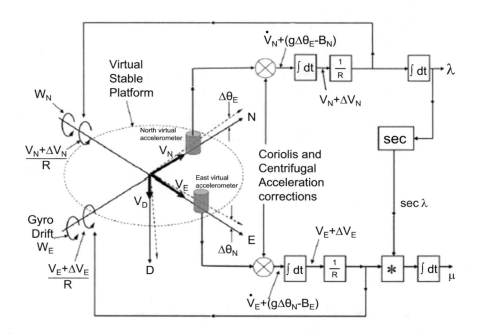

Fig. 6.9 Schuler tuned strap-down INS (note: The Earth rotation rates about the North and East axes are left out for clarity)

Fig. 6.10 Schuler tuning loop
(Earth rotation rate corrections not
shown for clarity)

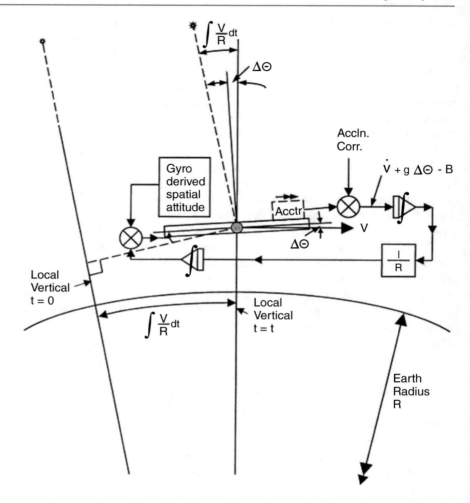

2. Gyro drift rates which introduce a direct error in the applied vehicle rates about the North and East axes of W_N and W_E, and have a major effect on the system accuracy, as will be explained.

These errors in turn produce tilt angle errors, $\Delta\theta_N$ and $\Delta\theta_E$, about the North and East axes. These introduce gravitational acceleration components, $g\Delta\theta_N$ and $g\Delta\theta_E$, into the derived accelerations along the East and North axes.

The resulting velocity errors, ΔV_E and ΔV_N, along the East and North axes are given by

$$\Delta V_E = \int (g\Delta\theta_N - B_E)dt \qquad (6.7)$$

$$\Delta V_N = \int (g\Delta\theta_E - B_N)dt \qquad (6.8)$$

These velocity errors result in vehicle rate errors $\Delta V_E/R$ and $\Delta V_N/R$ about the North and East axes. The sense of the vehicle rate errors in the Schuler loop is such as to slow down or speed up the applied vehicle rate to try to restore $\Delta\theta_N$ and $\Delta\theta_E$ to zero. In fact, due to the two integrations taking place

in the loop, the system will oscillate about the local vertical, as will be shown mathematically. A physical non-mathematical explanation of the Schuler loop is given at the end of this section.

The rate of tilt about the North axis is equal to ($\Delta V_E/R + W_N$). Hence

$$\Delta\dot\theta_N = -\frac{1}{R}\int (g\Delta\theta_N - B_E)dt + W_N \qquad (6.9)$$

(The minus sign in the $\Delta V_E/R$ term is because the feedback is negative.) Whence

$$\Delta\theta_N = -\frac{1}{R}\iint (g\Delta\theta_N - B_E)dtdt + \int W_N dt \qquad (6.10)$$

Differentiating twice and re-arranging gives

$$\left(D^2 + \frac{g}{R}\right)\Delta\theta_N = \frac{B_E}{R} + DW_N \qquad (6.11)$$

A similar equation for $\Delta\theta_E$ can be obtained with the forcing functions being B_N and W_E.

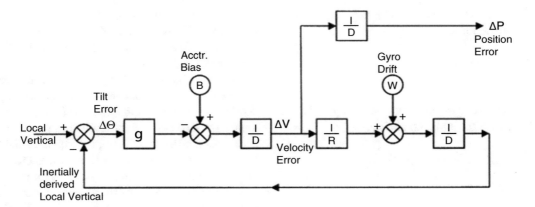

Fig. 6.11 Schuler tuned INS error model

Fig. 6.12 Error propagation due to accelerometer bias of half minute of arc

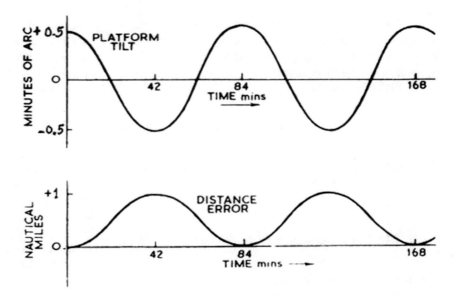

It is convenient to drop the suffices N and E in the above equations so that the solution can be applied to either axis.

$$\left(D^2 + \frac{g}{R}\right)\Delta\theta = \frac{B}{R} + DW \qquad (6.12)$$

The local vertical closed-loop tracking system produced by Schuler tuning is shown in block diagram form in Fig. 6.11 and provides an error propagation model of an INS.

The effects of accelerometer bias and gyro drift are considered separately for the sake of simplicity. The effects on the system of a combination of disturbing inputs of accelerometer bias and gyro drift can then be inferred.

1. *Effect of Accelerometer Bias*

Assume $W = 0$. The accelerometer bias can be equated to an initial tilt $= B/g$ radians. The solution of Eq. (6.12) for initial conditions

$$\Delta\theta(0) = B/g \quad \text{and} \quad \Delta\dot{\theta}(0) = 0$$

is

$$\Delta\theta = (B/g)\cos\omega_0 t \qquad (6.13)$$

where $\omega_0 = \sqrt{g/R}$ is the undamped natural frequency of Schuler loop.

The platform thus oscillates about the local vertical with amplitude B/g and a period is $2\pi\sqrt{R/g} = 84.4$ minutes.

The acceleration error is $g\Delta\theta$, the velocity error is $\int B\cos\omega_0 t \, dt$, and the distance error is $\int\int B\cos\omega_0 \, tdtdt$. Hence

$$\text{Distance error} = \frac{B}{\omega_0^2}(1 - \cos\omega_0 t) \qquad (6.14)$$

Figure 6.12 shows the effect of an accelerometer bias error equivalent to an initial tilt of one half minute of arc

$(1.45 \times 10^{-4}$ g). This results in an oscillatory error between zero and one nautical mile.

2. *Effect of Gyro Drift*

Assume $B = 0$ and gyro drift rate, W, is constant. The solution of Eq. (6.18) for initial conditions

$$\Delta\theta(0) = 0 \ \text{ and } \ \Delta\dot{\theta}(0) = W$$

is

$$\Delta\theta = \frac{W}{\omega_0} \sin \omega_0 t \tag{6.15}$$

The platform thus oscillates about the vertical as before with a 84.4-minute period. A gyro drift rate of 0.01°/hour results in a very small amplitude oscillation about the local vertical of 0.14 minutes of arc.

The velocity error is

$$\int \frac{gW}{\omega_0} \sin \omega_0 t \, dt = WR(1 - \cos \omega_0 t) \tag{6.16}$$

The velocity error thus oscillates between zero and 2WR.

The distance error is

$$\int WR(1 - \cos \omega_0 t) dt = WR\left(t - \frac{1}{\omega_0} \sin \omega_0 t\right) \tag{6.17}$$

The distance error is thus proportional to the time of flight, the small amplitude oscillatory component being swamped after a while.

A constant gyro drift rate of 0.01°/hour (0.6 minutes of arc per hour) will give an average velocity error of 0.6 knot, that is, a distance error which builds up at a rate of 0.6 NM/hour. It should be noted that this represents the error build up along the North and East axes, so that the radial error will be multiplied by $\sqrt{2}$ and hence will be equal to 1 NM/hour.

Figure 6.13 illustrates the error growth for a constant gyro drift of 0.01°/hour.

In practice, this is not a constant quantity and the distance error over a long period is proportional to the square root of the time of flight, if the variation in the gyro drift is random.

The very accurate vertical reference provided by Schuler tuning can be seen – even a gyro drift as high as 0.25°/hour results in a peak vertical error of only 3.5 minutes of arc.

A physical explanation of Schuler tuning is set out below, as it is important to have a physical appreciation, as well as a mathematical understanding. Referring to Fig. 6.14, suppose the platform is being rotated at the same rate as the local vertical, but is tilted downwards from the local vertical by a small angle, $\Delta\theta$.

At $t = 0$ the platform tilt results in the accelerometers measuring a small negative component of the gravitational acceleration as well as the true acceleration. (Accelerometer input axis tilted downwards results in a negative gravitational acceleration component being measured, as the accelerometer proof mass tends to move in the same direction as when the vehicle is being retarded, that is, negative acceleration.)

$t = 0$ *to* 21 *minutes*. As the aircraft flies over the Earth the negative acceleration error resulting from the platform tilt is integrated with time and causes the computed vehicle rate to become slower than the true vehicle rate. The platform is thus being rotated at a progressively slower rate than the local vertical thereby reducing the tilt error.

At $t = 21$ *minutes*, the tilt error is now zero but the platform is still being rotated at a slower rate than the local vertical so that the platform then starts to tilt the other way (i.e. upwards) after 21 minutes.

$t = 21$ *to* 42 *minutes*, the acceleration error due to the platform tilt is now positive so that the platform rate of rotation starts to increase.

At $t = 42$ *minutes*, the platform is now rotating at the same rate as the local vertical.

$t = 42$ *to* 63 *minutes*, after 42 minutes the platform 'vertical' is now starting to rotate at a faster rate than the true vertical and by 63 minutes the platform tilt is zero again. However, the platform is now being rotated at a faster rate than the local vertical.

$t = 63$ *to* 84 *minutes*, the tilt has changed sign so that the platform rate of rotation now starts to slow down until at 84 minutes it is rotating at the same rate as the local vertical.

The whole cycle is repeated over the next 84 minutes and so on (Table 6.1).

6.2.3 Platform Axes

The platform or inertial measuring axis frame commonly adopted for latitudes below 80° is an Earth referenced axis frame known as a 'local level, North slaved axis frame', or local North, East, Down (NED) axes (Fig. 5.26, Chap. 5).

Alternative co-ordinate systems are adopted for navigation over polar regions and these are discussed later.

In the case of a stable platform INS mechanisation, this enables the gimbal angles to provide a direct read out of the aircraft Euler angles, that is, the heading, pitch and bank angles.

Fig. 6.13 Error propagation due to gyro drift of 0.01°/hour

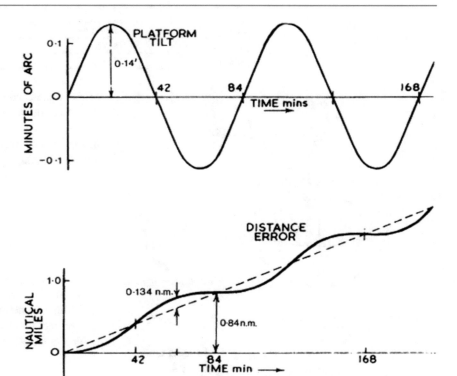

It should be noted that the term platform is used in a general sense and can denote a gimballed stable platform, or, the virtual stable platform maintained in the strap-down INS computer.

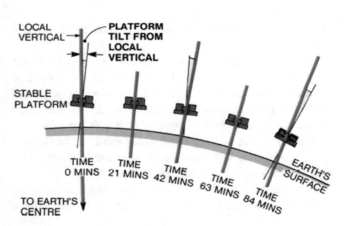

Fig. 6.14 Schuler oscillation

It should be noted that the quantity, R, in the vehicle rate terms is the distance of the aircraft from the Earth's centre and is equal to the Earth's radius, R_o, plus the aircraft's altitude, H, that is,

$$R = R_O + H \tag{6.18}$$

The value of R_o is taken as 6,378,137 m. Further correction must also be made to allow for the Earth being an ellipsoid and not a perfect sphere.

6.2.3.2 Acceleration Correction Terms

The NED axis frame is rotating with respect to an inertial axis frame and this introduces a further complication in deriving the aircraft's rate of change of velocity along the North and East axes. This is because the aircraft's linear motion is defined with respect to these axes which in turn are rotating with the rotation of the Earth.

Coriolis accelerations are introduced because of the linear motion with respect to a rotating axis frame, as the path in space is a curved one. Coriolis accelerations are named after the French mathematician who formulated the general principles for the study of moving bodies in a rotating frame of reference early in the nineteenth century.

Referring to the inset diagram on Fig. 6.15, the Coriolis acceleration experienced by a body moving with velocity, V, with respect to an axis frame which is rotating at an angular

6.2.3.1 Angular Rate Correction Terms

As explained earlier in Chap. 5, Sect. 5.3, it is necessary to rotate the gyro derived reference frame at the appropriate rates so that it stays aligned with the local NED axis frame.

The Earth's rate and vehicle rate correction terms are set out in Table 6.1, for reference.

Table 6.1 Angular rate correction terms

Rate term	North axis	East axis	Down axis
Earth's rate	$\Omega \cos \lambda$	0	$\Omega \sin \lambda$
Vehicle rate	V_E/R	$-V_N/R$	$-(V_E/R) \tan \lambda$

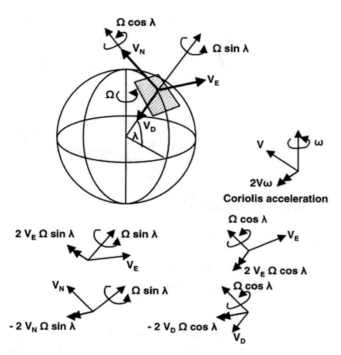

Fig. 6.15 Coriolis accelerations due to Earth's rotation

of the aircraft, so that it is necessary to correct the output for the gravitational acceleration, g.

The gravitational acceleration reduces as the altitude increases and follows an inverse square law. The value of g at an altitude, H, is given by

$$g = \frac{R_0^2}{(R_0 + H)^2} \cdot g_0 \qquad (6.19)$$

where g_0 is the value of the gravitational acceleration at the Earth's surface.

The local value of the gravitational acceleration, g_0, varies by a small amount with latitude. This is because of the centrifugal acceleration created by the Earth's rate of rotation and the fact that the Earth is an oblate spheroid.

The relationship is

$$g_0 = g_{equ} \left\{ \frac{1 + k \sin^2 \lambda}{\left(1 - e^2 \sin^2 \lambda\right)^{1/2}} \right\} \qquad (6.20)$$

where $g_{equ} = 9.7803267714$ m/s^2, $k = 0.00193185138639$, and $e^2 = 0.00669437999013$.

The rates of change of the aircraft velocity components along the NED axes, $\dot{V}_N$, $\dot{V}_E$, $\dot{V}_D$, are obtained by subtracting the acceleration corrections in Table 6.2 from the outputs of the North, East, Down accelerometers, a_N, a_E, a_D. (These are virtual accelerometers in the case of a strap-down INS.)

It is instructive to work out the magnitude of the Coriolis and centrifugal acceleration terms in order to assess their effect, if they were not corrected.

For example, the Coriolis acceleration experienced when moving in a straight line over the Earth's surface with a velocity of 600 knots (approximately 300 m/s) at a latitude of 52° is equal to $2 \times 300 \times 7.2717 \times 10^{-5} \sin 52°$ m/s^2, that is, 0.03438 m/s^2. ($\Omega = 15°$/hour $= 7.2717 \times 10^{-5}$ rads/s). This is equivalent to a tilt error of 0.03438/g rads, that is, 3.5 milli-radians, if uncorrected.

The magnitude of the centrifugal acceleration term $V_E^2 \tan \lambda/R$ at the same velocity and latitude is equal to 0.01806 m/s^2. This is equivalent to a tilt error of 1.84 milli-radians, if uncorrected.

The effect of an initial tilt error is analysed in the preceding section. Tilt errors in a good quality INS are of the order of 0.1 milli-radians (0.3 minute of arc approx), so that it can be seen that accurate compensation must be made for the Coriolis and centrifugal acceleration terms.

rate of ω rads/s is equal to $2V\omega$ and is mutually at right angles to the linear velocity and angular velocity vectors.

Referring to Fig. 6.15, the Coriolis acceleration components along the North, East, vertical axes due to the aircraft's linear velocity components V_N, V_E, V_D and the Earth's rotation rate components $\Omega \cos \lambda$ about the North Axis and $\Omega \sin \lambda$ about the vertical axis are as follows:

North axis	$2V_E\Omega \sin \lambda$
East axis	$-2V_N\Omega \sin \lambda - 2V_D\Omega \cos \lambda$
Vertical axis	$-2V_E\Omega \cos \lambda$

The aircraft is also turning in space as it flies over the Earth's surface, as the Earth is spherical, and this introduces centrifugal acceleration components. The centrifugal (strictly speaking centripetal) acceleration of a body moving with velocity V and turning at a rate ω rads/s is equal to $V\omega$.

The centrifugal acceleration components along the North, East and Down axes are set out in Table 6.2 together with the Coriolis acceleration components.

The Down (or vertical) axis accelerometer measures the gravitational acceleration as well as the vertical acceleration

Table 6.2 Acceleration correction terms

	North axis	East axis	Down axis
Acceleration component			
Coriolis	$2V_E \Omega \sin \lambda$	$-2V_N \Omega \sin \lambda - 2V_D \Omega \cos \lambda$	$-2V_E \Omega \cos \lambda$
Centrifugal	$(V_E^2 \tan \lambda - V_D V_N)/R$	$-(V_N V_E \tan \lambda + V_D V_E)/R$	$(V_N^2 + V_E^2)/R$
Gravitational			$\frac{R_0^2}{(R_0+H)^2} g_0$

6.2.4 Initial Alignment and Gyro Compassing

Inertial navigation can only be as accurate as the initial conditions which are set in. It is therefore essential to know the orientation of the accelerometer measuring axes with respect to the gravitational vector, the direction of true North, the initial position and the initial velocity components to very high accuracy.

The two basic references used to align an inertial system are the Earth's gravitational vector and the Earth's rotation vector.

The initial alignment process is basically the same in a stable platform and strap-down INS. The difference being that in a stable platform INS, the stable platform is physically rotated to bring it into alignment with the local NED axes by applying precession torques to the vertical and azimuth gyros on the platform. It is thus easier to visualise (literally). Whereas the strap-down system carries out the axis rotations within the system computer to create, in effect, a virtual stable platform as explained earlier.

The levelling operation takes place in two stages; a coarse levelling stage followed by a fine levelling stage using the horizontal accelerometer outputs. (In the case of a strap-down system, these are virtual horizontal accelerometers as the horizontal acceleration components are computed from the body mounted accelerometer outputs using the gyro derived attitude data.) These horizontal accelerometer outputs are directly proportional to the tilt angle from the horizontal of the accelerometer measuring axes when the aircraft is stationary on the ground. They also contain spurious accelerations and noise due to wind buffet, fuelling, crew and passengers moving about the aircraft, etc.

The coarse levelling of a stable platform INS is achieved by feeding the horizontal accelerometer outputs directly into the appropriate torque motors of the vertical gyro(s).

The fine levelling stage, which filters out the noise and spurious accelerations, is achieved by filtering the accelerometer outputs before feeding them into the vertical gyro torque motors. The filtering process is basically the same as in a strap-down INS, which is covered below. It relies on the fact that the integrated horizontal acceleration components, which give the horizontal velocity components, should be zero as the aircraft is stationary on the ground.

The accelerometers for an aircraft strap-down INS are generally mounted along the aircraft's principal axes so that the 'horizontal' accelerometers mounted along the forward and side-slip axes do not sense a large component of gravity. The pitch and bank angles of the aircraft are small as the aircraft is normally fairly level when stationary on the ground. The aircraft attitude integration process, using the incremental body angular rotations measured by the pitch, roll and yaw strap-down gyros, can be initialised by assuming the pitch and bank angles are both zero (if these are not known).

The fine levelling is carried out by using the fact that any tilt errors $\Delta\theta_N$ and $\Delta\theta_E$ about the computed North and East axes will couple gravitational acceleration components $g\Delta\theta_N$ and $g\Delta\theta_E$ into the East and North acceleration components derived from the accelerometers. The horizontal acceleration components are then integrated with respect to time to produce the horizontal velocity components of the aircraft. These horizontal velocity components should be zero as the aircraft is stationary on the ground. Any resulting horizontal velocity components that are measured are therefore fed back appropriately to correct the tilt and level the system.

The levelling loops are generally third-order loops using the integrals of the velocity errors as well as the velocity errors as control terms. The feedback gains are also varied.

Figure 6.16 shows the fine levelling and gyro compassing loops

A coarse azimuth alignment with respect to true North is made to within a degree or so using, say, a magnetic reference. The fine alignment to achieve the required accuracy is accomplished by the process of gyro compassing. During the gyro compassing phase the computed heading is adjusted until the component of the Earth's rotation sensed by the gyros about the East axis is zero.

As shown earlier, the components of the Earth's rate of rotation Ω about the North, East and Down axes at a latitude of λ are as follows:

North axis	$\Omega \cos \lambda$
East axis	0
Vertical axis	$\Omega \sin \lambda$

The inset diagram in Fig. 6.16 shows the components of the Earth's rotation rate which are sensed by the gyros when the derived NED axes are misaligned in the horizontal plane by an amount $\Delta\psi$ from true North. It can be seen that a rotation rate equal to $\Omega \cos \lambda . \sin \Delta\psi$ is measured about the nominal East pointing axis, that is, $\Omega \cos \lambda . \Delta\psi$, if $\Delta\psi$ is a small angle.

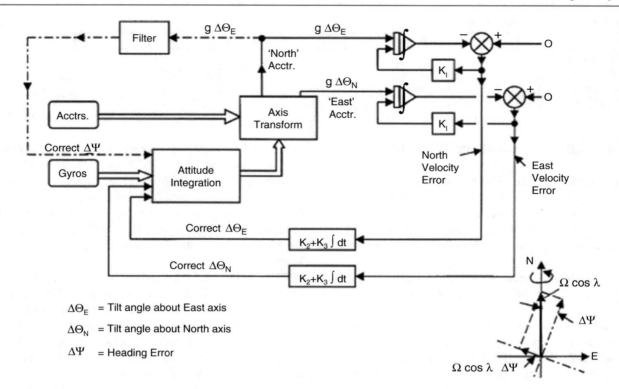

Fig. 6.16 Fine levelling and gyro compassing loops

The North pointing virtual accelerometer will hence be tilting away from the horizontal at a rate $\Omega \cos \lambda . \Delta\psi$ in the absence of any levelling or gyro compassing loops. (This is assuming the appropriate corrections are made to compensate for $\Omega \cos \lambda$ and $\Omega \sin \lambda$, respectively.)

A closed-loop gyro compassing system can be formed as shown in Fig. 6.16 using the suitably filtered North axis acceleration derived from the accelerometers and gyros; the acceleration being proportional to the tilt, $\Delta\theta_E$, about the East axis from the horizontal.

The gyro compassing loop adjusts the computed heading until the east component of the gyro angular rate measurement in the horizontal plane is nulled, the angular rate of rotation about the East axis being estimated from the summed East axis tilt correction

$$\Delta\theta_E = \int \Omega \cos \lambda . \Delta\psi dt)$$

The allowable gyro drift rate uncertainty can be determined from the accuracy required of the heading alignment. For example, if an accuracy of 0.1° is required for a latitude of 45°, then the component of the Earth's rate sensed at this latitude with a misalignment of 0.1° is equal to (0.1/57.3) sin 45° per hour, that is, 0.017 degrees per hour.

It can be seen that the magnitude of the component of Earth's rate to be sensed decreases with increasing latitude, so that gyro compassing is effectively restricted to latitudes below 80°.

The major factors which affect alignment accuracy and alignment times are as follows:

(a) Initial tilt
(b) Aircraft movements, for example, effect of wind gusts
(c) Accelerometer bias errors and gyro drift rates
(d) Change of the above quantities (c) with time as the system warms up
(e) Accelerometer resolution and gyro threshold

The loop gains in the levelling and gyro compassing loops are generally controlled by means of a Kalman filter to give an optimal alignment process. Typical alignment times are of the order of seven minutes for full accuracy IN performance.

Reduced alignment times are sometimes used and the system corrected to give full IN accuracy by subsequent position fixes using a position fixing navigation system (e.g. GPS).

6.2.5 Effect of Azimuth Gyro Drift

The effect of azimuth gyro drift is to cause position errors which build up with time and are a function of azimuth error, aircraft velocity and latitude.

From the previous section on gyro compassing it can be seen that an azimuth error $\Delta\psi$, will effectively inject a drift of $\Omega \cos \lambda . \Delta\psi$ about the East axis. An error of $0.2°$ for instance, would generate an effective drift of $0.033°$/hour (2 minutes of arc/hour approximately). This is injected into the Schuler loop (Fig. 6.13) as a disturbance $W = 0.033°$/hour and would give a northerly velocity error of 2 NM/hour. An azimuth gyro drift rate will give a distance error which builds up proportionately to the square of the time. This gives a latitude error which affects the accuracy of the other correction terms, for example, the meridian convergence term $V_E/R \cdot \tan \lambda$, causing still further error to build up with time. For this reason it is essential to have a good azimuth gyro performance.

6.2.6 Vertical Navigation Channel

It has been shown in Sect. 6.2.2 that Schuler tuning effectively constrains the error build up with time in the horizontal channels of the INS. Unfortunately, however, there is no such action in the vertical channel and small errors in computing the vertical acceleration correction terms are integrated with time – in fact, the vertical distance error builds up exponentially with time. As explained earlier, the principal correction terms for the vertical accelerometer channel are as follows:

(i) Gravity – gravity varies with altitude according to the inverse square law. The gravitational acceleration at an altitude H is given by

$$g = g_0 \frac{R_0^2}{(R_0 + H)^2}$$

where g_0 = gravitational acceleration at the earth's surface.

Errors in correcting for the variation in gravitational acceleration with height result in a vertical distance error which builds up as cosh function of time. Also, gravity varies over the Earth's surface and has anomalous values in certain places.

(ii) Centrifugal acceleration – a vehicle moving over the Earth's surface is describing a circular trajectory in space and so will experience a centrifugal acceleration $\left(V_E^2 + V_N^2\right)/R$ where V_E and V_N are the easterly and northerly velocity components of the vehicle, respectively.

(iii) Coriolis acceleration – the component of the Earth's rotation rate $\Omega \cos \lambda$ about the North axis is combined with the easterly velocity V_E and generates a Coriolis acceleration along the vertical axis which is equal to $2V_E\Omega \cos \lambda$.

It should be noted that there is also a small centrifugal acceleration due to the Earth's rotation which slightly modifies the local value of gravity.

Monitoring of the inertially derived height with some other source of vertical information such as pressure (or barometric) altitude from the air data system, or a radio altimeter is thus essential for long time operation.

The 'Standard Atmosphere' which is used to derive the pressure (or barometric) altitude from the static pressure measurement is explained in Chap. 7. Errors can arise in the altitude (or height) derived from the static pressure, however, due to the variations in the lapse rate, the height of the tropopause and in sea level temperature and pressure. Even though a datum is established before take-off, climbing or diving through a non-standard atmosphere (i.e. one which differs from the assumed standard) can result in errors in the height rate of up to 8% with corresponding errors in the height.

An inertial system is accurate over a short period provided it is correctly initialised and it also has an excellent dynamic response. The optimum mixing of inertially derived and barometrically derived height is thus able to compensate to a very considerable extent for the deficiencies present in the individual sources. The stages in deriving an optimum mixing system are briefly outlined below.

Consider first the simple second-order mixing system shown in bold lines in Fig. 6.17a. The baro-inertial feedback loop is capable of being opened and the gains K_1 and K_2 which determine the damping and response times, respectively, are also capable of being varied. The system response can be shown to be given by

$$H_I = \frac{D^2}{(D^2 + K_1 D + K_2)} \cdot H + \frac{(K_2 + K_1 D)}{(D^2 + K_1 D + K_2)}$$
$$\cdot H_P + \frac{1}{(D^2 + K_1 D + K_2)} . \Delta a_D \tag{6.21}$$

where H is the true height, H_P is the pressure altitude, Δa_D is the uncorrected vertical acceleration error and H_I is the baro/inertial mixed height.

The inertially derived component of H_I, that is,

$$\frac{D^2}{(D^2 + K_1 D + K_2)} \cdot H$$

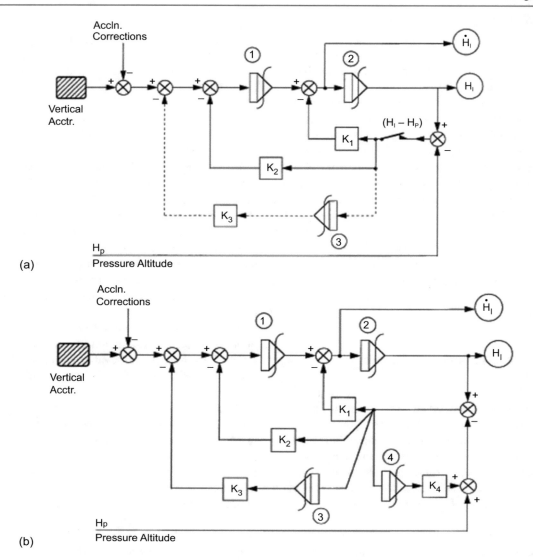

(a)

(b)

Fig. 6.17 (**a**) Baro-inertial mixing. Simple second-order baro-inertial mixing shown in bold lines. Integral term correction for inertial drift shown in dotted lines. (**b**) Fourth-order baro-inertial mixing (with correction for inertial channel drift and pressure altitude error)

couples the inertially measured height through a second-order high pass filter transfer function which responds to dynamic changes in height without lag, but 'DC blocks' the steady-state value.

The barometrically derived component of H_I, that is,

$$\frac{(K_2 + K_1 D)}{(D^2 + K_1 D + K_2)} \cdot H_P$$

couples the barometrically derived height through a low pass filter which smooths the pressure height signal and attenuates the noise present in the signal. The dynamic response is thus slugged by the low pass filter but the steady-state value of the pressure height is not affected. The combination of the two

components results in a fast response with a low noise content.

However, there is no drift compensation for the inertial channel. Assuming the system is trimmed before flight, the dominant error would probably be poor 'g' compensation at high altitudes due to the error in the height so that Δa_D will increase. This in turn will cause an increasing offset between H_P and H_I. In a closed-loop climb, the inertial height, H_I, will be forced to follow changes in pressure height, H_P, with possible resultant height rate errors of up to 8%. An open-loop climb, however, would result in the system becoming divergently unstable because of the uncompensated acceleration errors. The transient which would normally occur on closing the loop could be reduced by a 'fast re-set' phase in which the K_1 feedback gain is temporarily increased.

The next stage is to correct the inertial channel drift by the addition of an integral correction term $K_3 \int (H_P - H_I)dt$ which is fed to the input of the first integrator (1) as shown in the dotted lines on Fig. 6.17a. This would improve the open-loop (i.e. pure inertial) climb performance as the inertial channel would start in a well-trimmed state. However, even with a fast re-set the system cannot settle without a lengthy transient on re-closing the loop. This is because to force H_I to equal H_p requires additional drift compensation from integrator (3) to accommodate the new error in the 'g' correction following the open-loop climb.

This situation can be improved by adding the fourth integrator (4) as shown in Fig. 6.17b to correct the pressure height error as the inertial system will measure the changes in height over a short period accurately if correctly initialised. This eliminates the transient on re-closing the inertial loop and makes an attempt to operate in 'true' height.

Optimum performance over the flight envelope requires the gains K_1, K_2, K_3 and K_4 to be variable. This can be effectively achieved by means of a Kalman filter. With a Kalman filter, the terms in the covariance matrix which compares the errors in the system state variables, reflect the previous history of the system. For example, if a prolonged period of level flight had resulted in a very well-trimmed inertial system and hence correspondingly small values of variances associated with it, almost the whole discrepancy arising at the start of a climb would be attributed to an abnormal atmosphere. The Kalman filter can thus be designed to have a capability to learn the structure of the atmosphere and make short term use of the knowledge.

Mention has been made earlier of the importance of being able to derive the aircraft's velocity vector from the ground speed vector V_G output of the INS and the baro/inertially derived vertical velocity, $\dot{H}_I$. The magnitude of the velocity vector is equal to $\sqrt{V_G^2 + \dot{H}_I^2}$. The angle the velocity vector makes with the horizontal is equal to $\tan^{-1} \dot{H}_I/V_G$.

6.2.7   Choice of Navigation Co-ordinates

The choice of a navigation co-ordinate system reduces practically to either:

(a)   Spherical co-ordinates (latitude/longitude)
(b)   Direction cosines

A spherical co-ordinate system has two mathematical 'poles' or singularities where the value of one co-ordinate (the longitude) is indeterminate and near which the high rates of change of that co-ordinate make any system implementation impractical (longitude rate $\dot{\mu} = (V_E/R) \sec \lambda$ becomes infinite when $\lambda = 90°$). With the latitude/longitude co-ordinate

system (also known as the geographical or 'geodetic' system) these poles are the true North and South poles of the Earth. This co-ordinate system is satisfactory provided there is no requirement for navigation in the polar regions. It also provides the latitude/longitude information for display to the flight crew.

When polar navigational capability is required there are three possible ways to overcome the difficultly, namely:

(a)   A spherical co-ordinate system with the poles removed to some other regions
(b)   A 'unipolar' system
(c)   Direction cosines

The spherical co-ordinate system with transferred poles gives rise to added complications in the ellipsoidal corrections because of the asymmetry which results from the misalignment between the main co-ordinate axis and the axis of the geographic ellipsoid. More seriously, it is a solution of limited usefulness because the singularities still exist.

Space does not permit the 'unipolar' system to be discussed at length. It is a compromise solution which has just one singularity (which may be made the South pole if arctic navigation is required for instance). Again, it is not a general solution and has nothing to recommend it in preference to the third solution, namely, the direction cosine system.

The direction cosine system is the system generally used. In this system the co-ordinates of a point are the three direction cosines of its radius vector in an orthogonal, Earth centred, Earth-fixed reference frame (together with its height above the Earth's surface). The attitude of the platform may also be described by the direction cosines of its axes in the same reference frame. This system has no singularities. The basic computations are not significantly more involved than those for latitude/longitude computations and present no problems to the modern micro-processor based system. An additional separate computation is required to derive latitude and longitude for display purposes but this presents no significant problems.

6.2.8   Strap-Down IN System Computing

The basic computing flow diagram for a strap-down INS is shown in Fig. 6.18. A strap-down IN system carries out the same functions as a stable platform type. INS and many elements and functional areas are common to both systems. There are two crucial areas in the strap-down mechanisation. These are as follows:

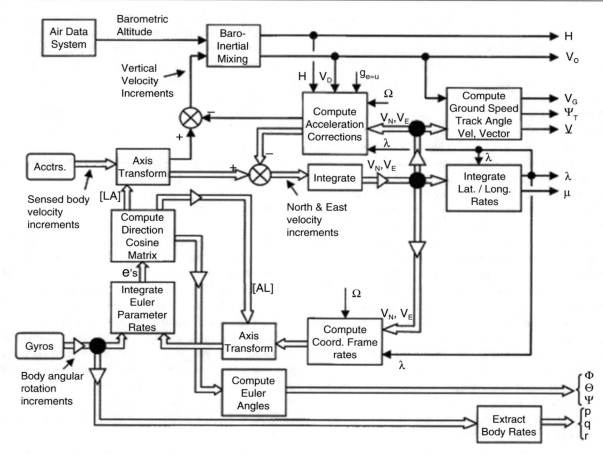

Fig. 6.18 Strap-down INS computing flow diagram

1. Attitude integration whereby the vehicle attitude is derived by an integration process from the body incremental angular rotations measured by the gyros.
2. Accelerometer resolution whereby the corrected outputs of the body mounted accelerometers are suitably resolved to produce the horizontal and vertical acceleration components of the aircraft.

The derivation of the aircraft attitude from the strap-down gyro outputs by continuously updating the four Euler symmetrical parameters each iteration period has been explained in Chap. 5 (Sect. 5.3.2.1). Very high accuracy is required in the attitude integration process, the integration period should be as short as possible and accurate integration algorithms must be used (e.g. Runge–Kutta algorithms). The ortho-normalisation of the transition matrix is essential using the constraint equation for the Euler parameters, $\left(e_0^2 + e_1^2 + e_2^2 + e_3^2 = 1\right)$.

The derivation of the equivalent (or virtual) horizontal and vertical accelerometer outputs from the body mounted accelerometers is as explained in Chap. 5, Sect. 5.3.2.2, using the Direction Cosine Matrix derived from the Euler parameters.

It should be noted that the velocity increments derived from the pulse rate outputs of the three body mounted accelerometers may need to be corrected to allow for vehicle rotation during the integration interval when the velocity increments are being accumulated. These corrections are known as Coriolis body rotation corrections and have been left out up to now for simplicity.

The computing system, after deriving the equivalent North, East, Down accelerometer outputs and transforming the co-ordinate frame rates from local NED axes to aircraft body axes, is then the same as for a stable platform mechanisation.

The implementation of a strap-down INS is shown in Fig. 6.19. The aim is to give the reader an appreciation of the small number of solid state modules required. These comprise the following:

1. The Inertial Measuring Unit (IMU) comprising three orthogonally mounted laser gyros and three orthogonally mounted accelerometers.
2. The Processor Module carrying out all the processing tasks just described, and also monitoring and self-test functions.

Fig. 6.19 Solid state modular implementation of a strap-down INS

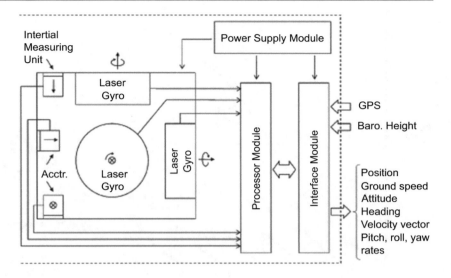

3. The Interface Module carrying out all the interfacing tasks.
4. The Power Supply Unit.

The complexity resides in the software and the processor and interface micro-chips.

The Honeywell *Laseref VI Micro Inertial Reference System* is an example of a current state of the art INS. The inset illustration and data are published by courtesy of Honeywell.

- Overall size: 6.5″ × 6.4″ × 6.4″
  (16.5 cm × 12.3 cm × 12.3 cm)
- Weight: 9.1 lbs (4.14 kg)
- Power consumption: 20 watts
- Reliability >50,000 hours

Position accuracy (unaided) is 2 NM/hour.

The inertial sensors comprise the Honeywell GG 1320 AN Digital Laser Gyro and the QA 2000 Q Flex Accelerometer, a closed-loop, torque balance accelerometer. The system incorporates GPS integration. Hybrid mode position and velocity accuracy is 12 m and 0.25 knots (0.13 m/s). GPS integration also enables IRS alignment in motion and even dispatch prior to navigation mode. This feature eliminates delays while waiting for the IRS to align.

6.3 Aided IN Systems and Kalman Filters

The time dependent error growth of an IN system makes it necessary that some form of aided navigation system using an alternative navigation source is introduced to correct the INS error build up on long distance flights. For example, as mentioned earlier, a good quality unaided INS of 1 NM/hour accuracy could have a positional error of 5 NM after a 5-hour flight.

A variety of navigation aids can be used for this purpose, for example GPS.

Consider, firstly, a simple position reset at each position fix as shown in Fig. 6.20. The error growth is limited but follows a saw tooth pattern, the amplitude depending on the period between updates and the magnitude of the velocity and tilt errors.

Now suppose the errors present in the INS, such as attitude errors, velocity errors, gyro drifts and accelerometer errors could be determined from the positional fix information. This is what a Kalman filter does and corrections can then be applied to the INS as shown in the block diagram in Fig. 6.21. The Kalman filter provides an optimum estimate of the IN system errors taking into account the errors present in the position fixing system. The resulting error propagation using a Kalman filter to estimate and correct the INS errors follows the pattern shown in Fig. 6.22 – a substantial improvement. An accurate velocity reference system, such as a Doppler radar velocity sensor, can also be used in conjunction with a Kalman filter to estimate and correct the INS errors. In fact, a number of navigational aids can all be combined with an INS in an optimal manner by means of a Kalman filter. The dissimilar nature of the error characteristics of an INS and the various position (and velocity) aids is exploited by the Kalman filter to achieve an

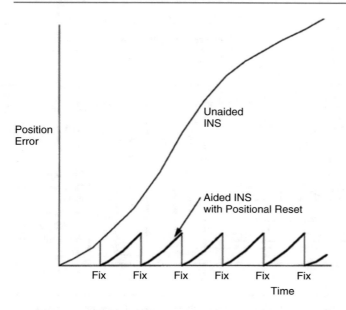

Fig. 6.20 Aided INS with simple positional reset

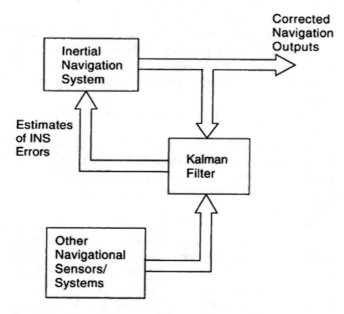

Fig. 6.21 Block diagram of aided IN system with Kalman filter

overall accuracy and performance which is better than the individual systems. The optimum blending of the individual system characteristics which can be achieved can be seen from the brief summary of the various navigation sources below. The complementary filtering of dissimilar systems using simple fixed gain mixing has already been discussed in Chap. 5 and also in the preceding section on baro/inertial mixing. To recap, the various navigation sources comprise:

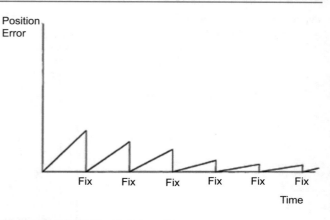

Fig. 6.22 Aided INS with Kalman filter

1. *Position Data*
   • GPS, VOR/DME, TACAN
   • Terrain reference navigation systems
   • Radar
   • Visual fixes (e.g. use of helmet mounted sight)
   • Astro (Stellar) navigation (using automatic star trackers)
2. *Velocity Data*
   • Doppler radar
   • GPS
3. *Altitude Data*
   • Barometric altitude from the air data computer
   • Radio altimeter

These sources provide good information on the average at low frequency but are subject to high frequency noise due to causes such as instrument noise, atmospheric effects, antenna oscillation and unlevel ground effects.

In contrast, IN systems provide good high frequency information content (above the Schuler frequency) despite vehicle motion. The low frequency information, however, is poor due to the inherent long-term drift characteristics, as already explained.

It should be stressed at this point that a Kalman filter can be used to provide an optimum estimate of the errors in any measuring system and its use is not confined to navigation systems, although it has been particularly effective in this field. Examples of other applications are as follows:

• Radar and infrared automatic tracking systems
• Fault detection and in monitoring of multiple (redundant) sensors
• Initial alignment and gyro compassing of an INS

The Kalman filter was first introduced in 1960 by Dr. Richard Kalman (see reference list at the end of this chapter). It is essentially an optimal, recursive data processing algorithm which processes sensor measurements to estimate the quantities of interest (states) of the system using the following:

1. A knowledge of the system and the measurement device dynamics
2. A statistical model of the system model uncertainties, noises, measurement errors
3. Initial condition information

The recursive nature of the filter, that is, using the same equations over and over again make it ideally suited to a digital computer. The filter only requires the last value of the state of the system it be stored and does not need the value of the old observations to be retained. This considerably reduces the amount of computer storage required.

The basic application of a Kalman filter to a mixed navigation system is shown in the flow diagram in Fig. 6.23. The filter contains an error model for all the systems involved, enabling the dynamic behaviour of the system errors to be modelled. The computer contains a current estimate for each term in the error model and this estimate, which is based on all previous measurements, is periodically updated. At the time of each new measurement, the difference in the outputs of the systems is predicted based on the current estimate of the errors in the systems. This difference between the predicted and actual measurements is then used to update each of the estimates of the errors through a set of weighting coefficients – the Kalman gains. The weighting coefficients are variables which are computed periodically in the system computer and are based on the assumed statistical error model for the errors. The configuration takes into account the past history of the system including the effects of previously applied information and of vehicle motions which affect the system errors.

A fundamental feature of the Kalman filter is that the error measurements made of one quantity (or set of similar quantities) can be used to improve the estimates of the other error quantities in the system. For example, the Kalman filtering technique generates an improvement in the INS velocity accuracy by virtue of the strong correlation between position error (the measured quantity) and velocity error, which is essentially the direct integral of the position error.

An introductory overview of Kalman filtering is set out in the following pages with the aim of explaining some of the essential processes and the terms used so that the interested reader can follow up the extensive literature on the subject (references are given at the end of this chapter). It should be noted that applications of Kalman filtering alone can occupy a fairly thick book – for instance Agardograph No. 139 'Theory and Application of Kalman Filtering' is a soft cover book nearly 2.5 cm (1 inch) thick.

An essential element in the Kalman filter is the *System Error Model* which models the dynamic behaviour of the system errors. The system dynamic behaviour can be represented by n linear differential equations, where n is the number of state variables in the system. In matrix form this becomes

$$\dot{\mathbf{X}} = \mathbf{A}\mathbf{X} + \mathbf{B}\mathbf{U} \qquad (6.22)$$

where $\mathbf{X}$ is the system state vector comprising n state variables, $\mathbf{A}$ is the coefficient or plant matrix, $\mathbf{B}$ is the driving matrix and $\mathbf{U}$ is the input state vector. (Bold letters denote matrices.)

In deriving the Kalman filter, $\mathbf{U}$ is assumed to be a vector of unbiased, white, Gaussian noise sequences.

The state equations for the errors in one axis of an IN system are derived below as an example. Referring to the simplified block diagram, Fig. 6.11, the position error is ΔP, the velocity error is ΔV, the tilt error is $\Delta\theta$, the gyro drift error is W and the accelerometer bias error is B.

The relationships between these variables are set out below:

$$\Delta\dot{P} = \Delta V \qquad (6.23)$$

$$\Delta\dot{V} = -g\Delta\theta + B \qquad (6.24)$$

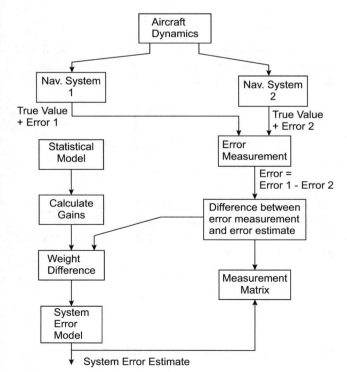

Fig. 6.23 Application of Kalman filter to mixed navigation systems

$$\Delta\dot\theta = \frac{1}{R}\Delta V + W \qquad (6.25)$$

These equations can be represented more compactly in matrix form as shown below.

$$\begin{bmatrix} \Delta\dot P \\ \Delta\dot V \\ \Delta\dot\theta \end{bmatrix} = \begin{bmatrix} 0 & 1 & 0 \\ 0 & 0 & -g \\ 0 & \dfrac{1}{R} & 0 \end{bmatrix} \begin{bmatrix} \Delta P \\ \Delta V \\ \Delta\theta \end{bmatrix}$$

$$+ \begin{bmatrix} 1 & 0 \\ 0 & 1 \end{bmatrix} \begin{bmatrix} B \\ W \end{bmatrix} \qquad (6.26)$$

That is,

$$\dot{\mathbf{X}} = \mathbf{A}\mathbf{X} + \mathbf{B}\mathbf{U}$$

As explained in Chap. 5, the state transition matrix $\boldsymbol\Phi_{\mathbf{n}}$ relates the system state vector at time nT, that is, $\mathbf{X_n}$, to the system state vector at time $(n + 1)T$, that is, $\mathbf{X_{(n+1)}}$ where T is the iteration period. The relationship is shown below:

$$\mathbf{X}_{(n+1)} = \boldsymbol\Phi_n \cdot \mathbf{X}_n \qquad (6.27)$$

Assuming a linear system

$$\boldsymbol\Phi_n = e^{\mathrm{AT}} = \mathbf{1} + \mathbf{A}T + \frac{\mathbf{A}T^2}{2} + \ldots \qquad (6.28)$$

If the iteration period T is short

$$\boldsymbol\Phi_n \approx \mathbf{1} + \mathbf{A}T + \mathbf{A}^2\frac{T^2}{2} \qquad (6.29)$$

For the above simple example

$$\boldsymbol\Phi_n = \begin{bmatrix} 1 & 0 & 0 \\ 0 & 1 & 0 \\ 0 & 0 & 1 \end{bmatrix} + \begin{bmatrix} 0 & 1 & 0 \\ 0 & 0 & -g \\ 0 & \dfrac{1}{R} & 0 \end{bmatrix} T$$

$$+ \begin{bmatrix} 0 & 1 & 0 \\ 0 & 0 & -g \\ 0 & \dfrac{1}{R} & 0 \end{bmatrix} \begin{bmatrix} 0 & 1 & 0 \\ 0 & 0 & -g \\ 0 & \dfrac{1}{R} & 0 \end{bmatrix} \frac{T^2}{2}$$

$$\boldsymbol\Phi_n = \begin{bmatrix} 1 & T & -gT^2/2 \\ 0 & (1 - gT^2/2R) & -gT \\ 0 & T/R & (1 - gT^2/2R) \end{bmatrix}$$

$$(6.30)$$

Hence

$$\Delta P_{(n+1)} = \Delta P_n + T\,\Delta V_n - gT^2/2 \cdot \Delta\theta_n \qquad (6.31)$$

From initial estimates of the uncertainties in ΔP, ΔV, $\Delta\theta$ at time, $t = 0$, the values at time $t = (n + 1)T$ can thus be derived by a step-by-step integration process using the transition matrix at each iteration.

As stated earlier this is a simplified example to illustrate the process of deriving a dynamic error model.

A typical Kalman filter INS error model consists of 17–22 error states. The 19 state model comprises the following error states: two horizontal position errors, two horizontal velocity errors, three attitude errors, three gyro biases, three gyro scale factor errors, three accelerometer bias errors, three accelerometer scale factor errors. The vertical channel is not modelled since it is unstable, as shown earlier, because of the gravity compensation with altitude which results in the positive feedback of altitude errors. The barometric altitude information from the air data system is combined with the INS vertical channel as shown earlier in a baro/inertial mixing loop. The GPS received error model typically comprises 12 states.

The measurement matrix, $\mathbf{H}$, is used to select the part or the component of the state vector $\mathbf{X}$ which is being measured. For example, suppose the state vector

$$\mathbf{X} = \begin{bmatrix} \Delta P \\ \Delta V \\ \Delta\theta \\ \Delta P_2 \end{bmatrix}$$

where ΔP, ΔV, $\Delta\theta$ are the inertial system errors and ΔP_2 is the error in the position reference system. To extract $(\Delta P - \Delta P_2)$ the measurement matrix $\mathbf{H} = [1\ 0\ 0\!-\!1]$

$$\mathbf{H} \cdot \mathbf{X} = \begin{bmatrix} 1 & 0 & 0 & -1 \end{bmatrix} \begin{bmatrix} \Delta P \\ \Delta V \\ \Delta\theta \\ \Delta P_2 \end{bmatrix} = \Delta P - \Delta P_2$$

The covariance matrix of the estimation errors is formed by multiplying the error state vector matrix by its transpose – the transpose of a matrix means interchanging the rows and columns of the matrix.

For example, matrix

$$\mathbf{M} = \begin{bmatrix} a & b \\ c & d \end{bmatrix}$$

The transpose of matrix $\mathbf{M}$ denoted $\mathbf{M^T}$ is thus

$$\mathbf{M}^{\mathrm{T}} = \begin{bmatrix} a & c \\ b & d \end{bmatrix}$$

As a simple example, the covariance matrix, P, of a state vector $\mathbf{X}$, comprising position, velocity and tilt errors, that is,

$$\mathbf{X} = \begin{bmatrix} \Delta P \\ \Delta V \\ \Delta \theta \end{bmatrix}$$

is thus

$$\mathbf{P} = \mathbf{X}\mathbf{X}^{\mathrm{T}} \tag{6.32}$$

$$\mathbf{P} = \begin{bmatrix} \Delta P \\ \Delta V \\ \Delta \theta \end{bmatrix} [\Delta P \ \Delta V \ \Delta \theta]$$

$$\mathbf{P} = \begin{bmatrix} \Delta P^2 & \Delta P \cdot \Delta V & \Delta P \cdot \Delta \theta \\ \Delta V \cdot \Delta P & \Delta V^2 & \Delta V \cdot \Delta \theta \\ \Delta \theta \cdot \Delta P & \Delta \theta \cdot \Delta V & \Delta \theta^2 \end{bmatrix}$$

It should be noted that the covariance matrix is symmetrical about a diagonal, with the diagonal elements comprising the mean square position, velocity and tilt errors, respectively. The off diagonal terms are cross-correlations between these same three quantities.

The covariance matrix changes with time as the initial errors propagate with time. It is modified with time by means of the transition matrix using the following relationship:

$$\mathbf{P}_{(n+1)} = \mathbf{\Phi}_n \cdot \mathbf{P}_n \cdot \mathbf{\Phi}_n^{\mathrm{T}} \tag{6.33}$$

($\mathbf{\Phi}_n^{\mathrm{T}}$ is the transpose of $\mathbf{\Phi}_n$).

The use of the Kalman filter equations is briefly set out below. The purpose of the Kalman filter is to provide an optimum estimate of the system state vector at iteration n. This is denoted by $\widehat{\mathbf{X}}_n$; the circumflex indicating a best estimate. The filter may be described in two stages. 'Extrapolation' indicates the period during which the filter simulates the action of the system between measurements. 'Update' occurs when a measurement is made on the system and is incorporated into the filter estimate. Quantities after extrapolation, immediately preceding the nth update are shown in the manner $\widehat{\mathbf{X}}_n(-)$, and those immediately succeeding that update as $\widehat{\mathbf{X}}_n(+)$. During extrapolation the filter estimate of the system state vector is modified according to the best available knowledge of the system dynamics

$$\widehat{\mathbf{X}}_n(-) = \mathbf{\Phi}_{n-1} \cdot \widehat{\mathbf{X}}_{n-1}(+) \tag{6.34}$$

The error covariance matrix is also extrapolated

$$\mathbf{P}_n(-) = Phi_{n-1} \cdot \mathbf{P}_{n-1}(+) \cdot \mathbf{\Phi}_{n-1}^{\mathrm{T}} + \mathbf{Q} \tag{6.35}$$

where $\mathbf{Q}$ is the covariance matrix of the random system disturbances.

During the filter update procedure, the difference between the actual measurement and a measurement made of the estimated state is weighted and used to modify the estimate state

$$\widehat{\mathbf{X}}_n(+) = \widehat{\mathbf{X}}_n(-) + \mathbf{K}_n \cdot \left[\mathbf{Z}_n - \mathbf{H}\widehat{\mathbf{X}}_n(-) \right] \tag{6.36}$$

$\mathbf{Z}_n$ is the actual measurement made of the state variables and $\mathbf{K}_n$ is the Kalman gain matrix.

The weighting factors, or Kalman gains, are calculated from the current estimate of the error covariance.

$$\mathbf{K}_n = \mathbf{P}_n(-) \cdot \mathbf{H}^{\mathrm{T}} \cdot \left[\mathbf{H} \cdot \mathbf{P}_n(-) \cdot \mathbf{H}^{\mathrm{T}} + \mathbf{R} \right] \tag{6.37}$$

where $\mathbf{R}$ is the measurement noise covariance matrix.

The error covariance matrix is then also updated

$$\mathbf{P}_n(+) = [\mathbf{1} - \mathbf{K}_n\mathbf{H}] \cdot \mathbf{P}_n(-) \tag{6.38}$$

6.4 Attitude Heading Reference Systems

6.4.1 Introduction

As explained in the introduction to this chapter, modern Attitude Heading Reference Systems are strap-down systems exploiting solid state gyros and accelerometers and are basically similar to modern strap-down IN systems. The major differences are in the accuracy of the inertial sensors and their consequent cost. There is no significant difference in reliability between the two systems as both exploit solid state implementation.

The key feature of Schuler tuning using very high accuracy gyros and accelerometers is the very high (and essential) vertical reference accuracy the system provides – around 0.3 minutes of arc for a good quality INS. This accuracy is not dependent on the aircraft's acceleration profile; a fundamental characteristic of Schuler tuning.

The lower limit in terms of gyro and accelerometer accuracies for Schuler tuning to be effective in providing an acceptable vertical reference accuracy in an AHRS system is

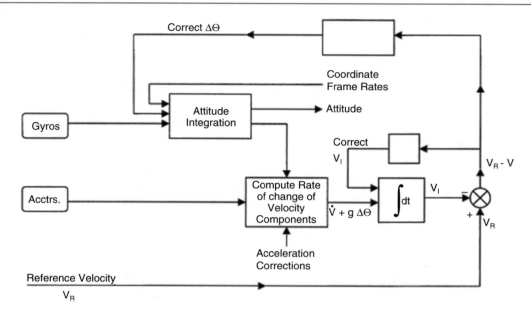

Fig. 6.24 Vertical monitoring using Doppler or air data/inertial velocity mixing

around 0.3°/hour gyro drift rate uncertainty and 500 µg (2 minutes of arc tilt) accelerometer bias uncertainty (Sect. 6.2.2, Figs. 6.12 and 6.13). It can be seen that such sensor errors would produce vertical errors of 4.2 minutes of arc and 2 minutes of arc, respectively, and the resulting stand-alone vertical accuracy would be around 0.1°.

An alternative method of monitoring the vertical reference which enables lower performance and hence lower cost gyros and accelerometers to be used in an AHRS is based on the use of an independent velocity source. The technique is known as *Doppler/inertial* or *air data/inertial* mixing depending on the source of the aircraft's velocity. The errors in the vertical reference resulting from the effects of accelerations during manoeuvres can be kept small with such a monitoring system, given an accurate velocity source. Small transient errors can be induced in the vertical reference due to changes in gyro drift rates or accelerometer bias errors. The gyros and accelerometers should thus be of reasonable accuracy although they can be up to two orders of magnitude below inertial standards.

The method uses the same techniques as used in IN systems to derive the aircraft velocity components from the AHRS accelerometers by integrating the suitably corrected accelerometer outputs with respect to time from known initial conditions.

The major source of error in the inertially derived velocity components arises from the vertical errors (or tilt errors) in the AHRS producing gravitational acceleration errors which are integrated with time. The inertially derived velocity components are, therefore, compared with the velocity components measured by the reference velocity system.

The velocity differences are then fed back to correct the vertical errors in the attitude reference and the inertial velocity errors. Figure 6.24 illustrates the basic concepts.

In the case of a Doppler velocity sensor, the Doppler and inertially derived velocities are compared along local NED axes, as the Doppler is a key part of the navigation system.

In the case of lower accuracy AHRS systems, the comparison of the air data and inertial velocity components is generally made along the aircraft body axes.

The mechanisation of an air data/inertial velocity mixing system to monitor the vertical reference of a strap-down AHRS is described below to show the following:

(a) The use of air data derived velocity as opposed to Doppler
(b) The use of aircraft body axes as a reference frame of axes
(c) The application to a lower accuracy AHRS with lower cost gyros in the few degrees/hour bias uncertainty performance bracket

The basic stages involved in the vertical monitoring system are set out below.

Derivation of velocity components from the air data system. The air data system provides outputs of true airspeed, V_T and the angle of attack/incidence, α, and the sideslip incidence angle, β.

The air data derived forward velocity, U_A, sideslip velocity, V_A, and vertical velocity, W_A, are computed from the air data system outputs V_T, α and β, the suffix A being used

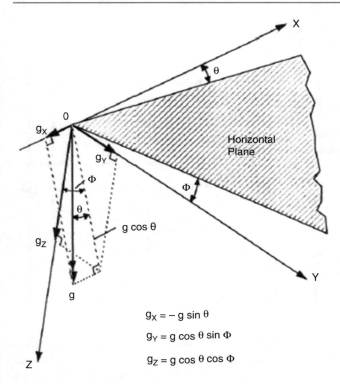

$$g_X = -g \sin \theta$$

$$g_Y = g \cos \theta \sin \Phi$$

$$g_Z = g \cos \theta \cos \Phi$$

Fig. 6.25 Resolution of gravity vector

henceforth to denote the source of U, V, W (i.e. $U = U_A$, $V = V_A$, $W = W_A$).

Inertial derivation of forward velocity, and side-slip velocity. The aircraft acceleration components along the body axes are derived from first principles in Chap. 3, Sect. 3.4.1.

Referring to Eqs. (3.13), (3.14), (3.15):

- Acceleration along forward axis, $OX = \dot{U} - Vr + Wq$.
- Acceleration along sideslip axis, $OY = \dot{V} + Ur - Wp$.
- Acceleration along vertical axis, $OZ = \dot{W} - Uq + Vp$.

Referring to Fig. 6.25 it can be seen that the gravitational acceleration components g_x, g_y, g_z along OX, OY, OZ, respectively, are

$$g_x = -g \sin q \quad (6.39)$$

$$g_y = g \cos q \sin F \quad (6.40)$$

$$g_z = g \cos q \cos F \quad (6.41)$$

The outputs of the forward, sideslip and vertical accelerometers are denoted by a_X, a_Y and a_Z, respectively

$$a_X = \dot{U} - Vr + Wq + g_x \quad (6.42)$$

$$a_Y = \dot{V} + Ur - Wp + g_y \quad (6.43)$$

$$a_Z = \dot{W} - Uq + Vp + g_z \quad (6.44)$$

The accelerometer bias errors are ignored at this stage for simplicity.

From Eqs. (6.42) and (6.43), it can be seen that $\dot{U}$ and $\dot{V}$ can be obtained by computing the centrifugal acceleration component terms $(-Vr + Wq)$ and $(Ur - Wp)$ and the gravitational acceleration component terms g_x and g_y and subtracting these computed components from the accelerometer outputs a_X and a_Y, respectively. The inertially derived velocity components U_I and V_I are then obtained by integrating $\dot{U}$ and $\dot{V}$ with respect to time.

The air data/inertial velocity mixing loops The overall air data/inertial velocity mixing loops are shown in Fig. 6.26. The computed centrifugal acceleration component terms $(-Vr + Wq)$ and $(Ur - Wp)$ are derived from the air data velocity components U_A, V_A, W_A and so are computed with respect to wind axes. The fact that the air mass is used as a reference frame and the wind axes are moving does not affect the values of these computed centrifugal acceleration terms provided the wind velocity is constant. Any acceleration of the reference frame due to changes in the wind velocity, however, can introduce errors. The assumption is made that the air mass accelerations generally appear as gusts of comparatively short duration which can be effectively smoothed by the air data/inertial mixing.

The gravitational acceleration components g_x and g_y are computed from the values of the pitch angle, θ and bank angle Φ, derived from the attitude computations using the incremental angular data from the strap-down gyros as explained earlier.

Integrating $\dot{U}$ and $\dot{V}$ with respect to time yields U_I and V_I and these values are subtracted from the air data derived values of U_A and V_A to yield $(U_A - U_I)$ and $(V_A - V_I)$.

The Coriolis acceleration components have been ignored for simplicity. (The air mass, of course, rotates with the Earth.) The major errors in deriving U_I and V_I are due to the tilt errors $\Delta\theta$ and $\Delta\Phi$ in deriving the pitch angle, θ and bank angle, Φ, from the gyro outputs, assuming the errors in the computed centrifugal accelerations are small. The accelerometer bias errors are also assumed to be small.

These tilt errors, $\Delta\theta$ and $\Delta\Phi$, result in acceleration errors $g\,\Delta\theta$ and $g\Delta\Phi$ (assuming $\Delta\theta$ and $\Delta\Phi$ are small angles) in the computation of g_x and g_y which are integrated with time causing a divergence between U_I and U_A and V_I and V_A unless corrected. The air data/inertial velocity errors $(U_A - U_I)$ and $(V_A - V_I)$ are thus fed back as shown in Fig. 6.26.

Fig. 6.26 Air data vertical
monitoring

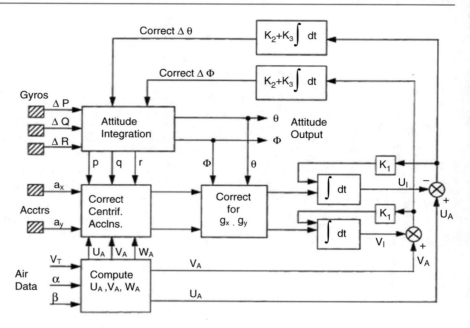

(a) To the U_I and V_I integrators to slave U_I to U_A and V_I to V_A.

(b) To the attitude computer to correct the pitch and bank angle tilt errors and reduce these to zero.

Proportional plus integral control, namely,

$$K_2 \,(\text{Velocity error}) + K_3 \int (\text{Velocity error})\, dt$$

is used in the feedback to the attitude computer to eliminate steady-state tilt errors due to gyro bias.

The values of the feedback gains, K_1, K_2 and K_3 are selected to ensure good damping of the closed loops and good overall accuracy. These loops are also used to align the AHRS when the aircraft is stationary on the ground as U_A and V_A are both zero. Any velocity errors result from initial tilt errors and so are fed back to correct the vertical reference. It should be noted that the aircraft is usually fairly level when on the ground so that the pitch and bank angles are generally small. The coarse values of pitch and bank angles used to initialise the attitude computation from the gyro incremental outputs can thus be set at zero and subsequently corrected by the alignment process using the accelerometer outputs.

The gyro derived attitude information is coupled in through a high pass filter with pass frequencies of the order of 0.003 Hz (period of 5.6 minutes approximately), or lower so that the short-term measurement of the attitude changes of the aircraft is provided mainly by the gyros. The attitude reference derived from the accelerometers is filtered by a low pass filter which smooths the noise present and attenuates short period fluctuations but retains the long-term accuracy of the gravitational reference.

It should be noted that there are considerable variations in the implementation and sophistication of the mixing systems used by strap-down AHRS manufacturers. The above description is a simplified outline only of the basic principles used; the corrections for Earth's rate and vehicle rate have also been omitted for clarity.

In the event of the loss of air data velocity, the AHRS reverts to a gravity monitoring system basically similar in principle to that of a simple standby vertical gyro unit (VGU). The computing power available, however, within the embedded processor in the solid state AHRS enables a more sophisticated levelling (or erection) loop with variable time constants to be implemented. The erection cut-out strategy during manoeuvres can also be made more adaptable.

6.4.2 Azimuth Monitoring Using a Magnetic Heading Reference

Gyro compassing is not suitable for an AHRS with lower accuracy gyros – alignment errors of 1° would result from using gyros with 0.1°/hour bias uncertainty. Typical AHRS gyros are in the 0.3°/hour to 5°/hour bias uncertainty bracket so that it can be seen that other methods must be used for initial alignment and subsequent monitoring of the aircraft heading. (The heading can be unmonitored for several hours if the gyro drift is less than 0.01°/hour.)

Magnetic monitoring using the Earth's magnetic field as the directional reference is generally used for the alignment and monitoring of the heading output of an AHRS using gyros which are not of inertial quality. This is a self-contained and simple system of relatively low cost and high reliability. Using modern magnetic sensors with computer compensation of the sensor errors together with an accurate vertical reference from the AHRS enables the heading errors to be constrained generally to less than 0.7° for latitudes up to about 60°. A magnetic heading reference is also used for coarse initial alignment of an INS with the fine alignment carried out by gyro compassing.

It should be noted that the Earth's magnetic poles are offset from the Earth's polar axis so that the direction of *magnetic North* differs from the direction of *true North*, the difference between the two is known as the *magnetic variation*. This is not constant and varies over the Earth's surface. Concentrations of iron ore in the Earth's crust, for instance, can cause significant changes (or 'magnetic anomalies') in the local magnetic variation. There are also long-term changes in the magnetic variation over several years. Magnetic variation has been accurately measured over large areas of the Earth's surface so that the magnetic heading can be corrected to give true heading by adding (or subtracting) the local magnetic variation, which can be stored within the system computer.

The presence of magnetic material on the aircraft causes local distortions in the Earth's magnetic field measured within the aircraft introducing a further error called 'magnetic deviation'. This is defined as the difference between the direction of the magnetic North measured in the aircraft and that at the same location in the absence of the aircraft. Extensive measurements are made with the aircraft on the ground to establish the magnitude of the errors due to the aircraft's internal magnetic environment.

The Earth's magnetic field is inclined at an angle to the horizontal in general. This angle is known as the *angle of dip* and is dependent on the latitude, being roughly 0° near the equator and 90° at the North and South magnetic poles. The magnetic heading is determined by measuring the direction of the horizontal component of the Earth's magnetic field. Errors in measuring the orientation of the magnetic sensors with respect to the vertical will result in a component of the vertical magnetic field being measured and will hence introduce an error in the magnetic heading. The vertical magnetic field component is equal to the horizontal magnetic field component multiplied by the tangent of the angle of dip and at high latitudes the vertical component can be several times greater than the horizontal component.

The magnetic heading error caused by a vertical error is thus 'geared up' by the tangent of the angle of dip. For example, an angle of dip of 68° (North America) results in a magnetic heading error which is 2.5 × *vertical error*. The importance of an accurate vertical reference can thus be seen. It should be noted that the magnetic heading reference becomes unusable at very high latitudes as the angle of dip approaches 90° and the horizontal component of the magnetic field approaches zero. The unmonitored gyro heading output must be relied upon for flight sectors over the polar regions.

The magnetic heading is measured by an instrument known as a *three axis fluxgate detector* in conjunction with an AHRS which provides the vertical reference for the fluxgate detectors. The system comprises three magnetic field sensors, or '*fluxgates*', which are mounted orthogonally in a block directly fixed to the airframe (i.e. 'strapped-down') with the sensor axes parallel to the principal axes of the aircraft. The location of the fluxgate triad is chosen to be as far away as practical from magnetic field sources (AC and DC) within the aircraft. Possible locations may be in one of the wings or in the fin, the small size and electrical output of the fluxgates giving a high degree of flexibility in the installation. The typical fluxgate detector consists of primary and secondary coils wound on a pair of soft iron cores. The primary is excited by an audio frequency sinusoidal current of sufficient amplitude to drive the cores into saturation at the current peaks. The secondary which is wound differentially about the two cores picks up the second harmonic content caused by the change in inductance at saturation. This output is approximately proportional to the magnetic field strength for applied static fields less than the saturation field of the cores. To improve the linearity of the output, the fluxgates are operated in a feedback mode. The output of the secondary is demodulated and fed back to the primary so that the net field in the cores is zero and the DC current in the primary is then a measure of the impressed magnetic field. The fluxgates, however, also measure any low frequency magnetic fields and not just the component of the Earth's magnetic field along the measuring axis of the fluxgate. The fluxgate outputs are thus in error from what is known as 'hard iron' effects due to unwanted AC magnetic fields and also from 'soft iron' effects that distort the Earth's magnetic field through the magnetic materials in the aircraft.

The components of the Earth's magnetic field with respect to the aircraft's principal axes are denoted H_X, H_Y and H_Z along the forward, sideslip and vertical axes, respectively (refer to Fig. 6.27).

These true components H_X, H_Y, H_Z of the Earth's magnetic field along each axis can be expressed in terms of a set of equations which relate the total measurements x_m, y_m, z_m of the three fluxgates. This is to account for axis misalignment and the 'hard iron' and 'soft iron' effects which are characteristic of the fluxgates and their environment in the aircraft.

This set of equations is of the form

$$H_X = a_1 x_m + b_1 y_m + c_1 z_m + d_1 \qquad (6.45)$$

$$H_Y = a_2 x_m + b_2 y_m + c_2 z_m + d_2 \qquad (6.46)$$

$$H_Z = a_3 x_m + b_3 y_m + c_3 z_m + d_3 \qquad (6.47)$$

Fig. 6.27 Components of the Earth's magnetic field

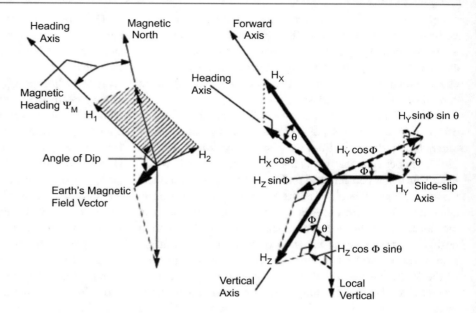

The coefficients a_1, a_2, a_3 and b_1, b_2, b_3 and c_1, c_2, c_3 and d_1, d_2, d_3 are measured and stored in the AHRS computer so that the fluxgate measurements can be corrected and the true components H_X, H_Y and H_Z of the Earth's magnetic field determined.

Referring to Fig. 6.27, the resolved components of the Earth's magnetic field in the horizontal plane along the heading axis, H_1, and at right angles to the heading axis, H_2, are given by

$$H_1 = H_X \cos \theta + H_Y \sin \Phi \ \sin \theta$$
$$+ H_Z \cos \Phi \ \sin \theta \qquad (6.48)$$

$$H_2 = H_Y \cos \ \Phi - H_Z \sin \ \Phi \qquad (6.49)$$

The horizontal component of the Earth's magnetic field, H_H, is therefore equal to

$$\sqrt{H_1^2 + H_2^2}$$

The aircraft's magnetic heading, Ψ_M, is thus given by

$$\Psi_M = \cos^{-1} \frac{H_1}{H_H} \quad \text{or} \quad \sin^{-1} \frac{H_2}{H_H} \qquad (6.50)$$

The computational processes in deriving the magnetic heading, Ψ_M, are shown in Fig. 6.27 and comprise the following:

1. Computation of the true components H_X, H_Y, H_Z of the Earth's magnetic field with respect to aircraft axes from the measured outputs of the three fluxgates x_m, y_m, z_m and the stored coefficients a_1, a_2, a_3, b_1, b_2, b_3, c_1, c_2, c_3, d_1, d_2, d_3 using Eqs. (6.45), (6.46) and (6.47).

2. Computation of the horizontal components of the Earth's magnetic field along the heading axis, H_1, and at right angles to the heading axis, H_2, using Eqs. (6.48) and (6.49).

3. Computation of the magnetic heading angle, Ψ_M, from

$$\Psi_M = \cos^{-1} \frac{H_1}{\sqrt{H_1^2 + H_2^2}} \quad \text{or} \quad \sin^{-1} \frac{H_2}{\sqrt{H_1^2 + H_2^2}}$$

The gyro/magnetic heading monitoring system is illustrated in Fig. 6.28, a basic second-order mixing system being shown for simplicity. (More complex mixing systems using a Kalman filter can be employed.) Referring to Fig. 6.28:

Ψ_G = Gyro derived heading from the strap-down AHRS

$$\Psi_G = \Psi + \int W \, dt$$

where Ψ is the true heading angle and W is gyro drift rate; $\Psi_{G/M}$ is the gyro/magnetic heading; K_1 is the mixing gain for $(\Psi_M - \Psi_{G/M})$ feedback and K_2 is the mixing gain for $\int (\Psi_M - \Psi_{G/M}) dt$ feedback.

From the block diagram it can be seen that

$$\Psi_{G/M} = \Psi_G + K_1 \int \left(\Psi_M - \Psi_{G/M} \right) dt$$

$$+ K_2 \ \int \int \left(\Psi_M - \Psi_{G/M} \right) dt \, dt \qquad (6.51)$$

Differentiating Eq. (6.51) twice and substituting $\dot{\Psi} + W$ for $\dot{\Psi}_G$ gives

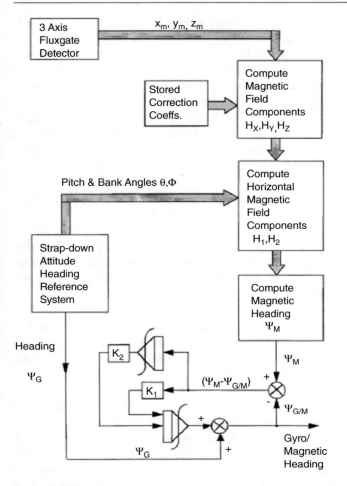

Fig. 6.28 Magnetic monitoring of gyro heading

$$\left(D^2 + K_1 D + K_2\right)\Psi_{G/M} = (K_2 + K_1 D)\Psi_M + D^2\Psi + DW \tag{6.52}$$

The dynamic response of the system is that of a simple second-order system with an undamped natural frequency, $\omega_0 = \sqrt{K_2}$ and damping ratio, $\zeta = K_1/2\sqrt{K_2}$. The integral feedback gain, K_2, determines the undamped natural frequency and the proportional feedback gain, K_1 determines the damping. K_1 and K_2 are determined by the value of the gyro bias uncertainty, W, and the need to have a well damped closed-loop response with ζ equal to or near to 1.

From Eq. (6.52):

$$\Psi_{G/M} = \frac{\omega_0^2\left(1 + \frac{2\zeta}{\omega_0}D\right)}{\left(D^2 + 2\zeta\,\omega_0 D + \omega_0^2\right)}\cdot\Psi_M + \frac{D^2}{\left(D^2 + 2\zeta\,\omega_0 + \omega_0^2\right)}\cdot\Psi$$
$$+ \frac{D^2}{\left(D^2 + 2\zeta\,\omega_0 D + \omega_0^2\right)}\cdot W \tag{6.53}$$

Ignoring the gyro bias uncertainty term W, as it has zero contribution to $\Psi_{G/M}$ if W is constant.

$$\Psi_{G/M} = F_1(D)\Psi_M + F_2(D)\Psi \tag{6.54}$$

where

$$F_1(D) = \frac{\omega_0^2\left(1 + \frac{2\zeta}{\omega_0}D\right)}{\left(D^2 + 2\zeta\omega_0 D + \omega_0^2\right)} \text{ and } F_2(D)$$
$$= \frac{D^2}{\left(D^2 + 2\zeta\omega_0 D + {}_0^2\right)}$$

The magnetic heading, Ψ_M, component of $\Psi_{G/M}$ is coupled through a low pass filter of transfer function $F_1(D)$. This greatly attenuates the magnetic noise content of the Ψ_M output and short-term transient errors such as would be caused by flexure of the structure at the fluxgate sensor location under manoeuvre loads.

The gyro measured component of $\Psi_{G/M}$ is coupled through a high pass, or, 'wash-out' filter of transfer function $F_2(D)$.

This enables the changes in heading to be measured without lag but ensures that the steady gyro bias, W, is 'DC blocked' (or 'washed-out').

The complementary filtering thus enables the 'best of both worlds' to be achieved. The basic accuracy and repeatability of the magnetic heading sensor is retained but any noise and transient errors during manoeuvres are heavily smoothed and filtered. The excellent dynamic response of the gyro system is retained and the gyro bias is DC blocked. There are also no steady-state magnetic heading errors due to gyro bias.

Figure 6.29 illustrates the system response to a constant rate turn (assuming a near critically damped system).

6.5 GPS: Global Positioning System

6.5.1 Introduction

GPS is basically a radio navigation system which derives the user's position from the radio signals transmitted from a number of orbiting satellites.

The fundamental difference between GPS and earlier radio navigation systems, such as LORAN-C (now no longer in use), is simply the geometry of propagation from ground-based transmitters compared with space borne transmitters. An orbiting satellite transmitter can provide line of sight propagation over vast areas of the world. This avoids the inevitable trade-offs of less accuracy for greater range which are inherent with systems using ground-based

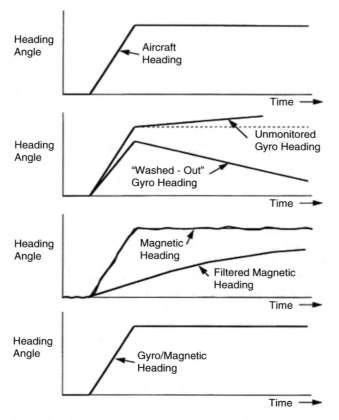

Fig. 6.29 Complementary filtered gyro/magnetic heading response to constant rate of turn

transmitters. The satellite signals also penetrate the ionosphere rather than being reflected by it so that the difficulties encountered with sky waves are avoided.

GPS provides a superior navigation capability to all previous radio navigation systems. For these reasons and also space constraints, coverage of radio navigation systems has been confined to GPS.

Satellite navigation can be said to have started with the successful launching by the Russians of the world's first orbiting satellite, SPUTNIK 1 in October 1957. The development of the first satellite navigation system TRANSIT 1 was triggered by observations made on the radio signals transmitted from SPUTNIK 1 and was initiated at the end of 1958. TRANSIT 1 resulted in a worldwide navigation system which has been in continuous operation since 1964.

GPS started as a series of preliminary system concept studies and system design studies in the late 1960s. A Phase 1 'Concept and Validation Programme' was carried out from 1973 to 1979 followed by a Phase 2 'Full Scale Development and System Test Programme' from 1979 to 1985. The Phase 3 'Production and Deployment Programme' was initiated in 1985. Twelve development satellites were used to develop and prove the system and the first production satellite was launched in February 1989. Some delay to the program was incurred by the space shuttle CHALLENGER disaster in 1986, as it had been originally intended to insert all the production standard GPS satellites into orbit using the space shuttle. A DELTA 2 launch vehicle was used subsequently and deployment of the 24 production standard GPS satellites was completed in the late 1990s.

A very similar satellite navigation system called GLONASS has been developed by the Russians. Although designed for military applications (like GPS), it is available but is not being used elsewhere at the present time (2010).

6.5.2 GPS System Description

The overall GPS system comprises three segments, namely, the space segment, the control segment and the user segment and is shown schematically in Fig. 6.30. The three segments are briefly summarised below.

Space Segment This comprises 24 GPS satellites placed in six orbital planes at 55° to the equator in geo-synchronous orbits at 20,000 km above the Earth. The orbit tracks over the Earth, forming an 'egg beater' type pattern.

Twenty-one satellites are required for full worldwide coverage and three satellites act as orbiting spares.

The GPS satellites use two frequency transmissions, L1 at 1575.42 MHz and L2 at 1227.6 MHz for transmitting the digitally encoded navigation message data at 50 Hz modulation on both the L1 and L2 channels. The navigation message data will be explained in more detail in the next section but basically comprises the satellite orbital position parameters, clock correction parameters and health information for itself and the other satellites, and the almanac data for all the satellites.

Spread spectrum techniques are used on both the L1 and L2 frequency channels. The L1 carrier is modulated by a 1.023 MHz clock rate pseudo-random code known as the Coarse/Acquisition (C/A) code; a different C/A code is assigned to each satellite. A quadrature carrier component of the L1 signal is modulated by the Precise (P) code which uses ten times the clock rate of the C/A code.

The L2 transmission is modulated by the P code only and enables corrections to be made for ionospheric delay uncertainties, the dual frequency transmission enabling these corrections to be derived.

It should be noted that, until fairly recently, the GPS accuracy available to civil users was deliberately degraded to 100 m. The full accuracy of 16 m could only be obtained by military users with access to the P code on the L2 transmission, which was encrypted. This restriction was removed in 2000.

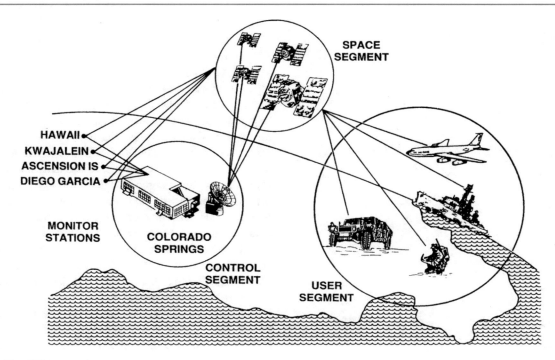

Fig. 6.30 The GPS system

Control Segment This comprises a Master Control Station at Colorado Springs in the USA and five monitor stations located worldwide. The control segment is operated by the United States Department of Defense (DoD). The control segment tracks the satellites and predicts their future orbital position data and the required satellite clock correction parameters, and updates each satellite on the uplink as it goes overhead.

The GPS full system accuracy is only available when the operational control system is functioning properly and navigation messages are uploaded on a daily basis. The GPS satellites are, however, designed to function with the control system inoperable for a period of 180 days with gradually degraded accuracy. This gives the GPS system a high degree of robustness.

User Segment The user segment equipment as mentioned earlier is entirely passive and comprises a GPS receiver. A very wide variety of compact, light weight and inexpensive GPS receivers are now available, all using the same basic concepts.

The user system operation is very briefly as follows. The operator first enters the estimated present position and the time. The GPS receiver then starts to search for and track satellites. The data coming in identifies the satellite number, locates the satellite in space and establishes the system time. As will be explained in the next section the GPS receiver needs to track the signals from at least four satellites to determine the user's position.

As mentioned in the introduction to this chapter, the user's 3D position is determined to an accuracy of 16 m RMS, 3D velocity to 0.1 m/s RMS by measuring the Doppler shifts, and time to within 100 ns (1 sigma).

6.5.3   Basic Principles of GPS

The basic principle of position determination using the GPS system is to measure the spherical ranges of the user from a minimum of four GPS satellites. The orbital positions of these satellites relative to the Earth are known to extremely high accuracy and each satellite transmits its orbital position data.

Each satellite transmits a signal which is modulated with the C/A pseudo-random code in a manner which allows the time of transmission to be recovered.

The spherical range of the user from the individual transmitting satellite can be determined by measuring the time delay for the satellite transmission to reach the user. Multiplying the time delay by the velocity of light then gives the spherical range, R, of the user from the transmitting satellite. The user's position hence lies on the surface of a sphere of radius, R, as shown in Fig. 6.31.

The system depends on precise time measurements and requires atomic clock reference standards. The need for extremely high accuracy in the time measurement can be

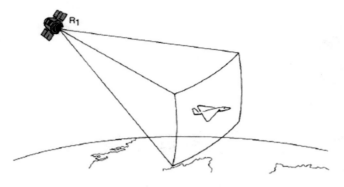

Fig. 6.31 GPS spherical ranging

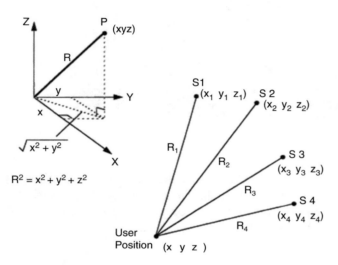

Fig. 6.32 User–satellite geometry

Fig. 6.33 GPS satellite waveforms with perfect satellite clocks

seen from the fact that a 10 ns (10^{-8} seconds) time error results in a distance error of 3 m, as the velocity of light is 3×10^8 m/s.

Each GPS satellite carries an atomic clock which provides the time reference for the satellite data transmission. Assume for the moment that this time is perfect – the corrections required will be explained shortly. Given a perfect time reference in the user equipment, measurement of the spherical ranges of three satellites would be sufficient to determine the user's position. The user's equipment, however, has a crystal clock time reference which introduces a time bias in the measurement of the transit times of the satellite transmissions. The measurement of the time delay is thus made up of two components. The first component is the transit time of the ranging signal and the second component is the time offset between the transmitter clock and the receiver clock due to the non-synchronisation of the clocks.

Measuring the spherical ranges from four satellites as shown in Fig. 6.32 enables the user's position to be determined and yields four equations containing the four unknowns, namely, the three position co-ordinates of the user and the time bias in the user's clock. The position co-ordinates of the user can thus be determined together with very accurate time information. Figure 6.33 shows the data transmission waveforms and illustrates the user time bias ΔT, and the time delays Δ_{t1}, Δ_{t2}, Δ_{t3} and Δ_{t4} for the signals transmitted from the satellites to reach the user.

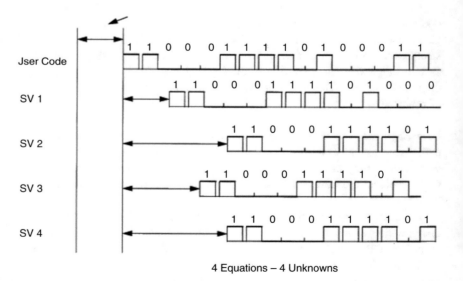

4 Equations – 4 Unknowns

Four pseudo ranges R_{1_p}, R_{2_p}, R_{3_p}, R_{4_p} to the four satellites S1, S2, S3, S4 can be determined, namely,

$$R_{1_p} = c\Delta t_1$$
$$R_{2_p} = c\Delta t_2$$
$$R_{3_p} = c\Delta t_3 \qquad (6.55)$$
$$R_{4_p} = c\Delta t_4$$

Let the range equivalent of the user's clock offset be T, that is,

$$T = c_\Delta T$$

Hence, from basic 3D co-ordinate geometry (see Fig. 6.31)

$$R_1 = \left[(X - X_1)^2 + (Y - Y_1)^2 + (Z - Z_1)^2\right]^{1/2}$$
$$= R_{1_p} - T \qquad (6.56)$$

$$R_2 = \left[(X - X_2)^2 + (Y - Y_2)^2 + (Z - Z_2)^2\right]^{1/2}$$
$$= R_{2_p} - T \qquad (6.57)$$

$$R_3 = \left[(X - X_3)^2 + (Y - Y_3)^2 + (Z - Z_3)^2\right]^{1/2}$$
$$= R_{3_p} - T \qquad (6.58)$$

$$R_4 = \left[(X - X_4)^2 + (Y - Y_4)^2 + (Z - Z_4)^2\right]^{1/2}$$
$$= R_{4_p} - T \qquad (6.59)$$

when R_1, R_2, R_3, R_4 are the actual ranges from the user's position to the four satellites S1, S2, S3, S4 and the co-ordinates of these satellites are $(X_1\ Y_1\ Z_1)$, $(X_2\ Y_2\ Z_2)$, $(X_3\ Y_3\ Z_3)$, $(X_4\ Y_4\ Z_4)$, respectively.

These four equations with four unknowns can thus be solved and yield the user's position co-ordinates (X, Y, Z) and the user's time offset, ΔT.

The assumption of perfect satellite clocks made initially in the discussion, however, is not a valid one and in fact the clocks are slowly but steadily drifting away from each other. The satellite clocks are, therefore, mathematically synchronised to a defined GPS Master time which is maintained at the Master Control Station. This GPS Master time is continuously monitored and related to the Universal Time Co-ordinate (UTC) maintained by the United States Naval Observatory.

Each satellite time is related to GPS time by a mathematical expression and the user corrects the satellite time to the GPS time using the equation

$$t = t_{s/c} - \Delta t_{s/c} \qquad (6.60)$$

where t is the GPS time in seconds, $t_{s/c}$ is the effective satellite time at signal transmission in seconds and $\Delta t_{s/c}$ is the time offset between the satellite and GPS master time.

The time offset is $\Delta t_{s/c}$ computed from the following equation:

$$\Delta t_{s/c} = a_0 + a_1\left(t - t_{o/c}\right) + a_2\left(t - t_{o/c}\right)2 + \Delta t_r \qquad (6.61)$$

where a_0, a_1 and a_2 are polynomial coefficients representing the phase offset, frequency offset and ageing term of the satellite clock with respect to the GPS master time and Δt_r is the relativistic term (seconds). The parameter t is the GPS time and $t_{o/c}$ is the *epoch time* at which the polynomial coefficients are referenced and generally $t_{o/c}$ is chosen at the mid-point of the fit interval. The polynomial coefficients a_0, a_1, a_2 are estimated by the control segment for each satellite clock and periodically uplinked to the satellite.

These coefficients are transmitted together with the satellite orbital position data, termed the *Ephemeris parameters*, to the navigation user equipment as navigation messages. The clock corrections for the four satellites are designated τ_1, τ_2, τ_3 and τ_4.

The spherical ranges R_1, R_2, R_3 and R_4 are thus given by

$$R_1 = c(\Delta t_1 + \Delta T - \tau_1)$$
$$R_2 = c(\Delta t_2 + \Delta T - \tau_2)$$
$$R_3 = c(\Delta t_3 + \Delta T - \tau_3) \qquad (6.62)$$
$$R_4 = c(\Delta t_4 + \Delta T - \tau_4)$$

All satellite clocks are mathematically synchronised to the GPS master time by means of these clock correction terms. The error in synchronisation will grow, however, if the polynomial coefficients a_0, a_1 and a_2 are not updated periodically.

The navigation user requires the Ephemeris parameters, that is, the instantaneous position data of the GPS satellites which are being used for range measurement, as well as the clock parameters in order to compute the user's position. The Ephemeris parameters defining the satellite orbital position data with respect to Earth reference axes comprise 16 parameters in all. Figure 6.34 illustrates the satellite–Earth geometry and the definition of the orbit parameters. The control segment processes the tracking data acquired from the monitor stations to generate the orbit estimates for the GPS satellites. The predicted estimates of the satellite position co-ordinates are generated by integrating the equations of motion of the GPS satellites. These Cartesian position co-ordinates are then fitted mathematically over a specified interval of time to compute the Ephemeris parameters.

Fig. 6.34 Satellite-Earth
geometry. Ephemeris parameter
definition

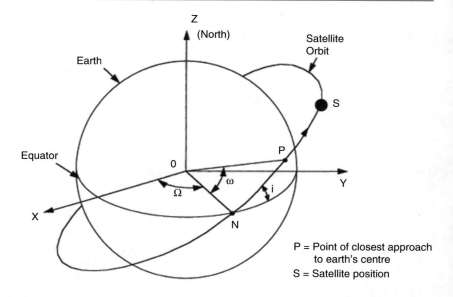

P = Point of closest approach
to earth's centre
S = Satellite position

Both clock and Ephemeris parameters are down linked to the user at 50 bits per second data rate modulated in both C/A and P code (Y code) navigation signals. The navigation message uses a basic format consisting of a 1500 bit long frame made up of five sub-frames, each sub-frame being 300 bits long. Sub-frames 4 and 5 are sub-commuted 25 times each so that a compete data message takes a transmission of 25 full frames. Sub-frame 1 contains the clock parameters and sub-frames 2 and 3 the Ephemeris parameters for the satellite. Sub-frames 1, 2 and 3 are repeated every 30 seconds so that it is possible for the user to update the clock and Ephemeris parameters every 30 seconds. Sub-frames 4 and 5 have each 25 pages so that these sub-frames are repeated only once in 12.5 minutes.

In addition to the Ephemeris and clock parameters of the satellite, each satellite transmits almanac data of all satellites to the user. This is primarily to facilitate satellite acquisition and to compute what is known as the *geometric dilution of precision* (GDOP) values to assist the selection of satellites to achieve better accuracy. The optimum geometry occurs when one satellite is at the user's zenith (directly overhead) and at least three other satellites are evenly spaced around the user's horizon. Conversely, a large error would occur if the satellites were clustered together, so that their lines of sight tended towards being parallel. Health data for all the satellites is also transmitted. The almanac data and satellite health data are contained in sub-frames 4 and 5. Ionospheric data for a single frequency user and the conversion parameters from GPS time to UTC are contained in sub-frame 4.

The navigation equations are basically non-linear as can be seen in Eqs. (6.56), (6.57), (6.58) and (6.59), but can be linearised about nominal values for their solution by applying Taylor's series approximations.

6.5.4 Solution of Navigation Equations

Let X_u, Y_u, Z_u denote the nominal (a priori best estimate) values of the user's co-ordinates X, Y, Z and clock offset range equivalent, T.

Let ΔX, ΔY, ΔZ and δT denote the corrections which need to be to these nominal values, that is,

$$X = X_u + \Delta X; \quad Y = Y_u + \Delta Y; \quad Z = Z_u + \Delta Z; \quad T = T_u + \delta t$$

The actual range from the ith satellite ($i = 1, 2, 3, 4 ...$) is given by

$$R_i = \left[(X - X_i)^2 + (Y - Y_i)^2 + (Z - Z_i)^2 \right]^{1/2} \quad (6.63)$$

where X_i, Y_i, Z_i are the position co-ordinates of the ith satellite.

Applying Taylor's series expansion about the nominal values

$$R_i \approx \left[(X_u - X_i)^2 + (Y_u - Y_i)^2 + (Z_u - Z_i)^2 \right]^{1/2} + h_{iX}\Delta X + h_{iY}\Delta Y + h_{iZ}\Delta Z \quad (6.64)$$

where

$$h_{iX} = \frac{\partial R_i}{\partial X}; h_{iY} = \frac{\partial R_i}{\partial Y}; \; h_{iZ} = \frac{\partial R_i}{\partial Z}$$

evaluated at $X = X_u$, $Y = Y_u$, $Z = Z_u$.

Partially differentiating Eq. (6.63) with respect to X gives

$$\frac{\partial R_i}{\partial X}$$

$$= \frac{1}{2}\left[(X-X_i)^2 + (Y-Y_i)^2 + (Z-Z_i)^2\right]^{-1/2}.2(X-X_i) \tag{6.65}$$

whence

$$h_{iX} = (X_u - X_i)/R_{iu}$$
$$h_{iY} = (Y_u - Y_i)/R_{iu} \tag{6.66}$$
$$h_{iZ} = (Z_u - Z_i)/R_{iu}$$

where

$$R_{iu} = \left[(X_u - X_i)^2 + (Y_u - Y_i)^2 + (Z_u - Z_i)^2\right]^{1/2} \tag{6.67}$$

whence

$$R_i = R_{iu} + h_{iX}\Delta X + h_{iY}\Delta Y + h_{iZ}\Delta Z \tag{6.68}$$

The range error of the ith satellite, ΔR_I, is given by

$$\Delta R_i = R_{ip} - T - R_i \tag{6.69}$$

whence substituting $T = T_u + \delta T$

$$R_{ip} - T_u - R_{iu} = \Delta R_i = h_{iX}\Delta X + h_{iY}\Delta Y + h_{iZ}\Delta Z + \delta T \tag{6.70}$$

whence

$$\begin{bmatrix} \Delta R_1 \\ \Delta R_2 \\ \Delta R_3 \\ \Delta R_4 \end{bmatrix} = \begin{bmatrix} h_{1X} & h_{1Y} & h_{1Z} & 1 \\ h_{2X} & h_{2Y} & h_{2Z} & 1 \\ h_{3X} & h_{3Y} & h_{3Z} & 1 \\ h_{4X} & h_{4Y} & h_{4Z} & 1 \end{bmatrix}\begin{bmatrix} \Delta X \\ \Delta Y \\ \Delta Z \\ \delta T \end{bmatrix} \tag{6.71}$$

This linear matrix equation is solved to yield ΔX, ΔY, ΔZ, δT and these values are used to correct the nominal values of the user's position and the clock offset. The process is repeated until ΔX, ΔY, ΔZ, δT become negligible.

It should be noted that the measurement uncertainties and errors have been left out of the equations at this stage for simplicity.

A sequential processing of the measurements is used in dynamic user applications because better estimates can be obtained by utilising the previous estimates and their associated uncertainties. Kalman filters are often used. The matrix of the partial derivatives h_{iX}, h_{iY}, h_{iZ} is generally referred to as the **H** matrix. It should be noted that h_{iX}, h_{iY}, h_{iZ} are the direction cosines of the ith satellite range vector and the **H** matrix can be used to assess the GDOP (Geometric Dilution of Precision).

6.5.5 Integration of GPS and INS

GPS and INS are wholly complementary and their information can be combined to the mutual benefit of both systems. For example:

- Calibration and correction of INS errors – the GPS enables very accurate calibration and correction of the INS errors in flight by means of a Kalman filter.
- The INS can smooth out the step change in the GPS position output which can occur when switching to another satellite because of the change in inherent errors.
- Jamming resistance – like any radio system, GPS can be jammed, albeit over a local area, although it can be given a high degree of resistance to jamming. The INS, having had its errors previously corrected by the Kalman filter, is able to provide accurate navigation information when the aircraft is flying over areas subjected to severe jamming.
- Antenna obscuration – GPS is a line of sight system and it is possible for the GPS antenna to be obstructed by the terrain or aircraft structure during manoeuvres.
- Antenna location corrections – the GPS derived position is valid at the antenna and needs to be corrected for reference to the INS location. The INS provides attitude information which together with the lever arm constants enables this correction to be made.

6.5.6 Differential GPS

6.5.6.1 Introduction
As explained in the preceding section the horizontal position accuracy available to all GPS users (civil and military) is now 16 m. This was not the case, however, until 2000 when the restriction of 'Selective Availability' was removed.

Concerns about potential enemies using GPS to deliver missiles and other weapons against the USA had led to a policy of accuracy denial, generally known as Selective Availability. The GPS ground stations deliberately introduced satellite timing errors to reduce the positioning accuracy available to civil users to a horizontal positioning accuracy of 100 m to a 95% probability level. This was deemed adequate for general navigation use, but in practice it did not satisfy the accuracy or integrity requirements for land or hydrographic surveying, coastal navigation or airborne navigation. It should be noted that even the 16 m accuracy, now available, is insufficiently accurate for many applications. For example, positioning of off-shore oil drilling rigs or automatic landing in the case of airborne applications.

A supplementary navigation method known as *Differential GPS* (DGPS) has therefore been developed to improve the positioning accuracy for the growing number of civil applications.

DGPS can be defined as:

> The positioning of a mobile station in real-time by corrected (and possibly Doppler or phase smoothed) GPS pseudo ranges. The corrections are determined at a static 'reference station' and transmitted to the mobile station. A monitor station may be part of the system, as a quality check on the reference station transmissions.

The success of DGPS can be seen from its application to new markets such as locating land vehicles used by the emergency services. Successful trials for automatic landings and taxiway guidance have also been conducted. It is now widely used in land and hydrographic surveying applications.

6.5.6.2 Basic Principles

The basic principle underlying DGPS is the fact that the errors experienced by two receivers simultaneously tracking a satellite at two locations fairly close to each other will largely be common to both receivers.

The basic differential GPS concept is illustrated in Fig. 6.35. The position of the stationary GPS Reference Station is known to very high accuracy so that the satellite ranges can be very accurately determined, knowing the satellite ephemeris data. The errors in the pseudo-range measurements can then be derived and the required corrections computed and transmitted to the user's receiver over a radio link. The errors present in a GPS system are illustrated schematically in Fig. 6.36 and are briefly discussed below.

GPS satellite clocks GPS satellites are equipped with very accurate atomic clocks and corrections are made via the Ground Stations, as explained in the preceding section. Even so, very small timing errors are present and so contribute to the overall position uncertainty.

Selective Availability deliberately introduced noise equivalent to around 30 m in the individual satellite clock signals.

Satellite ephemeris errors The satellite position is the starting point for all the positioning computations, so that errors in the Ephemeris data directly affect the system accuracy. GPS satellites are injected into very high orbits and so are relatively free from the perturbing effects of the Earth's upper atmosphere. Even so, they still drift slightly from their predicted orbits and so contribute to the system error.

Atmospheric errors Radio waves slow down slightly from the speed of light in vacuo as they travel through the ionosphere and the Earth's atmosphere. This is due to the charged

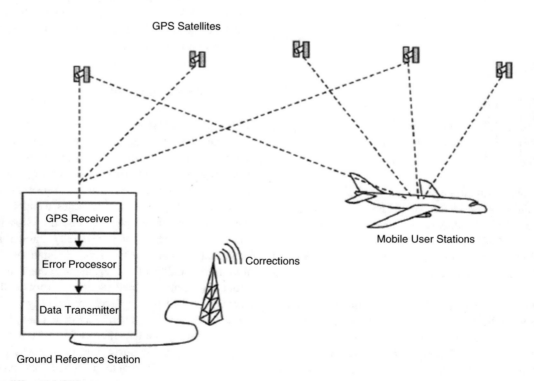

Fig. 6.35 The differential GPS concept

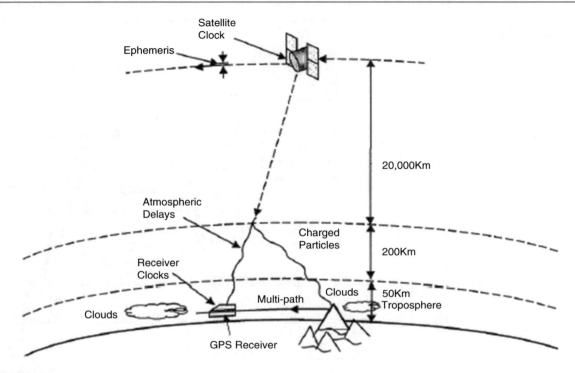

Fig. 6.36   GPS error sources

particles in the ionosphere and the water vapour and neutral gases present in the troposphere. These delays translate directly into a position error.

The use of different frequencies in the L1 and L2 transmissions enables a significant correction to be made for ionospheric delays. (It should be appreciated that this facility was not available to civil users prior to 2000.)

A predicted correction factor for the Earth's atmosphere path is made in the receiver but this is based on a statistical model and there are inevitable residual errors present.

Multi-path errors   The GPS satellite signal is received by the direct line of sight (LOS) path, but the signal may also be received as the result of reflections off local obstructions. The reflected signals arrive slightly delayed from the direct LOS signal and are termed multi-path signals. The resulting noise is called multi-path error. The multi-path errors experienced in a moving receiver occur in a random fashion which results in a noise like pseudo-range error. The problem is considerably alleviated by the use of an early-late delay-lock loop and suitable filtering techniques.

Receiver clock   Internal noise in the GPS receiver clock introduces a small error.

DGPS enables most of the above errors to be counteracted, as they are common to both the Reference Station receiver and the user receiver. The exceptions are

Table 6.3   Summary of GPS error sources

Typical error budget (in metres)		
Per satellite accuracy	Standard GPS	Differential GPS
Satellite clocks	1.5	0
Orbit errors	2.5	0
Ionosphere	5.0	0.4
Troposphere	0.5	0.2
Selective availability[a]	30.0	0
Receiver noise	0.3	0.3
Multi-path (reflections)	0.6	0.6
Typical positioning accuracy		
Horizontal	50.0	1.3
Vertical	78.0	2.0
3D	93.0	2.8

[a]*Note*: Selective Availability error is shown to demonstrate the effectiveness of the DGPS technique, although the Selective Availability restriction has now been removed

the multi-path and receiver errors as these are strictly local phenomena.

The dramatic improvement in accuracy effected by DGPS can be seen in Table 6.3.

Separation distances between the Ground Reference Station and the mobile user can be up to 300–500 km and separations of 1000 km or beyond are not unknown.

Figure 6.37 is a simplified functional diagram of a generic differential GPS system.

Radio links can be realised in one of a number of forms to provide a suitable narrow band communication channel to the

Fig. 6.37 Simplified functional diagram of a generic differential GPS system

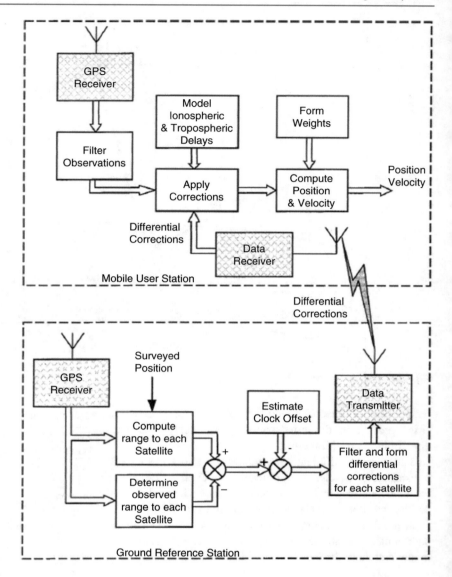

mobile receiver. Dedicated radio channels within the MF, HF, VHF or UHF bands are used. Frequency diversity and error checking codes are used to provide protection against data corruption caused by the vagaries of propagation.

The satellite pseudo-range measurement in the user's GPS receiver is carried out by the correlation of the received satellite signal with the receiver generated replica of the known satellite C/A code using a code tracking loop. The accuracy of the time measurement by the C/A code tracking loop is inevitably ultimately limited by noise sources such as multi-path reception.

GPS receivers also employ carrier tracking loops to monitor the Doppler shifted carrier component of the satellite navigation signals. The Doppler shift is proportional to the radial velocity of the receiver relative to the satellite. The velocity of the user vehicle is computed from these Doppler shifts.

The Ground Reference GPS Station in the DGPS system uses these Doppler frequency shift measurements to improve the accuracy of the pseudo-range measurements. Integrating the Doppler shift over an interval provides a measurement of the change in satellite-to-receiver range. The range itself cannot be measured because the integer number of carrier cycles is not known. The information extracted from a carrier tracking loop can take the form of a Doppler observation or a continuously integrated Doppler observation more commonly called a *carrier phase* observation. The multi-path and noise induced errors found in carrier phase observations are negligible compared to those on C/A code observations. The accuracy improvement is commensurate with the rates of the L1 code wavelength and carrier wavelength, 290 m and 0.19 m, respectively. Carrier phase filtering can be carried out using Kalman filtering techniques.

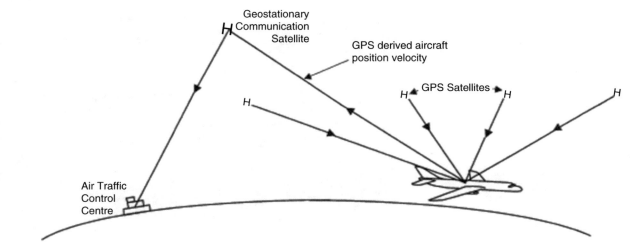

Fig. 6.38 Remote air traffic control

6.5.7   Future Augmented Satellite Navigation Systems

The advent of satellite navigation systems and satellite communication links has provided new capabilities for aircraft precision navigation, particularly in civil operations.

Providing the integrity and accuracy requirements can be met, satellite navigation systems are able to support all phases of flight including all-weather precision approaches to airports not equipped with ILS (or MLS) installations.

In conjunction with satellite communication links, they can also prove the capability for remote air traffic control as shown in Fig. 6.38.

Successful concept proving trials were, in fact, conducted by the UK Air Traffic Control authorities in conjunction with British Airways around 1996. The trials used the on-board GPS receivers and SAT COM radios in a British Airways Boeing 747 airliner to monitor the aircraft flight paths on normal commercial flights to the West Indies. Detail changes in the aircraft flight path over the West Indies were accurately monitored from the UK, over 3000 miles away. The potential for more flexible air traffic control systems can be seen and it is only a matter of time before they are introduced.

Some reservations exist as to whether the integrity of the present GPS systems is sufficiently high to meet the navigation integrity requirements in safety critical phases of the flight and adverse weather conditions. Although the probability of GPS receivers producing erroneous position data is very low, there have been recorded instances of erroneous GPS position data in flight.

There have also been reservations about total reliance on GPS, as it is a military system which is completely under the control of the US military command, although it is freely available to any user. The accuracy available to civil users during the 1990s was also limited to 100 m by the policy of Selective Availability, as already explained, and this was inadequate for precision approaches.

An augmented satellite navigation system provided by additional satellites under international civil control was therefore proposed and studied in detail by the European civil authorities from the late 1990s. The additional ranging signals and monitoring will enable the integrity requirements to be met and will also provide increased accuracy.

The go-ahead for the new system, known as the Galileo system, was announced by the participating countries in the European Union in March 2002. The Galileo system will be inter-operable with GPS from a user perspective and will comprise up to 40 orbiting satellites. The system will provide a positional accuracy of the order of 1 m worldwide when it comes into operation. The Galileo satellite navigation system entered service in 2016.

The development of differential GPS has enabled Ground Reference Stations to monitor the quality of the satellite transmissions. They provide an additional check on the GPS system integrity for users within a 500–1000 mile radius of these stations and enable differential corrections to be transmitted to these users. Increased accuracy is obtained with errors in the few metre bracket, depending on the range from the Ground Station (as explained in the preceding section).

A GPS satellite augmentation system is being developed in the USA under the auspices of the FAA called the Wide Area Augmentation System, WAAS.

The WAAS system is shown in Fig. 6.39. This will provide increased integrity by monitoring the GPS satellite transmissions from a network of monitor ground stations in

Fig. 6.39 Wide area augmentation system, WAAS, concept

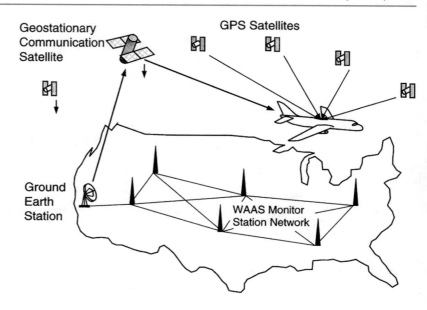

the USA and increase the system accuracy by transmitting the differential corrections over communication satellite radio links. User position accuracy should be within 3 m enabling precision approaches to be accomplished in Cat 2 visibility conditions.

Extensive trials of a WAAS evaluation system have been conducted by the FAA over the last decade, with very promising results. It appears very likely that the full system will be built and become operational over the next decade.

6.6　Terrain Reference Navigation

6.6.1　Introduction

Terrain reference navigation (TRN) is a generic classification which covers any technique of aided navigation which relies on correlating terrain measurement data from a suitable terrain sensor (or sensors) with the data stored in a digital map database. The system is operated in conjunction with a DR navigation system; the position fixes derived by the TRN system are then used to update and correct the DR system errors by means of a Kalman filter. TRN systems can only be used over land and require a very accurate database. The latter may need to be derived from satellite data when accurate map data are not available. Terrain reference navigation systems can be divided into three basic types:

1. *Terrain contour navigation (TCN)*. In this system, the terrain profile is measured by a radio altimeter and is matched with the stored terrain profile in the vicinity of

the aircraft's position as previously estimated by the DR system.

2. *Terrain characteristic matching (TCM)*. This type of system can use a variety of sensors, for example, radar or radiometric sensors to sense changes in the terrain characteristics below the aircraft. For example, these terrain characteristics include flying over lakes, rivers, roads, woods, buildings, etc. The data can then be processed to detect the edges or boundaries where the terrain characteristics change abruptly. The detected edges of the overflown features can then be matched with the stored terrain features in the vicinity of the aircraft's estimated position.

Both TCN and TCM can be used in a complementary manner, TCN providing very good positional fixes where there are reasonable terrain contour variations. Over very flat terrain the TCN accuracy is degraded; however, the TCM system can then provide very good positional fixes from the detected edges of specific terrain features.

3. *Scene matching area correlation (SMAC)*. This is also known as digital scene matching area correlation (DSMAC) in the USA. These systems generally use an infrared imaging sensor to locate and lock on to specific recognisable landmarks or set of features at known positions on the route. An area correlation technique is used to match the processed image data with the stored feature data so that the system will track and lock on to the feature. The aircraft position must be known fairly accurately in the first place using, say, TCN.

The order of magnitude accuracies which can be achieved with these systems are as follows:

TCN	Around 50 m
TCM	Around 10 to 20 m
SMAC	Around 1 to 2 m

It should be noted that while the basic concepts of these TRN systems are relatively straightforward, there is a very considerable mathematical content involved in explaining the detail implementation of these systems. For example, statistics, probability theory, correlation techniques, state estimation methods and Kalman filters. Space constraints have, therefore, limited the coverage of TRN systems to a brief overview only.

6.6.2 Terrain Contour Navigation

The basic concepts of terrain contour navigation are illustrated in Fig. 6.40. Three types of measurement data are required by the system:

1. A sequence of measurements of the ground clearance (i.e. height above ground level) of the aircraft. This is normally obtained by sampling the output of a radio altimeter, possibly with some additional pre-filtering. A horizontal sampling interval of about 100 m is typically used, but this is not critical.
2. Data from a barometric or baro-inertial height system is required to measure any vertical motion of the aircraft between successive ground clearance measurements.
3. Some form of DR navigation system, for example, an INS or Doppler radar plus heading reference is required to measure the relative horizontal positions of the ground clearance measurements.

The essence of TCN is to use this data to reconstruct the terrain profile under the aircraft's track. A digital map of terrain height is then searched to find a matching profile in the vicinity of the aircraft's position as previously estimated. This can then be used as the basis of an update to the estimated position.

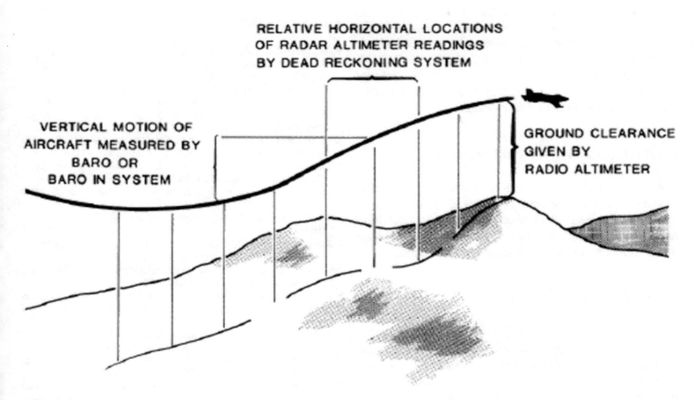

Fig. 6.40 Terrain reference system

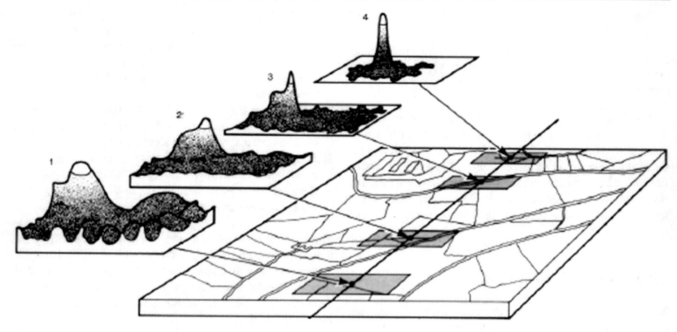

Fig. 6.41 Terrain characteristic matching – edge detection navigation

There are a considerable number of TCN implementations which have been developed in the USA and Europe to provide very accurate navigation systems such as TERCOM, TERPAC, SITAN, CAROTE, TERPROM, SPARTAN. Space constraints limit further coverage and the reader is referred to a number of papers which have been published on TRN systems in the Further Reading.

6.6.3 Terrain Characteristic Matching

The edge detection of specific terrain features has been mentioned earlier in this section. The basis of the technique is shown in Fig. 6.41 which shows a sequence of 'position likelihood' distributions which indicate how the position estimate improves as feature boundaries are detected.

6.6.4 Civil Exploitation of TRN

It should be noted that while TRN systems have been used in exclusively military applications to date, they can clearly be exploited in civil aircraft as a very accurate self-contained navigation aid particularly in the terminal area. Accurate radio altimeters are installed in most civil aircraft together with accurate DR navigation systems such as IN systems. The TRN computer and terrain database is a modest equipment increment with current electronic technology.

Further Reading

Agardograph No. 139.: Theory and Application of Kalman Filtering

Andreas, R.D., Hostetler, L.D., Beckmann, C.: Continuous Kalman Updating of an Inertial Navigation System Using Terrain Measurements: NAECON, pp. 1263–1270 (1978)

Chen, G., Chui, C.K.: Kalman Filtering with Real-Time Applications Springer Series in Information Sciences, vol. 17. Springer, New York, USA (1991)

Daly, P.: Navstar GPS and GLONASS: Global satellite navigation systems. Electron. Commun. Eng. J., IEE. **5**, 349–357 (1993)

Farrell, J.L.: Integrated Aircraft Navigation. Academic press, Cambridge, Massachusetts, USA (1976)

Forssell, B.: Radio Navigation Systems. Prentice Hall, New Jersey, USA (1991)

Gelling, L.A.: The global positioning system. IEEE Spectr., 36–47 (1993)

Hofmana-Wellanhot, B., Lichtenegger, H., Collins, J.: GPS Theory and Practice. Springer, New York, USA (1992)

Hostetler, L.D.: Optimal terrain-aided navigation systems. In: AIAA Guidance and Control Conference (1978)

Kalman, R.E.: A new approach to linear filtering and prediction problems. J. Basic Eng. Trans. Am. Soc. Mech. Eng. 82(1):35-45 (1960)

Kalman, R.E., Bucy, R.S.: New results in linear filtering and prediction theory. J. Basic Eng. Trans. Am. Soc. Mech. Eng. **83**, 95 (1961)

Morgan Owen, G.J., Johnston, G.T.: Differential GPS positioning. Electron. Commun. Eng. J., IEE. **7**, 11–21 (1995)

Schlee, F.H., Toda, N.F., Islam, M.A., Standish, C.J.: Use of an external cascaded Kalman filter to improve the performance of a global positioning system (GPS) inertial navigator. In: Proceedings of the NAECON '88 Conference, Dayton, Ohio, 23–27 May 1988

Siouris, G.M.: Aereospace Avionic Systems. Academic press, Cambridge, Massachusetts, USA (1993)

Skarman, E.: Kalman filter for terrain aided navigation. In: Conference on Remotely Piloted Vehicles, Bristol, UK (1979)

Air Data and Air Data Systems

7.1   Introduction

Air data systems provide accurate information on quantities such as pressure altitude, vertical speed, calibrated airspeed, true airspeed, Mach number, static air temperature and air density ratio. This information is essential for the pilot to fly the aircraft safely and is also required by a number of key avionic sub-systems which enable the pilot to carry out the mission. It is thus one of the key avionic systems in its own right and forms part of the essential core of avionic sub-systems required in all modern aircraft, civil or military.

This chapter explains the importance of air data information and shows how the fundamental physical laws are derived and the air data quantities are generated by an air data computing system.

7.2   Air Data Information and Its Use

7.2.1   Air Data Measurement

The air data quantities, such as pressure altitude, vertical speed, calibrated airspeed, true airspeed, Mach number, are derived from three basic measurements by sensors connected to probes which measure the following:

- *Total (or Pitot) pressure*
- *Static pressure*
- *Total (or indicated) air temperature*

Figure 7.1 illustrates a basic air data system.

The *total pressure*, P_T, is measured by means of an absolute pressure sensor (or transducer) connected to a Pitot tube facing the moving airstream. This measures the *impact pressure*, Q_C, that is the pressure exerted to bring the moving airstream to rest relative to the Pitot tube plus the *static pressure*, P_S, of the free airstream, that is, $P_T = Q_C + P_S$.

The *static pressure* of the free airstream, P_S, is measured by an absolute pressure transducer connected to a suitable orifice located where the surface pressure is nearly the same as the pressure of the surrounding atmosphere.

High-performance military aircraft generally have a combined Pitot/static probe which extends out in front of the aircraft so as to be as far away as practicable from aerodynamic interference effects and shock waves generated by the aircraft structure. Some civil transport aircraft have Pitot probes with separate static pressure orifices located in the fuselage generally somewhere between the nose and the wing. The exact location of the static pressure orifices (and the Pitot tubes or probes) is determined by experience and experimentation. However, many civil transport aircraft (e.g. Boeing 747) have side mounted ('L') type Pitot static probes. The trend on new military aircraft is to use flush port systems for stealth reasons (protruding probes give a significant radar return). It is interesting to note that the Pitot tube, which is universally used on aircraft (and wind tunnels) for measuring airspeed because of its simplicity and effectiveness, was invented over 250 years ago by the French mathematician and scientist Henri Pitot for measuring the flow of water in rivers and canals – one invention which has stood the test of time.

From the measurements of static pressure, P_S, and total pressure, P_T, it is possible to derive the following quantities:

1. *Pressure altitude*, H_P. This is derived from the static pressure, P_S, measurement by assuming a 'standard atmosphere'.
2. *Vertical speed*, $\dot{H}_P$. This is basically derived by differentiating P_S.
3. *Calibrated airspeed*, V_C. This is derived directly from the impact pressure, Q_C, which is in turn derived from the difference between the total and static pressures ($Q_C = P_T$ P_S).

R. P. G. Collinson, *Introduction to Avionics Systems*, https://doi.org/10.1007/978-3-031-29215-6_7

Fig. 7.1 Basic air data system

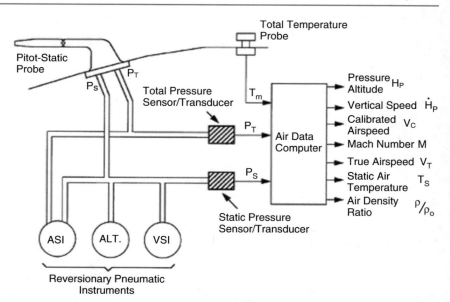

4. *Mach number, M.* This is the ratio of the true airspeed, V_T, to the local speed of sound, A, that is, $M = V_T/A$, and is derived directly from the ratio of the total pressure to the static pressure, P_T/P_S. (True airspeed is defined as the speed of the aircraft relative to the air.)

The third measurement, namely that of the *measured (or indicated) air temperature*, T_m, is made by means of a temperature sensor installed in a probe in the airstream. This gives a measure of the free airstream temperature, T_S, plus the kinetic rise in temperature due to the air being brought partly, or wholly, to rest relative to the temperature sensing probe. The temperature assuming the air is brought totally to rest (i.e. recovery ratio = 1) is known as the *total air temperature*, T_T.

The computation of the aircraft's Mach number, M, together with the known recovery ratio of the probe (which allows for the air not being brought wholly to rest) enables the correction factor for the kinetic heating effect to be derived to convert the measured (or indicated) air temperature to the free airstream or *static air temperature*, T_S. The static air temperature so derived then enables the local speed of sound, A, to be determined as this is dependent only on the air temperature. *True airspeed*, V_T can then be readily computed, namely, $V_T = MA$. Air density ratio, ρ/ρ_0 can then be computed from P_S and T_S (ρ is air density and ρ_0 is air density at standard sea level conditions).

7.2.2 The Air Data Quantities and Their Importance

The use and importance of the air data quantities of pressure altitude, vertical speed (rate of climb/descent), calibrated airspeed, Mach number, and true airspeed by the pilot and the key avionic sub-systems are discussed below.

7.2.2.1 Air Data Information for the Pilot

The pilot is presented with displays of the above air data quantities, all of which are very important at various phases of the flight or mission. However, the two basic quantities which are fundamental for the piloting of any aircraft from a light aircraft to a supersonic fighter are the *pressure altitude* and the *calibrated airspeed*. Pressure altitude is the height of the aircraft above sea level derived from the measurement of the static pressure assuming a standard atmosphere. Calibrated airspeed is the speed which, under standard sea level conditions, would give the same impact pressure as that measured on the aircraft. The altimeter displaying pressure altitude, and the calibrated (or indicated) airspeed display thus form part of the classic 'T' layout of vital instrument displays centred around the artificial horizon display. This layout has been largely retained for the modern electronic displays as can be seen in Fig. 2.37.

The use of the air data information by the pilot is discussed below.

- *Calibrated airspeed* – The reason for the importance of calibrated airspeed information is that it provides a direct measure of the impact pressure, by definition, and the impact pressure together with the angle of incidence determines the aerodynamically generated lift and drag forces and moments acting on the aircraft. (The angle of incidence is the angle between the direction of the airflow and a datum line through the aerofoil section of the wing or control surface.) These aerodynamic forces and moments in turn determine the aircraft's ability to fly and manoeuvre, its controllability and response and its performance in terms of speed, range, operating height, etc.

The impact pressure is a function of the true airspeed, V_T and the air density, ρ, and at low airspeeds up to about 100 m/sec (200 knots) where compressibility effects can be neglected is equal to $1/2\rho V_T^2$. (This relationship no longer holds as the speed increases and compressibility effects increase, and the impact pressure becomes a function of Mach number as well.) The air density is directly related to the altitude so that to maintain the same lift force at high altitudes as at sea level requires an increase in the true airspeed in order to produce the same impact pressure. Hence, the critical speeds which affect the aircraft's behaviour, controllability or safety are specified in terms of calibrated airspeed as this is independent of the air density variation with altitude or temperature. Such critical speeds include the *rotation speed* for take-off, the *stalling speed* and the *not to exceed speed* in a dive when the aerodynamic forces and moments exerted during the pull-out would approach the structural limits of the airframe or the controllability limits would be reached.

It should be noted that the quantity *indicated airspeed* is frequently used in this context. The indicated airspeed is basically the same quantity as calibrated airspeed but includes the pressure error present in the Pitot/static installation and the instrument errors present in a simple mechanical type of airspeed indicator (ASI) instrument. (Calibrated airspeed is derived by the air data computer using very much more accurate pressure sensors and the inherent pressure errors in the Pitot/static probe installation can be compensated by the computer.)

- *Pressure altitude* – Accurate measurement of the aircraft's altitude is essential for the control of the flight path in the vertical plane. For instance, to maintain adequate clearance of mountains and hills, under conditions of poor visibility, flying in cloud or at night. Altitude and airspeed are also vital displays during the approach and landing. The air traffic control (ATC) authorities also require very accurate measurement of the pressure altitude for air traffic control to ensure safe vertical separation in busy airways. Pressure altitude is therefore automatically reported to the ATC ground control by the *ATC transponder* as will be explained in the next section. The ATC authorities also require that the reported pressure altitude must be the same as that displayed on the pilot's altimeter display.
- *True airspeed* – This information is displayed to the pilot for navigation purposes.
- *Mach number* – As the aircraft speed increases and approaches the speed of sound, or exceeds it in the case of a supersonic aircraft, there is a large increase in drag, the lift characteristics change and the pitching moment characteristics change due to compressibility effects. The

performance and controllability of the aircraft are dependent on the aircraft's Mach number in this high-speed regimen. Accurate information on the aircraft's Mach number is thus an essential display for the pilot. It is also essential information for other aircraft sub-systems which are discussed later.

- *Vertical speed or rate of climb/descent* – A display of vertical speed or rate of climb/descent is also required by the pilot and this quantity is generated within the air data computer by differentiating the static pressure. Rate of descent is particularly important during a *ground controlled approach* (GCA) where the pilot will set up a given rate of descent (and speed) in the approach to the airfield. The vertical speed indicator (VSI) display is also used during a turn to detect any tendency to lose height, the pilot applying appropriate corrective movements to the control column or 'stick' to hold a constant height turn.
- *Angle of incidence* – The importance of the angle of incidence has already been mentioned. Generally, the lift force from the wings increases fairly linearly with increasing incidence angle up to near the maximum permissible incidence angle at which point the airflow starts to break away and the further increase would result in the wing stalling with the consequent sudden loss of lift. Airflow sensors to measure the angle of incidence are thus frequently installed so that the pilot can monitor the situation and ensure the critical value is not reached. (It should be noted that the term angle of attack is generally used in the USA for the angle of incidence.)

7.2.2.2 Air Data for Key Sub-systems

The key sub-systems requiring air data information and their use of this information are briefly described below.

- *Air traffic control transponder* – Pressure altitude is supplied to the air traffic control (ATC) transponder for automatic reporting to the air traffic ground control system. The ATC authorities specify the flight levels which aircraft must maintain in 'controlled airspace' in terms of pressure altitude and these are set so that there is a minimum of 1000 ft vertical separation between aircraft flying in the vicinity of each other. As stated earlier, pressure altitude is derived from the measurement of the static pressure and the assumption of a 'standard atmosphere' which enables a unique mathematical law to be derived relating altitude to the static pressure. Pressure altitude, however, can differ from the true altitude because of day-to-day variations from the standard atmosphere. These differences are small at low altitudes if the *ground pressure correction* is applied but can be much larger at higher altitudes.

This difference between pressure altitude and true altitude does not matter from an air traffic control standpoint providing the pressure altitude measurements are sufficiently accurate. This is because the difference is common to all the pressure altitude measurements made in the vicinity of each other. For example, suppose aircraft A is given a flight level of 33,000 ft to fly by the ATC authorities and aircraft B is given a flight level of 34,000 ft. Let the difference from the true altitude be Δ_H so that ignoring any errors in the pressure altitude measurements in the two aircraft, the true altitude of aircraft A will be $(33,000 + \Delta H)$ and that of aircraft B will be $(34,000 + \Delta H)$. The required vertical separation of 1000 ft is maintained. The cruising altitudes of jet airliners are typically in the 29,000–45,000 ft band and it will be shown in Sect. 7.3 that the ATC 1000 ft separation levels place very stringent accuracy requirements on the Static Pressure Sensors.

- *Flight control systems* – Calibrated airspeed and pressure altitude information is required by the flight control system (FCS). This is to enable automatic adjustment to be made to the gains (or 'gearings') of the FCS with airspeed and height to compensate for the wide variation in control effectiveness and aircraft response over the flight envelope. This is frequently described as 'air data gain scheduling'. This is covered in more detail in Chaps. 3 and 4. Adequate redundancy must be provided in the system to ensure failure survival in the case of a fly-by-wire flight control system. This may involve three or more independent air data computing systems.

- *Autopilot system* – A number of autopilot control modes require air data information, for example, 'height acquire/hold', 'Mach number acquire/hold' and 'airspeed acquire/hold' (auto-throttle system). Air data gain scheduling may also be required by the autopilot control loops. Autopilots are covered in Chap. 8.

- *Navigation system* – Pressure altitude and true airspeed are required by the navigation system. Pressure altitude is required for navigation in the vertical plane. It can be combined (or mixed) with the inertially derived information from the inertial navigation system (INS) to provide vertical velocity and altitude information which is superior to either source on its own. (The technique is referred to as 'Barometric/Inertial' mixing and is covered in Chap. 6.)

The aircraft's velocity vector is derived from the vector sum of the barometric/inertial vertical velocity and the horizontal velocity (or ground speed vector) output from the INS. The velocity vector information is used for guidance and control, for example, HUD, flight director system and weapon aiming system.

Pressure altitude is also essential information for a terrain reference navigation (TRN) system as it measures the vertical motion of the aircraft and so enables the ground profile to be derived from the radio altimeter measurements. TRN systems operate by correlating the radar altimeter measurements of the ground profile with a stored topographical map elevation database and hence establishing the aircraft position and correcting the dead reckoning position estimate. (It is interesting to note that the standard deviation from the horizontal of an isobar (line joining points of equal pressure) over a distance of 1 nautical mile at sea level is 1 foot.)

True airspeed is required for dead reckoning (DR) navigation, DR position being computed from a knowledge of the true airspeed, aircraft heading and forecast/estimated wind velocity. True airspeed information is also used to compute the wind velocity vector using the ground speed information from the INS. Other sources of navigational information, for example, GPS or Doppler, can also be used to derive wind velocity from the true airspeed information. Information on wind velocity is required by the flight management system (FMS) as will be explained. Wind velocity information is also required for weapon aiming in the case of a military aircraft.

- *Flight management system* – The flight management system requires information on all the air data quantities: pressure altitude, vertical speed, Mach number, static air temperature, true airspeed and calibrated airspeed.
Air data information is essential for the FMS to maintain the aircraft on the most fuel-efficient flight path and for achieving 4D flight management (3D position and time). FMS systems are covered in Chap. 8.

- *Engine control systems* – Height and calibrated airspeed information are required by the engine control systems.

The flow of air data information from the air data computing system to the key avionic sub-systems is shown in Fig. 7.2. Modern aircraft systems use a time multiplexed digital data bus system, such as MIL STD 1553B on a military aircraft, or ARINC 629 in the case of a civil aircraft to transmit the data and interconnect the system. The degree of redundancy has been omitted for clarity: for instance, civil aircraft have at least two independent air data computing systems.

7.3 Derivation of Air Data Laws and Relationships

The object of this section is to show how the mathematical laws relating altitude, air density, Mach number, calibrated airspeed, free airstream temperature and true airspeed are

Fig. 7.2 Flow of air data to key avionic sub-systems (redundancy omitted for clarity)

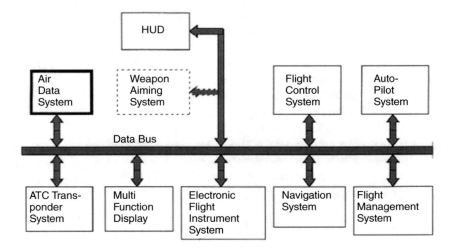

derived from the basic air data measurements of static pressure, total pressure and total (or indicated) air temperature. The units used for the air data quantities are explained in Chap. 1, Sect. 1.3.

It should be noted that the *standard atmospheric pressure* is the atmospheric pressure at sea level under standard temperature conditions of 288.15°K (15 °C) and is equal to 101.325 kN/m² (kPa) or 1013.25 mb (also equal to 29.9213 inches of mercury).

7.3.1 Altitude–Static Pressure Relationship

By making certain assumptions regarding a 'standard' atmosphere, it is possible to express the altitude above sea level at any point in the Earth's atmosphere as a single-valued function of the atmospheric pressure at that point. These assumptions concern the chemical constitution of the atmosphere, initial atmospheric conditions at sea level and the temperature distribution throughout the atmosphere.

These assumed conditions have been agreed internationally and enable a mathematical law to be established relating static pressure and altitude so that by measuring the pressure of the free airstream the pressure (or barometric) altitude can be derived.

These assumptions of a standard atmosphere are based on statistical data and the pressure altitude so derived can differ from the true altitude above sea level by several thousand feet. However, as explained earlier, provided the static pressure is measured accurately and the computation errors are small all aircraft flying at a particular pressure altitude will be flying at the same altitude and the fact that the pressure altitude does not coincide with the true altitude does not matter from an air traffic control aspect. A standard atmosphere also enables flight tests, wind tunnel results and general aircraft design and performance parameters to be related to a common reference.

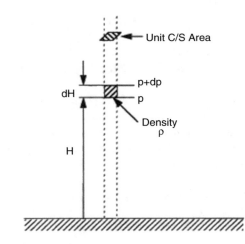

Fig. 7.3 Static pressure and altitude relationship

The pressure altitude–static pressure relationship is derived as follows: referring to Fig. 7.3, the change in pressure, dp, of air of density, ρ, resulting from a small change in height, dH, is derived from equating the forces acting in the vertical plane on an elemental volume of air from which

$$-dp = \rho g \, dH \qquad (7.1)$$

where g is the gravitational constant.

From the gas law

$$p = \rho R_a T \qquad (7.2)$$

where T is air temperature (°K) and R_a is the gas constant for unit mass of dry air.

Combining Eqs. (7.1) and (7.2) yields

$$-\frac{dp}{P} = \frac{g}{R_a T} dH \qquad (7.3)$$

T can be expressed as a function of H only, by making the following assumptions which are based on statistical data.

1. The temperature at sea level, T_0, and the pressure at sea level, P_{S0}, are assumed to be constant.
2. The temperature decreases linearly with increasing height until a height known as the *tropopause* is reached above which height the temperature remains constant until the *stratopause* height is reached. The region below the tropopause height is known as the *troposphere*.

   The law relating temperature, T, at altitude, H, up to the tropopause height is

$$T = T_0 - LH$$

   where L is the temperature lapse rate.
3. The temperature above the tropopause height stays constant at a value T_{T*} until a height known as the *stratopause* is reached. (The asterisk $*$ is used to distinguish the tropopause temperature T_{T*} from the total temperature T_T.)

   The region between the tropopause and stratopause heights is known as the *stratosphere*.
4. At heights above the stratopause height, the temperature starts to increase linearly with height, this region being known as the *chemosphere*.

The temperature in the chemosphere is given by

$$T = T_{T*} + L(H - H_S)$$

where L is the temperature rise rate and H_S is the height of stratopause.

The altitude–pressure law is thus made up of three parts covering the troposphere, stratosphere and chemosphere regions, respectively.

The gravitational acceleration, g at an altitude, H differs from the value at the Earth's surface, g_0 and obeys an inverse square law, as explained earlier, namely,

$$g = \frac{R_0^2}{(R_0 + H)^2} g_0 \qquad (7.4)$$

where R is the radius of the Earth.

The effect of this variation in deriving the pressure altitude is very small over the normal operating altitudes and it is convenient to assume g is constant at the sea level value g_0 thereby simplifying the integration of Eq. (7.4). The value of altitude so derived is known as the *geopotential altitude*. This small difference does not matter from the point of view defining a pressure altitude which is a single-valued function of the static pressure.

(a) *Troposphere region* $T = T_0 - \text{LH}$

Substituting for T and assuming g is constant and equal to g_0 in Eq. (7.3) and integrating both sides gives

$$-\int_{P_{s0}}^{P_s} \frac{1}{p}\, dp = \frac{g_0}{R_a} \int_0^H \frac{1}{(T_0 - \text{LH})}\, dH$$

where P_S0 is the value of p at sea level, $H = 0$ and P_S is the value of p at altitude H.

From which

$$\log_e \frac{P_S}{P_{S0}} = \frac{g_0}{LR_a} \log_e \frac{(T_0 - \text{LH})}{T_0}$$

Hence

$$P_S = P_{S0} \left(1 - \frac{L}{T_0} H\right)^{g_0/LR_a} \qquad (7.5)$$

and

$$H = \frac{T_0}{L} \left[1 - \left(\frac{P_S}{P_{S0}}\right)^{LR_a/g_0}\right] \qquad (7.6)$$

(b) *Stratosphere region* $T = T_{T*}$

Substituting T_{T*} for T and g_0 for g in Eq. (7.3) and integrating both sides gives

$$-\int_{P_{sT}}^{P_s} \frac{1}{p}\, dp = \frac{g_0}{R_a T_{T*}} \int_{H_T}^H dH$$

where P_{ST} is the pressure at tropopause altitude, H_T. Whence

$$P_S = P_{ST}\, e^{-(g_0/R_a T_{T*})(H - H_T)} \qquad (7.7)$$

and

$$H = H_T + \frac{R_a T_{T*}}{g_0} \log_e \frac{P_{ST}}{P_S} \qquad (7.8)$$

(c) *Chemosphere region*

The static pressure versus altitude relationship can be derived by a similar process by substituting $T = T_{T*} + L(H -$

Table 7.1 Pressure–altitude law constants

Constant	Standard atmosphere value
Pressure at sea level, P_{S0}	101.325 kPa (1013.25 mb)
Temperature at sea level, T_0	288.15° K
Troposphere lapse rate, L	6.5×10^{-3} °C/m
Tropopause height, H_T	11,000 m (36,089.24 ft)
Tropopause temperature, T_{T*}	216.65° K (-56.5 °C)
Stratopause height, H_S	20,000 m (65,617 ft)
Chemosphere rise rate, L	1.0×10^{-3} °C/m
Chemosphere height limit	32,004 m (105,000 ft)
g_0	9.80665 m/sec^2
R_a	287.0529 Joules/° K/kg
g_0/LR_a (Troposphere)	5.255879
$g_0/R_a T$	1.576885×10^{-4} m^{-1}
g_0/LR_a (Chemosphere)	34.163215

H_S) and $g = g_0$ in Eq. (7.3) and integrating both sides between the appropriate limits. Whence

$$P_S = P_{SS}\left[1 + \frac{L}{T_{T*}}(H - H_S)\right]^{-g_0/R_a L} \qquad (7.9)$$

and

$$H = H_S + \frac{T_{T*}}{L}\left[\left(\frac{P_{SS}}{P_S}\right)^{R_a L/g_0} - 1\right] \qquad (7.10)$$

The values of the constants used in the pressure–altitude law are set out in Table 7.1 in SI units with the alternative units in brackets.

The symbol H_P is used henceforth to denote pressure altitude.

The formulae relating P_S and H are set out below and are derived by substituting the appropriate values in the table in Eqs. (7.5), (7.7) and (7.9), respectively, and H_P for H.

(a) *Troposphere region*: -914.4 to 11,000 m (-3000 to 36,089 ft)

$$P_S = 1,013.25\left(1 - 2.25577 \times 10^{-5} H_P\right)^{5.255879} \text{ mb} \qquad (7.11)$$

(b) *Stratosphere region*: 11,000 to 20,000 m (36,089 to 65,617 ft)

$$P_S = 226.32\, e^{-1.576885 \times 10^{-4}(H_P - 11,000)} \text{ mb} \qquad (7.12)$$

(c) *Chemosphere region*: 20,000 to 32,004 m (65,617 to 105,000 ft)

$$P_S = 54.7482\left[1 + 4.61574 \times 10^{-1}(H_P - 20,000)\right]^{-34.163215} \text{ mb} \qquad (7.13)$$

H_P is expressed in metres. The conversion factor to convert feet to metres is 1 ft $= 0.3048$ m. The chemosphere region formulae have been omitted for brevity as already mentioned.

The static pressure–altitude relationship computed from the above formulae is plotted in Fig. 7.4. It can be shown that the difference between the pressure, or geopotential, altitude, H_P, derived from the above formulae, which assume $g = g_0$, and the geometric altitude, H_G, obtained by allowing for the variation of g with altitude is given to a close approximation over normal altitudes by H_G^2/R where R is the radius of Earth $= 6356.8$ km at latitude $45°$. As mentioned earlier, the difference is small over normal altitudes, although it follows a square law. For example, the difference is 19 m (62 ft) at 11,000 m (36,089 ft).

7.3.2 Variation of Ground Pressure

Errors due to the variation in the ground pressure from the assumed standard value are taken out of the altimeter reading by setting a scale to a given pressure. This action affects the zero point and so alters the altimeter reading over the whole of its range by a height corresponding to the pressure set as determined by the altitude–pressure law.

The pressure that is set for any given occasion varies with the type of altitude indication to be used. If the standard ground level pressure (1013.25 mb) is set then the altimeter will read pressure altitude (or barometric altitude). If the 'QFE' system is used, the pressure at ground level (not necessarily sea level) is set and the altimeter reads height above ground at this point if the atmosphere is assumed standard. Thus if the QFE for the airfield is set, the altimeter will read zero on touch down (assuming an accurate instrument system).

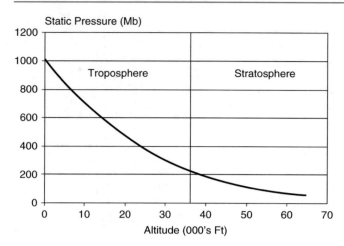

Fig. 7.4 Static pressure versus altitude

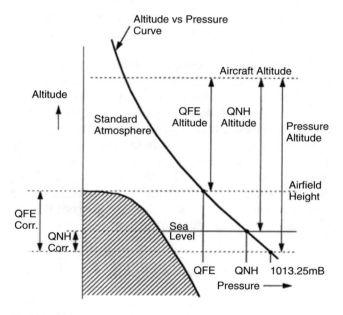

Fig. 7.5 Altimeter ground pressure adjustments

If the QNH system is used the pressure at mean sea level (often computed) for the region concerned is set and the altimeter reads height above sea level, if the atmosphere is assumed standard.

The different modes of height indication are shown in Fig. 7.5.

7.3.3   Air Density Versus Altitude Relationship

The relationship between air density, ρ, and altitude, H, is derived from the equation relating P_S and H and using the gas law

$$P_S = \rho R_a T$$

to eliminate P_S.

For example, in the troposphere

$$P_S = P_{S0}\left(1 - \frac{L}{T_0}H\right)^{g_0/LR_a}$$

$T = T_0 - L.H.$ Hence

$$\frac{\rho}{\rho_0} = \left(1 - \frac{L}{T_0}H\right)^{(g_0/LR_a)-1} \tag{7.14}$$

where ρ_0 is the density at standard sea level conditions $= P_{S0}/R_aT_0$. The ratio ρ/ρ_0 is referred to as the *density reduction factor*.

Example Compute the air density at an altitude of 10,000 m (30,480 ft) given $\rho_0 = 1.225$ kg/m^3 ($g_0/LR_a = 5.255879$)

$$\text{Density at } 10,000 \text{ m} = 1.225\left(1 - \frac{6.5}{1000} \times \frac{10000}{288.15}\right)^{5.255879-}$$

$$= 0.4127 \text{ kg/m}^3$$

In practice, air density ratio ρ/ρ_0 when required is normally computed from the static pressure, P_S, and the measured (or indicated) air temperature, T_m using the following relationships: static air temperature, $T_S = T_m/(1 + r0.2\,M^2)$ (Sect. 7.3.8) and

$$\frac{\rho}{\rho_0} = \frac{P_S}{P_{S0}} \cdot \frac{T_0}{T_S}$$

from which

$$\frac{\rho}{\rho_0} = \frac{P_S}{P_{S0}} \cdot \frac{T_0\left(1 + r0.2\,M^2\right)}{T_m} \tag{7.15}$$

7.3.4   Speed of Sound

The derivation of the formulae for the speed of sound, A, is set out briefly below as it is fundamental to the derivation of Mach number, calibrated airspeed, static air temperature and true airspeed from the measurements of total pressure, static pressure and indicated air temperature.

Consider first a stream tube of air of unit cross-sectional area, a, through which a pressure wave is being transmitted with velocity V, travelling from right to left. Imagine now, the

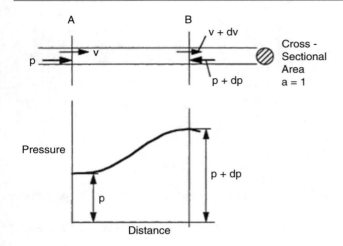

Fig. 7.6 Pressure wave

equivalent situation where the pressure wave is stationary and the air is moving with velocity V and travelling from left to right as shown in Fig. 7.6.

Consider a section at A where the pressure is p and air velocity V.

Let the section at B represent the adjacent portion of the stream tube where the pressure of the air increases by dp in passing through the pressure wave and let the velocity change by dV and the density by $d\rho$.

Equating the mass flows at A and B

$$\rho a V = (\rho + d\rho)a(V + dV)$$

Neglecting second-order terms

$$-V d\rho = \rho dV \qquad (7.16)$$

Force acting on stream tube of air between sections A and B

$$= \text{Change of momentum per second}$$
$$= \text{Mass per second} \times \text{change in velocity}$$
$$pa - (p + dp)a = \rho a V dV$$

Hence

$$dp = -\rho V dV \qquad (7.17)$$

Combining Eqs. (7.16) and (7.17) yields

$$dp/d\rho = V^2 \qquad (7.18)$$

The transmission of a pressure wave closely approximates an adiabatic process because the pressure change takes place very suddenly and consequently there is no time for any appreciable interchange of heat. The law for an adiabatic process is

$$p = K\rho^\gamma \qquad (7.19)$$

where

$$\gamma = \frac{\text{Specific heat of gas at constant pressure } c_p}{\text{Specific heat of gas at constant volume } c_v}$$

($\gamma = 1.4$ for air) and $K = $ constant.

$$\frac{dp}{d\rho} = K\gamma\rho^{\gamma-1} = \frac{\gamma p}{\rho} \qquad (7.20)$$

Equating Eqs. (7.18) and (7.20). Let A be the speed of sound ($= V$)

$$A = \sqrt{\frac{\gamma p}{\rho}} \qquad (7.21)$$

Substitute $p = \rho R_a T$ in Eq. (7.21)

$$A = \sqrt{\gamma R_a T} \qquad (7.22)$$

Hence, the speed of sound depends only on the temperature of the air.

The speed of sound at sea level, A_0, under standard pressure and temperature conditions is thus equal to

$$\sqrt{1.4 \times 287.0529 \times 288.15}$$

That is, $A_0 = 340.294$ m/s.

The speed of sound reduces with increasing altitude as the temperature decreases until the tropopause height is reached. Thereafter the speed stays constant within the stratosphere height band as the temperature is constant.

The variation in the troposphere region is derived from the relationship

$$A = A_0\sqrt{1 - \frac{L}{T_0}H}$$

The variation from sea level to 65617 ft is set out in Table 7.2.

7.3.5   Pressure–Speed Relationships

The relationship between total pressure, P_T, static pressure, P_S, true airspeed, V_T, and the local speed of sound is explained from the first principles in this section as it is fundamental to the derivation of Mach number, M, and

Table 7.2 Speed of sound variation with height

H	Speed of sound (A)	
0 ft	340.3 m/s	(661.5 knots)
10,000 ft	328.4 m/s	(637.4 knots)
20,000 ft	316.0 m/s	(614.3 knots)
30,000 ft	303.2 m/s	(589.4 knots)
36,089 ft	295.1 m/s	(573.6 knots)
65,617 ft	295.1 m/s	(573.6 knots)

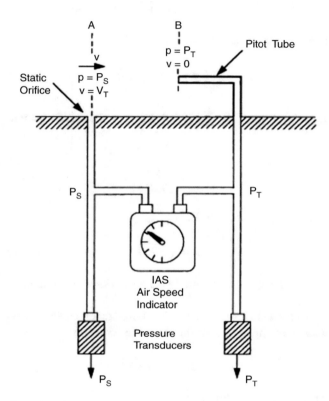

Fig. 7.7 Impact pressure measurement

calibrated airspeed, V_C. These are covered in Sects. 7.3.6 and 7.3.7, respectively.

As stated earlier, knowledge of the Mach number enables the correction for the kinetic heating effect to be determined so that the static air temperature, T_S, can be derived from the measured air temperature, T_m, as will be shown in Sect. 7.3.8. True airspeed can then be derived from the Mach number and the static air temperature as will be shown in Sect. 7.3.9.

(a) *Subsonic speeds*

The total and static pressure measurement system is shown schematically in Fig. 7.7. Consider first the case of low airspeeds below $M = 0.3$, where the air can be considered to be incompressible and the density therefore is constant.

The momentum equation for the airflow is

$$dp + \rho V dV = 0$$

In the free airstream $p = P_S$ and $V = V_T$. At the probe face $p = P_T$ and $V = 0$.

Integrating the momentum equation between these limits

$$\int_{P_S}^{P_T} dp + \rho \int_{V_T}^{0} V \, dV = 0 \tag{7.23}$$

Hence

$$P_T - P_S = \frac{1}{2}\rho V_T^2 \tag{7.24}$$

(Equation (7.24) is Bernoulli's equation.)

$$V_T = \sqrt{\frac{2}{\rho}} \cdot \sqrt{P_T - P_S} \,(\text{at low airspeeds}) \tag{7.25}$$

However, air is a compressible fluid and the density is not constant. The change in density due to the high impact pressures resulting from high airspeeds must therefore be taken into account. Assuming adiabatic flow, the relationship between pressure and density is

$$P = K\rho^\gamma$$

From which

$$\rho = \frac{1}{K^{1/\gamma}} P^{1/\gamma} \tag{7.26}$$

Substituting for ρ in the momentum equation $dp + \rho V \, dV = 0$

$$dp + \frac{1}{K^{1/\gamma}} P^{1/\gamma} V dV = 0 \tag{7.27}$$

In the free airstream $p = P_S$ and $V = V_T$. At the probe face $p = P_T$ and $V = 0$.

Rearranging Eq. (7.27) and integrating between these limits

$$\int_{P_S}^{P_T} p^{1/\gamma} dp + \frac{1}{K^{1/\gamma}} \int_{V_T}^{0} V \, dV = 0 \qquad (7.28)$$

$$\frac{\gamma}{\gamma - 1} \left[P_T^{(\gamma - 1)/\gamma} - P_S^{(\gamma - 1)/\gamma} \right] = \frac{1}{K^{1/\gamma}} \cdot \frac{V_T^2}{2} \qquad (7.29)$$

From Eq. (7.26)

$$K^{1/\gamma} = \frac{P_S^{1/\gamma}}{\rho}$$

Substituting for $K^{1/\gamma}$ in Eq. (7.29) and rearranging gives

$$\frac{P_T}{P_S} = \left[1 + \frac{(\gamma - 1)}{2} \cdot \frac{\rho}{\gamma P_S} \cdot V_T^2 \right]^{\gamma/(\gamma - 1)}$$

$(\gamma P_S)/\rho = A^2$ and putting $\gamma = 1.4$ in the above equation gives

$$\frac{P_T}{P_S} = \left[1 + 0.2 \frac{V_T^2}{A^2} \right]^{3.5} \qquad (7.30)$$

and

$$Q_C = P_S \left[\left(1 + 0.2 \frac{V_T^2}{A^2} \right)^{3.5} - 1 \right] \qquad (7.31)$$

(b) *Supersonic speeds*

The aerodynamic theory involved in deriving the relationship between P_T/P_S and Mach number (V_T/A) at supersonic speeds ($M > 1$) is beyond the scope of this book. The formula relating pressure ratio P_T/P_S and Mach number derived by Rayleigh is therefore set out below without further explanation. This can be found in suitable aerodynamic textbooks for those readers wishing to know more.

$$\frac{P_T}{P_S} = \frac{\left[\frac{(\gamma + 1)}{2} \left(\frac{V_T}{A} \right)^2 \right]^{\gamma/(\gamma - 1)}}{\left[\frac{2\gamma}{(\gamma + 1)} \left(\frac{V_T}{A} \right)^2 - \frac{(\gamma - 1)}{(\gamma + 1)} \right]^{1/(\gamma - 1)}} \qquad (7.32)$$

Substituting $\gamma = 1.4$, Eq. (7.32) becomes

$$\frac{P_T}{P_S} = \frac{166.92 \left(\frac{V_T}{A} \right)^7}{\left[7 \left(\frac{V_T}{A} \right)^2 - 1 \right]^{2.5}} \qquad (7.33)$$

$$Q_C = P_S \left[\frac{166.92 \left(\frac{V_T}{A} \right)^7}{\left[7 \left(\frac{V_T}{A} \right)^2 - 1 \right]^{2.5}} - 1 \right] \qquad (7.34)$$

7.3.6 Mach Number

Mach number can be derived from Eqs. (7.30) and (7.33) for the subsonic and supersonic regions, respectively, by substituting $M = V_T/A$

(a) *Subsonic speeds*

From Eq. (7.30)

$$\frac{P_T}{P_S} = (1 + 0.2 M^2)^{3.5} \qquad (7.35)$$

(b) *Supersonic speeds*

From Eq. (7.33)

$$\frac{P_T}{P_S} = \frac{166.92 M^7}{(7 M^2 - 1)^{2.5}} \qquad (7.36)$$

The pressure ratio, P_T/P_S, is plotted against the Mach number, M, in Fig. 7.8 for the subsonic and supersonic regimens.

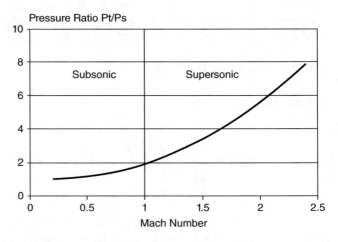

Fig. 7.8 Pressure ratio versus Mach number

7.3.7 Calibrated Airspeed

Calibrated airspeed, V_C, can be derived directly from Eqs. (7.31) and (7.34) for the subsonic and supersonic cases, respectively, by substituting for sea level conditions, that is $P_S = P_{S0}$ and $V_T = V_C$ (by definition).

(a) *Subsonic speeds* $(V_C \leq A_0)$

Referring to Eq. (7.31)

$$Q_C = P_{S0} \left\{ \left[1 + 0.2 \left(\frac{V_C}{A_0} \right)^2 \right]^{3.5} - 1 \right\} \qquad (7.37)$$

A simpler approximate expression relating Q_C and V_C can be obtained by applying the binomial expansion to the inner factor

$$\left[1 + 0.2 \left(\frac{V_C}{A_0} \right)^2 \right]^{3.5}$$

and noting $A_0^2 = (\gamma P_{S0})/\rho_0$. The expression obtained is

$$Q_C \approx 1/2 \rho_0 V_c^2 \left[1 + \frac{1}{4} \left(\frac{V_C}{A_0} \right)^2 \right] \qquad (7.38)$$

The error in this expression is about 1.25% when $V_C = A_0$ (661.5 knots). It should be noted that the exact formula in Eq. (7.37) is used to compute calibrated airspeed. The approximate formula (7.38) is useful, however, for appreciating the effects of increasing airspeed and Mach number. At low airspeeds, it can be seen that $Q_C \approx 1/2 \rho_0 V_C^2$.

(b) *Supersonic speeds* $(V_C > A_0)$

Referring to Eq. (7.34)

$$Q_C = P_{S0} \left[\frac{166.92 \left(\frac{V_C}{A_0} \right)^7}{\left[7 \left(\frac{V_C}{A_0} \right)^2 - 1 \right]^{2.5}} - 1 \right] \qquad (7.39)$$

The impact pressure, Q_C, is plotted against calibrated airspeed, V_C, in Fig. 7.9 for $V_C \leq A_0$ and $V_C > A_0$ speeds.

Note in Fig. 7.9 the very low impact pressure of 16.3 mb at a calibrated airspeed of 51.5 m/s (100 knots) and the very high impact pressure of 1456 mb at a calibrated airspeed of 412 m/s (800 knots), approximately $M = 1.2$ at sea level.

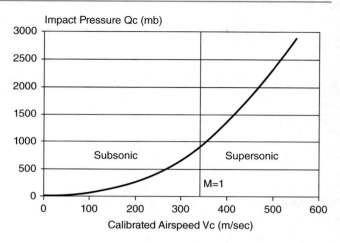

Fig. 7.9 Impact pressure versus calibrated airspeed

7.3.8 Static Air Temperature

As already mentioned, the temperature sensed by a thermometer probe in the airstream is the free airstream temperature plus the kinetic rise in temperature due to the air being brought partly or wholly to rest relative to the sensing probe.

The kinetic rise in temperature can be obtained by the application of Bernoulli's equation to compressible flow and assuming the pressure changes are adiabatic.

For the unit mass of air

$$\frac{P_1}{\rho_1} + \frac{1}{2} V_1^2 + E_1 = \frac{P_2}{\rho_2} + \frac{1}{2} V_2^2 + E_2 \qquad (7.40)$$

where P_1, ρ_1, V_1, E_1 and P_2, ρ_2, V_2, E_2 represent pressure, density, velocity, temperature and internal energy at two points in a streamline flow, namely, in the free airstream and at the probe.

$$\frac{P_1}{\rho_1} = R_a T_1 \quad \text{and} \quad \frac{P_2}{\rho_2} = R_a T_2 \quad \text{(gas law)}$$

In the free airstream, $V_1 = V_T$ and $T_1 = T_S$. At a stagnation point at the probe $V_2 = 0$ and $T_2 = T_T$.

Substituting these values in Eq. (7.40) gives

$$\frac{1}{2} V_T^2 = (E_2 - E_1) + R_a (T_T - T_S) \qquad (7.41)$$

The change in internal energy becomes heat and is given by

$$E_2 - E_1 = Jc_v (T_T - T_S) \qquad (7.42)$$

where J is the mechanical equivalent of heat (Joule's constant) and c_v is the specific heat of air at constant volume. Hence, Eq. (7.41) becomes

$$\frac{1}{2}V_T^2 = (Jc_v + R_a)(T_T - T_S) \qquad (7.43)$$

From thermodynamic theory

$$R_a = J(c_p - c_v) \qquad (7.44)$$

where c_p is the specific heat of air at constant pressure.

Combining Eqs. (7.43) and (7.44) yields

$$\frac{1}{2}V_T^2 = \frac{R_a c_p}{(c_p - c_v)}(T_T - T_S)$$

Rearranging and substituting $\gamma = c_p/c_v$ yields

$$T_T - T_S = \frac{(\gamma - 1)}{2} \cdot \frac{1}{\gamma R_a} \cdot V_T^2 \qquad (7.45)$$

$A^2 = \gamma R_a T_S$ (Eq. (7.21)).

Substituting A^2/T_S for γR_a and putting $\gamma = 1.4$ in Eq. (7.47) yields

$$T_S = \frac{T_T}{(1 + 0.2M^2)} \qquad (7.46)$$

A constant, r, known as the recovery factor, is generally introduced into the above equation, its value being dependent on the temperature probe installation, namely,

$$T_S = \frac{T_m}{(1 + r0.2M^2)} \qquad (7.47)$$

where T_m is the measured (or indicated) air temperature.

At a stagnation point $r = 1.0$ and $T_m = T_T$. Although r is normally assumed to be constant, it can vary slightly due to change of heat transfer from the probe with altitude; significant variations can also occur flying through rain or cloud.

7.3.9 True Airspeed

The computation of the temperature of the free airstream, T_S, enables the local speed of sound, A, to be established as $A = \sqrt{\gamma R_a T_S}$.

True airspeed, V_T can then be obtained from the Mach number.

$$V_T = MA = M\sqrt{\gamma R_a T}$$

Hence

$$V_T = \sqrt{\gamma R_a} \cdot M \cdot \sqrt{\frac{T_m}{(1 + r0.2M^2)}} \qquad (7.48)$$

$$V_T = 20.0468M\sqrt{\frac{T_m}{(1 + r0.2M^2)}} \ \text{m/sec} \qquad (7.49)$$

(This can be readily converted to knots using the conversion factor 1 m/s = 1.9425 knots.)

The air data formulae are set out in Table 7.3.

7.3.10 Pressure Error

The pressures measured at the Pitot head and the static orifice are not exactly equal to the true values. The error in the total pressure measured at the Pitot probe is usually very small provided the incidence angles of the probe to the airspeed vector are small. The error in the static pressure measurement, however, can be significant and is due to disturbance of the airflow near the detecting orifices by the pressure head itself and by the aircraft structure. To minimise the latter effect, the Pitot static probe is often mounted in front of the foremost part of the aircraft in which case disturbance due to the aircraft disappears above the speed of sound. The normal error pattern for this type of installation is for the error to build up to a maximum at $M = 1$ and for it then suddenly to drop to a low value at supersonic speeds.

The pressure error for the aircraft installation is determined by extensive flight testing and in general is a function of Mach number and altitude. This function is often referred to as the *static source error correction* (SSEC) and can be a relatively complex non-linear function around $M = 1$. The SSEC is generated by the air data computer and corrections are applied to the appropriate outputs.

7.4 Air Data Sensors and Computing

7.4.1 Introduction

The key air data sensors, as already mentioned, comprise two pressure sensors and a temperature sensor.

The temperature sensor generally comprises a simple resistance bridge with one arm of the bridge consisting of a resistive element exposed to the airstream, the resistance of this element being a function of temperature. The siting of the probe and establishing the recovery factor, r (Sect. 7.3.8), are the key factors of the sensor. In terms of technology, the temperature sensor is a relatively simple and straightforward device which fairly readily meets the system accuracy requirements. The pressure sensors, however, merit considerable discussion because of their very high accuracy requirements and thus influence on the overall system accuracy, long-term stability, reliability and overall cost.

Table 7.3 Air data formulae

Quantity	Computational formulae
Geopotential pressure altitude	(a) Troposphere 0–11,000 m (0–36,089 ft)
H_P metres	$P_s = 101.325(1-2.25577 \times 10^{-5} H_P)^{5.225879}$ kPa
	(b) Stratosphere 11,000–20,000 m (36,089–65,617 ft)
	$PS = 22.632 \, e^{-1.576885 \times 10-4(HP - 11,000)}$ kPa
Air density ratio	
$\frac{\rho}{\rho_0}$	$\frac{\rho}{\rho_0} = \frac{P_S}{0.35164 T_S}$
Mach number	(a) Subsonic speeds ($M \leq 1$)
M	$\frac{P_T}{P_S} = \left(1 + 0.2M^2\right)^{3.5}$
	(b) Supersonic speeds ($M > 1$)
	$\frac{P_T}{P_S} = \frac{166.92M^7}{\left[7M^2 - 1\right]^{2.5}}$
Calibrated airspeed	(a) $V_C \leq A_0$
V_C m/s	$Q_C = 101.325 \left[\left[1 + 0.2\left(\frac{V_C}{340.294}\right)^2\right]^{3.5} - 1\right]$ kPa
	(b) $V_C > A_0$
	$Q_C = 101.325 \left[\frac{166.92\left(\frac{V_C}{340.294}\right)^7}{\left[7\left(\frac{V_C}{340.204}\right)^2 - 1\right]^{2.5}} - 1\right]$ kPa
Static air temperature	
$T_S \,°K$	$T_S = \frac{T_m}{1 + r0.2M^2} \,°K$
True airspeed	
V_T m/s	$V_T = 20.0468 M \sqrt{T_S}$ m/s

A very wide variety of air data pressure sensors has been developed by different companies and organisations/technical establishments worldwide. You name an effect which is a function of pressure and some organisation has developed (or tried to develop) a pressure sensor exploiting it.

Air data pressure sensors require an extremely high accuracy (which will be explained shortly) and involve a long expensive development to establish and qualify a producible, competitive device. Like most sensors, they not only sense the quantity being measured but they can also be affected by the following:

- Temperature changes
- Vibration
- Shock
- Acceleration
- Humidity

The art of sensor design is to minimise and if possible eliminate these effects on the sensor.

7.4.2 Air Data System Pressure Sensors

7.4.2.1 Accuracy Requirements

(a) *Static pressure sensor*

Typical air data static pressure sensors have a full-scale pressure range of 0 to 130 kPa (0–1300 mb) and a minimum non-derangement pressure of 390 kPa (3.9 atmospheres) to cover mishandling in testing, etc. They are required to operate over a temperature range of $-60\,°C$ to $+90\,°C$ (or higher).

The effect of a small error in static pressure measurement and the resulting altitude error can be derived by rearranging Eq. (7.3), namely,

$$dH = -\frac{R_a T}{g} \frac{1}{P} dp$$

In the troposphere

$$dH = -\frac{R_a T_0}{g_0} \left(1 - \frac{L}{T_0} H\right) \frac{1}{P_S} dP_S \tag{7.50}$$

In the stratosphere

$$dH = -\frac{R_a T_{T^*}}{g_0} \cdot \frac{1}{P_S} dP_S \tag{7.51}$$

The effect of a 100 Pa (1 mb) error in static pressure measurement at sea level is thus equal to

$$\frac{287.0529 \times 288.15}{9.80665} \times \frac{1}{1013.25} \times 1 = 8.32\,\text{m}(27.3\,\text{ft})$$

However, the effect of a 100 Pa (1 mb) error in static pressure measurement at an altitude of 13,000 m (42,650 ft) when the static pressure is 16.5 kPa (165 mb) is equal to

$$\frac{287.0529 \times 216.65}{9.80665} \times \frac{1}{165} \times 1 = 38.43 \, \text{m} \, (126 \, \text{ft})$$

The need for accuracy in the static pressure measurement at the high altitudes used in cruise flight can be seen bearing in mind ATC height level separations of 1000 ft. A major part of the error budget in the static pressure measurement is the inherent pressure error due to the installation and location of the static pressure orifices. This means that the pressure sensors may not be measuring the true static pressure. This *pressure error*, or *position error* as it is sometimes called, has to be determined experimentally and is a function of Mach number and incidence. The uncertainty in this pressure error can be of the order of 100–150 Pa (1–1.5 mb) so that the error contribution from the pressure sensor should be as small as possible, say less than 30 Pa (0.3 mb). This would give a worst-case overall error of 150 + 30 = 180 Pa (1.8 mb) at 13,000 m altitude (42,650 ft) in the measurement of the static pressure which corresponds to an error in the pressure altitude of 69 m (227 ft) of which the static pressure sensor contribution is less than 12 m (38 ft).

The errors in static pressure measurement over the low altitude pressure range should also be less than 30 Pa (0.3 mb). This ensures the altimeter will read the true altitude on landing within 3.3 m (11 ft approximately) providing the correct ground pressure has been set in and is known to an accuracy of 0.1 mb.

Modern technology air data pressure sensors generally meet or better these figures and achieve accuracies which approach those of the pressure standards used in calibration. It should be noted that an accuracy of 10 Pa (0.1 mb) in 100 kPa (1000 mb) corresponds to an accuracy of 0.01%.

The pressure sensors must also have a very high resolution in order to be able to generate a rate of change of altitude $\left(\dot{H}\right)$ output with the required dynamic range.

A typical figure aimed for height resolution is better than 0.03 m (0.1 ft) at sea level which corresponds to a pressure resolution in the region of 0.1 Pa (0.001 mb).

(b) *Total pressure sensor*

A typical total pressure sensor has a full-scale pressure range of 0–260 kPa (0–2600 mb). The figure of 260 kPa corresponds to a calibrated airspeed of 426 m/s (828 knots). Minimum non-derangement pressure is of the order of three times full-scale pressure to cover mishandling in testing, etc.

The accuracy requirements for the total pressure sensor are every bit as exacting as those required from the static pressure sensor. Using a total pressure sensor and a static pressure

sensor to measure impact pressure ($P_T - P_S$) requires very high accuracy from both sensors as it involves taking the difference of two large comparable numbers at low airspeeds as the impact pressure is small.

The effect of errors in measuring the impact pressure at low airspeeds can be derived as follows:

$$Q_C = 1/2\rho V_T^2$$

Hence

$$dQ_C = \rho V_T dV_T$$

Consider the requirement for the measurement of airspeed to an accuracy of 0.5 m/s (approx. 1 knot) at an approach speed of 50 m/s (100 knots approx.).

$$dQ_C = 1.225 \times 50 \times 0.5 = 30.6 \, \text{Pa}$$

The maximum error in each sensor must therefore be less than 15 Pa (0.15 mb) to give the required accuracy at 50 m/s. The accuracy requirements become even more exacting as the airspeed is reduced below 50 m/s.

7.4.2.2 Pressure Sensor Technology

Two basic types of pressure sensor have now become well established in modern digital air data systems. Although there are other types of pressure sensor in service, attention has been concentrated on these two types as they account for most of the modern systems. The two main types can be divided into the following:

(a) Vibrating pressure sensors
(b) Solid state capsule pressure sensors

(a) *Vibrating pressure sensors*

The basic concept of this family of sensors is to sense the input pressure by the change it produces in the natural resonant frequency of a vibrating mechanical system.

The output of the sensor is thus a frequency which is directly related to the pressure being measured. This frequency can be easily measured digitally with very high resolution and accuracy without the need for precision analogue to digital conversion circuitry. This confers a significant advantage as it enables a very simple and very accurate interface to be achieved with a micro-processor for the subsequent air data computation.

The vibrating cylinder sensor is shown schematically in Fig. 7.10. The pressure sensing element consists of a thin-walled cylinder with the input pressure acting on the inside of the cylinder and with the outside at zero vacuum reference pressure. The cylinder is maintained in a hoop mode of

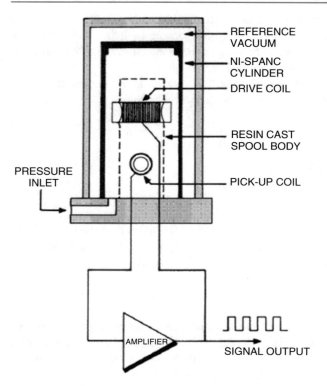

Fig. 7.10 Vibrating pressure sensor schematic. (By courtesy of Schlumberger Industries)

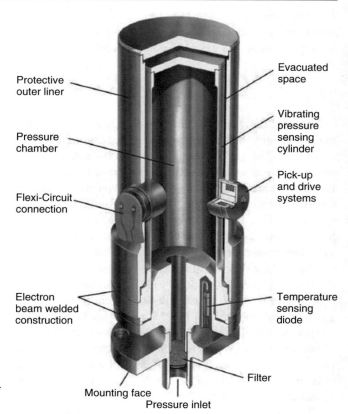

Fig. 7.11 Vibrating pressure sensor Type 3088

vibration by making it part of a feedback oscillator by sensing the cylinder wall displacement, processing and amplifying the signal and feeding it back to a suitable force-producing device. Electromagnetic drive and pick-off coils are used so that there is no contact with the vibrating cylinder.

The vibrating cylinder sensor is also density sensitive as the air adjacent to the cylinder wall effectively forms part of the mass of the cylinder so that a change in air density produces a small change in output frequency. This change, however, is much smaller than that due to pressure changes. The air density and pressure are directly related as shown in Sect. 7.2 so that this effect can be allowed for when the sensor is calibrated.

The small changes in density due to temperature variations together with the very small changes in the modulus of elasticity of the cylinder material with temperature can be compensated by measuring the sensor temperature and applying the appropriate corrections.

This type of sensor is extremely rugged, the very high Q of the cylinder conferring a very low susceptibility to vibration and shock, and the absence of moving parts enables it to withstand very high acceleration environments. The inherent simplicity and very low power consumption enable a very high reliability to be achieved together with very high accuracy and long-term stability.

Acknowledgement is made to Schlumberger industries for permission to publish the illustration in Fig. 7.11 of their

Type 3088 pressure sensor. Schlumberger industries have developed the vibrating cylinder type of pressure sensor to an outstanding performance and reliability level. Their pressure sensors have achieved very wide-scale usage in air data systems manufactured by several major avionics companies for large numbers of military and civil aircraft worldwide.

(b) *Solid state capsule pressure sensors*

This type of pressure sensor consists essentially of a capsule with a relatively thin diaphragm which deflects under the input pressure. They are fabricated from materials such as silicon, quartz, fused silica, or special ceramics. Figure 7.12 illustrates the basic construction. These materials exhibit virtually 'perfect mechanical properties' with negligible hysteresis and very stable and highly linear stress/strain characteristics. The modulus of elasticity variation with temperature is very stable and can be readily compensated by measuring the sensor temperature.

The deflection of the diaphragm is linear with input pressure but is also very small and a number of techniques are used to measure this deflection. These will be discussed later.

Semi-conductor technology is used in the fabrication of these sensors and this together with the absence of moving

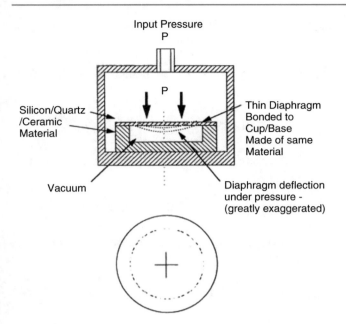

Input Pressure
P

Silicon/Quartz /Ceramic Material

P

Thin Diaphragm Bonded to Cup/Base Made of same Material

Vacuum

Diaphragm deflection under pressure - (greatly exaggerated)

Fig. 7.12 'Solid state' capsule pressure transducer

parts has led to the description 'solid state'. The technology also enables very small sensors to be fabricated with excellent repeatability because of the semi-conductor processes used.

The deflection of the solid state capsule under pressure is typically only 25–50 μm full scale, or less. The techniques which have been adopted to measure this very small deflection are briefly described below.

(i) *Integral strain gauges.* Piezo-resistive networks (or bridges) are ion implanted at the edge of a thin silicon diaphragm. Application of pressure causes the diaphragm to deflect thereby deforming the crystal lattice structure of the silicon which in turn causes the resistance of the piezo-resistive elements to change. The piezo-resistive elements are connected in a Wheatstone's bridge configuration as shown in Fig. 7.13 with two opposite resistive elements of the bridge located radially and the other two located tangentially on the diaphragm. The applied input pressure causes the resistance of the radial elements to decrease and the tangential elements to increase, or vice versa, depending whether the pressure is increasing or decreasing and so unbalances the Wheatstone's bridge. The output of the bridge is proportional to both pressure and temperature as the modulus of elasticity of silicon is temperature dependent. A temperature-sensitive resistive element is therefore incorporated into the diaphragm to measure the temperature so that the temperature-dependent errors can be corrected in the subsequent processing of the sensor output. (These temperature-dependent errors are very stable and repeatable.) A precision analogue to digital

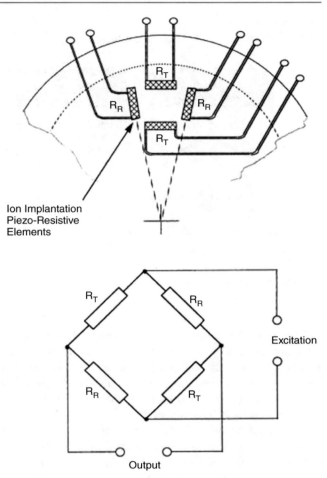

Ion Implantation Piezo-Resistive Elements

R_T
R_R
R_R
R_T

R_T
R_R
R_R
R_T

Excitation

Output

Fig. 7.13 Strain gauge configuration

(A to D) conversion circuit is required to convert the sensor analogue voltage output to a digital output. The supporting electronics including the individual device characterisation, A to D conversion and digital interface to the 'outside world' can all be incorporated in the sensor with modern micro-electronic technology. This technology is sometimes referred to as 'smart sensor' technology.

(ii) *Capacitive pick-off.* An alternative technique is to deposit a metallic film on an area at the centre of the diaphragm to form a capacitive element (or 'pick-off') whose capacitance changes as the diaphragm deflects under pressure. This forms part of a capacitance bridge network. Figure 7.14 shows the device construction. A correctly designed capacitive pick-off bridge combination can have extremely high resolution and can detect the incredibly small changes in capacitance resulting from minute deflections of the diaphragm. (It is noteworthy that the highest resolution yet achieved in an angular or linear displacement sensor has been achieved with a capacitive pick-off, leaving optical measuring techniques orders of magnitude behind.)

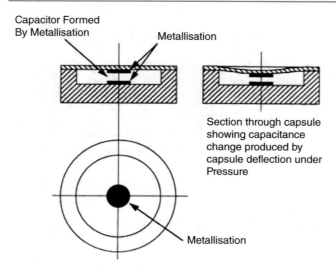

Section through capsule showing capacitance change produced by capsule deflection under Pressure

Fig. 7.14 Capacitance pick-off

This type of pressure capsule is fabricated from quartz, fused silica or special ceramic materials. The variation of the modulus of elasticity with temperature in these materials is less than with silicon. An integral temperature sensor is incorporated in the capsule to enable the temperature-dependent errors to be compensated.

The capacitive pick-off and bridge configuration must be designed so that it is not affected by changes in the stray capacitances inherent in the lead out wires and screening, etc., as these changes can be comparable or even exceed the very small changes due to the input pressure.

A precision A to D conversion is required to provide the required digital output.

It should be noted that the solid state capsule types of pressure sensors have an inherent sensitivity to acceleration (or 'g') because of the mass of the diaphragm. The sensitivity is small and can generally be ignored if the pressure sensors are mounted so that their sensitive axes are orthogonal to the normal acceleration vector.

The types of pressure sensor described in (a) and (b) are all used in modern digital air data systems. In a highly competitive market, all meet the requirements. As the saying goes – 'You pays your money and you takes your pick'.

7.4.3 Air Data Computation

A flow diagram of the air data computation processes is shown in Fig. 7.15. The computations are briefly described below.

7.4.3.1 Pressure Altitude
The static pressure can be computed at suitable increments of altitude from −914.4 m (−3000 ft) to 32,004 m (105,000 ft

approx.) and the data stored in a table look-up store using the appropriate formulae relating static pressure and altitude.

(a) Troposphere −914 m to 11,000 m (−3000 ft to 36,089 ft)

$$P_S = 1,013.25\left(1 - 2.25577 \times 10^{-5} H_P\right)^{5.255879}$$

(b) Stratosphere 11,000 m to 20,000 m (36,089 ft to 65,617 ft)

$$P_S = 226.32\, e^{-1.576885 \times 10^{-4}(H_P - 11,000)}$$

(c) Chemosphere 20,000 m to 32,000 m (65,617 ft to 105,000 ft)

$$P_S = 54.7482\left[1 + 4.61574 \times 10^{-6}(H_P - 20,000)\right]^{-34.163215}$$

The appropriate pressure altitude corresponding to the measured static pressure can then be derived from the table look-up store in conjunction with a suitable interpolation algorithm.

7.4.3.2 Vertical Speed, $\dot{H}_P$
The vertical speed or rate of change of altitude, $\dot{H}_P$, is derived from the rate of change of static pressure, $\dot{P}_s$, using the basic formulae relating the differentials dH and dP_S in Eq. (7.3), namely,

$$dH = -\frac{R_a T}{g_0}\frac{1}{P_s}dP_S$$

Hence

$$\dot{H}_P = -\frac{R_a T_S}{g_0}\frac{1}{P_S}\dot{P}_S$$

This can be computed from the actual measured and corrected air temperature value or from the standard atmosphere temperature. The latter is the usual case so that H_P is purely a function of P_S.

(a) Troposphere

$$\dot{H}_P = 8434.51\left(1 - 2.25577 \times 10^{-5} H_P\right)\frac{1}{P_S}\dot{P}_S$$

(b) Stratosphere

$$\dot{H}_P = 6341.62\frac{1}{P_S}\dot{P}_S$$

Fig. 7.15 Air data computation flow diagram

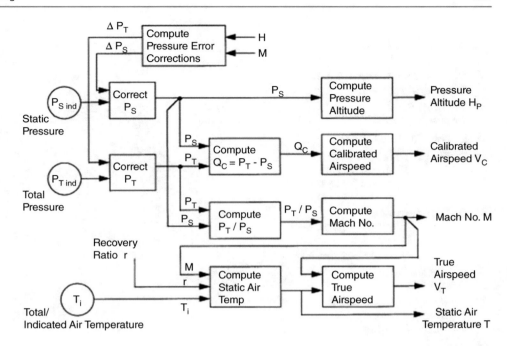

Deriving $\dot{H}_P$ by first differentiating P_S to generate $\dot{P}_S$ and then multiplying essentially by $1/P_S$ enables a better resolution to be obtained with fewer computational delays.

7.4.3.3 Mach Number
The pressure ratio, P_T/P_S, can be computed at suitable increments of Mach number and the results are stored in a table look-up store using the appropriate formulae.

(a) Subsonic speeds

$$\frac{P_T}{P_S} = \left(1 + 0.2M^2\right)^{3.5}$$

(b) Supersonic speeds

$$\frac{P_T}{P_S} = \frac{166.92M^2}{\left(7M^2 - 1\right)^{2.5}}$$

The actual pressure ratio, P_T/P_S, is then computed from the measured total pressure P_T and static pressure P_S. The appropriate Mach number corresponding to this computed pressure ratio is then derived from the table look-up store in conjunction with a suitable interpolation algorithm.

7.4.3.4 Calibrated Airspeed
The impact pressure, Q_C, can be computed at suitable increments over the range of calibrated airspeeds, say 25 m/s (50 knots) to 400 m/s (800 knots) and the results are stored in a table look-up store using the appropriate formulae.

(a) $V_C \leq 340.3$ m/s (661.5 knots)

$$Q_C = 101.325\left[1 + 0.2\left(\frac{V_C}{340.294}\right)^{3.5} - 1\right] \text{kN/m}^2$$

(b) $V_C \geq 340.3$ m/s (661.5 knots)

$$Q_C = 101.325\left[\frac{166.92\left(\frac{V_C}{340.294}\right)^7}{\left[7\left(\frac{V_C}{340.294}\right)^2 - 1\right]^{2.5}} - 1\right]$$

Q_C is derived by subtracting the measured total and static pressures. The appropriately calibrated airspeed corresponding to this impact pressure is then derived from the table look-up store in conjunction with a suitable interpolation algorithm.

7.4.3.5 Static Air Temperature
The static air temperature, T_S, is derived by computing the correction factor $1/(1 + r0.2\ M^2)$ and multiplying the measured (indicated) air temperature, T_m, by this correction factor, namely,

MAIN FEATURES :

- **Size: 4.69x3.96x4.86 in (excluding vane)**
- **Weight: 6.6lbs (including vane)**
- **Reliability: 6,000 hrs**
- **Outputs: Total Pressure
  Static Pressure
  Dynamic Pressure
  Local Flow Angle**
- **Output formats:
  MIL-STD-1553B
  RS-232C
  (Maintenance link)**
- **+-20v DC supplied from FCC**
- **115v 400Hz supplied from A/C Power Bus for vane de-icing**
- **Probe and Transducer Modules**
- **Electronic processing and interfacing modules**

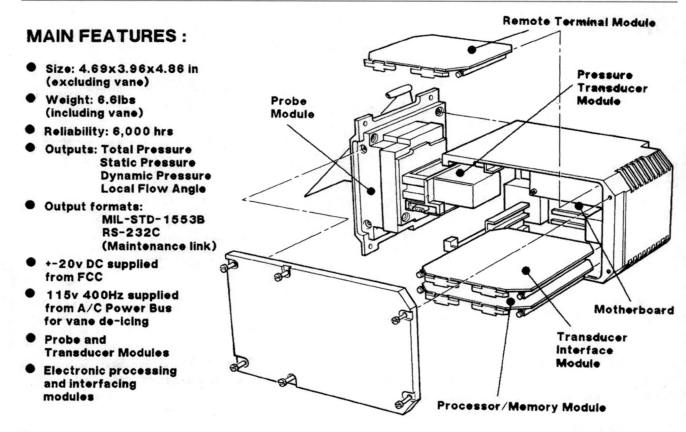

Fig. 7.16 Integrated air data transducer assembly. (By courtesy of BAE Systems)

$$T_S = \frac{T_m}{\left(1 + r0.2M^2\right)}$$

7.4.3.6 True Airspeed

The true airspeed, V_T, is derived from the computed Mach number and the computed static air temperature, T_S, namely,

$$V_T = 20.0468 \, M\sqrt{T_S}$$

The air data computations can be carried out by any modern 16-bit microprocessor, the software to carry out this task being relatively straightforward. Computations are typically carried out at an iteration rate of 20 Hz, although FBW flight control systems may require a faster iteration rate – up to 80 Hz.

A more exacting software task is the generation of the software to perform all the 'built in test' (BIT) and enable fault diagnosis to module level to be achieved.

7.4.4   Angle of Incidence Sensors

The importance of angle of incidence information has already been briefly explained. Knowledge of the angle of incidence can become particularly important at the extremes of the flight envelope when high values of incidence are reached and the limiting conditions approached. Many aircraft have incidence sensors installed so that the pilot can ensure the aircraft does not exceed the maximum permissible angle of incidence. Incidence sensors are also essential sensors for a fly-by-wire (FBW) flight control system, as mentioned earlier, and a redundant, failure survival system of sensors is required. It may also be necessary to measure the sideslip incidence angle, β.

A typical incidence sensor comprises a small pivoted vane suitably located on the aircraft fuselage near the nose of the aircraft. The vane is supported in low friction bearings so that it can align itself with the incident airflow under the action of the aerodynamic forces acting on it, like a weather vane. The angle of the vane relative to the fuselage datum and hence the angle of incidence is measured by a suitable angular position pick-off such as a synchro resolver (the output of which can be readily digitised).

It should be noted that an incidence sensor measures 'indicated angle of incidence' which must be converted to 'true angle of incidence' by an appropriate correction formula which is a function of the Mach number.

It is also possible to locate the Pitot probe on the vane together with the static pressure orifices to form an integrated unit which together with the appropriate sensors measures angle of incidence, total pressure and static pressure. This type of integrated Pitot probe/incidence vane is being adopted on several new high-performance aircraft as it minimises incidence contamination effects on the pressure measurements at high angles of incidence as well as providing a compact integrated solution. The integrated air data transducer, ADT, is illustrated in Fig. 4.5, Chap. 4. Figure 7.16 is an 'exploded view' showing the ADT assembly.

Further Reading

British Standard 2G199: Tables relating altitudes airspeed and Mach numbers for use in aeronautical instrument design and calibration, 1984

Dommasch, D.D., Sherby, S.S., Connolly, T.F.: Airplane Aerodynamics. Pitman, New York, USA (1967)

Autopilots and Flight Management Systems

8

8.1 Introduction

Autopilots and flight management systems (FMS) have been grouped together in one chapter as modern aircraft requires a very close degree of integration between the two systems, particularly in their respective control laws.

Much of the background to the understanding of autopilots has been covered in earlier chapters. For instance, the dynamics of aircraft control is covered in Chap. 3 and fly-by-wire flight control systems in Chap. 4, including pitch rate and roll rate manoeuvre command control systems. Chapter 4 also covers the methods used to achieve the very high integrity and safety required to implement FBW flight control through the use of redundancy. Similar levels of safety and integrity are required in some autopilot modes, for example, automatic landing and automatic terrain following (T/F). The digital implementation of flight control systems is also covered.

Some of the background to flight management systems has also been covered in previous chapters; for example, air data and air data systems in Chap. 7, navigation systems in Chap. 6 and displays and man-machine interaction in Chap. 2.

The basic function of the autopilot is to control the flight of the aircraft and maintain it on a pre-determined path in space without any action being required by the pilot. (Once the pilot has selected the appropriate control mode(s) of the autopilot.) The autopilot can thus relieve the pilot from the fatigue and tedium of having to maintain continuous control of the aircraft's flight path on a long duration flight so the pilot can concentrate on other tasks and the management of the mission.

A well-designed autopilot system which is properly integrated with the aircraft flight control system can achieve a faster response and maintain a more precise flight path than the pilot. Even more important, the autopilot response is always consistent whereas a pilot's response can be affected by fatigue and work load and stress. The autopilot is thus able to provide a very precise control of the aircraft's flight path for such applications as fully automatic landing in very poor, or even zero visibility conditions. In the case of a military strike aircraft, the autopilot in conjunction with a T/F guidance system can provide an all-weather automatic terrain following capability. This enables the aircraft to fly at high speed (around 600 knots) at very low altitude (200 ft or less) automatically following the terrain profile to stay below the radar horizon of enemy radars. Maximum advantage of terrain screening can be taken to minimise the risk of detection and alerting the enemy's defences.

The basic autopilot modes are covered in the next section. These include such facilities as automatic coupling to the various radio navigation systems such as VOR and the approach aids at the airport or airfield such as ILS and MLS. Flight path guidance derived from the aircraft's GPS system is also coming into increased use. The autopilot then steers the aircraft to stay on the path defined by the radio navigation aid. The autopilot can also be coupled to the flight management system which then provides the steering commands to the autopilot to fly the aircraft on the optimum flight path determined by the FMS from the flight plan input by the pilot.

The autopilot is thus an essential equipment for most military and civil aircraft, including helicopters. The advent of the micro-processor has also enabled relatively sophisticated and affordable autopilots to be installed in large numbers of general aviation type aircraft.

The prime role of the flight management system is to assist the pilot in managing the flight in an optimum manner by automating as many of the tasks as appropriate to reduce the pilot workload. The FMS thus performs a number of functions, such as the following:

- Automatic navigation and guidance including '4D' navigation
- Presentation of information
- Management of aircraft systems

© The Author(s), under exclusive license to Springer Nature Switzerland AG 2023
R. P. G. Collinson, *Introduction to Avionics Systems*, Methods in Molecular Biology 2687,
https://doi.org/10.1007/978-3-031-29215-6_8

- Efficient management of fuel
- Reduction of operating costs

The broad concepts and operation of an FMS are covered in Sect. 8.3. It should be appreciated that the detailed implementation of an FMS is a complex subject and can involve over 100 man-years of software engineering effort and very extensive (and expensive) flight trials before certification of the system can be obtained from the regulatory authorities. It is only possible, therefore, because of space constraints to give an overview of the subject.

It should also be pointed out that an FMS has an equally important role in a military aircraft. Accurate adherence to an optimum flight path and the ability to keep a rendezvous at a particular position and time for, say, flight refuelling or to join up with other co-operating aircraft are clearly very important requirements.

8.2 Autopilots

8.2.1 Basic Principles

The basic loop through which the autopilot controls the aircraft's flight path is shown in the block diagram in Fig. 8.1. The autopilot exercises a guidance function in the outer loop and generates commands to the inner flight control loop. These commands are generally attitude commands which operate the aircraft's control surfaces through a closed-loop control system so that the aircraft rotates about the pitch and roll axes until the measured pitch and bank angles are equal to the commanded angles. The changes in the aircraft's pitch and bank angles then cause the aircraft flight path to change through the flight path kinematics.

For example, to correct a vertical deviation from the desired flight path, the aircraft's pitch attitude is controlled to increase or decrease the angular inclination of the flight path vector to the horizontal. The resulting vertical velocity component thus causes the aircraft to climb or dive so as to correct the vertical displacement from the desired flight path.

To correct a lateral displacement from the desired flight path requires the aircraft to bank in order to turn and produce a controlled change in the heading so as to correct the error.

The pitch attitude control loop and the heading control loop, with its inner loop commanding the aircraft bank angle, are thus fundamental inner loops in most autopilot control modes.

The outer autopilot loop is thus essentially a slower, longer period control loop compared with the inner flight control loops which are faster, shorter period loops.

8.2.2 Height Control

Height is controlled by altering the pitch attitude of the aircraft, as just explained. The basic height control autopilot loop is shown in Fig. 8.2.

The pitch rate command inner loop provided by the pitch rate gyro feedback enables a fast and well damped response to be achieved by the pitch attitude command autopilot loop. As mentioned earlier, an FBW pitch rate command flight control system greatly assists the autopilot integration.

The pitch attitude command loop response is much faster than the height loop response – fractions of a second compared with several seconds. The open-loop transfer function of the height loop in fact approaches that of a simple integrator at frequencies appreciably below the bandwidth of the pitch attitude command loop. The height error gain, K_H (or gearing) is chosen so that the frequency where the open-loop gain is equal to unity (0 dB) is well below the bandwidth of the pitch attitude loop to ensure a stable and well damped height loop response. (The design of autopilot loops is explained in more detail in the worked example in the next section.)

The pitch attitude loop bandwidth thus determines the bandwidth of the height loop so that the importance of achieving a fast pitch attitude response can be seen.

The transfer function of the flight path kinematics is derived as follows. Vertical component of the aircraft's velocity vector is

Fig. 8.1 Autopilot loop

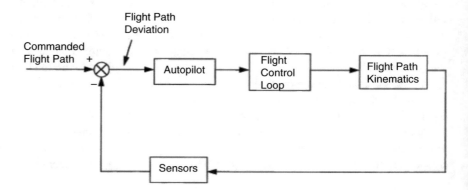

Fig. 8.2 Height control (heavy lines show FBW pitch rate command loop). Pitch rate gyro and AHRS are shown separately for clarity. In a modern aircraft, a strap-down AHRS/INS provides both θ and q outputs

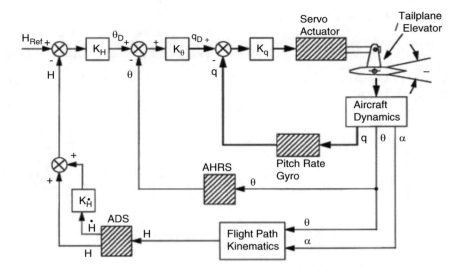

$$V_T \sin \theta_F = \dot{H}$$

where θ_F is the flight path angle, that is, the inclination of the velocity vector V_T to the horizontal.

$$\theta_F = \theta - \alpha$$

where θ is the aircraft pitch angle and α is the angle of incidence, $V_T \approx U$, aircraft forward velocity and θ_F is assumed to be a small angle. Hence

$$\dot{H} \approx U(\theta - \alpha) \qquad (8.1)$$

Thus

$$H = \int U(\theta - \alpha)dt \qquad (8.2)$$

8.2.3 Heading Control Autopilot

The function of the heading control mode of the autopilot is to automatically steer the aircraft along a particular set direction. As explained in Chap. 3, Sect. 3.5, aircraft bank to turn, and assuming a perfectly co-ordinated banked turn (i.e. no sideslip). Centripetal acceleration is $g \tan \Phi = U\dot{\Psi}$, where Φ is the aircraft bank angle, U is the aircraft forward velocity and $\dot{\Psi}$ is the rate of change of heading.

Hence, for small bank angles

$$\dot{\Psi} = \frac{g}{U}\Phi \qquad (8.3)$$

The basic control law used to control the aircraft heading is thus to demand a bank angle, Φ_D, which is proportional to

the heading error, $\Psi_E = (\Psi_{COM} - \Psi)$, where Ψ_{COM} is the commanded heading angle and Ψ the heading angle.

$$\Phi_D = K_\Psi \cdot \Psi_E \qquad (8.4)$$

where K_ψ is the heading error gain.

The heading control loop is shown in more detail in the block diagram in Fig. 8.3. The inner bank angle command loop should have as high a bandwidth as practical as well as being well damped in order to achieve a 'tight' heading control with good stability margins. This will be brought out in the worked example which follows.

Roll rate command FBW enables a fast, well damped bank angle control system to be achieved.

As explained earlier in Chap. 3, the dynamic behaviour of the lateral control system of an aircraft is complicated by the inherent cross-coupling of rolling and yawing rates and sideslip velocity on motion about the roll and yaw axes.

The dynamic behaviour of the aircraft is thus represented by a set of matrix equations as shown in Chap. 3 (Sect. 3.4.5).

However, a good appreciation of the aircraft and autopilot behaviour can be obtained by assuming pure rolling motion and neglecting the effects of cross-coupling between the roll and yaw axes. This can also be justified to some extent by assuming effective auto-stabilisation loops which minimise the cross-coupling effects.

Considering the bank angle demand loop, the aerodynamic transfer function relating roll rate, p, to aileron angle, ξ, is derived in Chap. 3, Sect. 3.6.1 which shows

$$\frac{p}{\xi} = \frac{L_\xi}{L_p} \cdot \frac{1}{1 + T_R D} \qquad (8.5)$$

Fig. 8.3 Heading control loop (heavy lines show FBW roll rate command loop). Roll rate gyro and AHRS are shown separately for clarity. In a modern aircraft, a strap-down AHRS/INS provides both p and Φ outputs

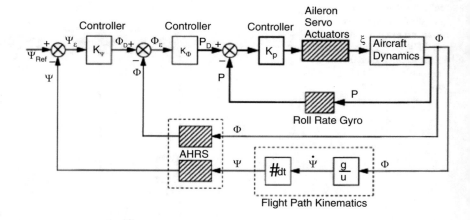

Fig. 8.4 Block diagram of bank angle demand loop

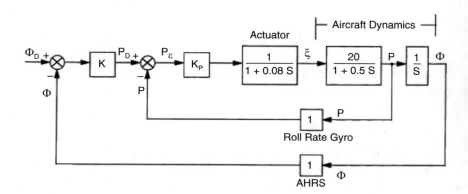

where T_R is the roll rate response time constant ($= I_x/L_p$) and L_ξ is the rolling moment derivative due to aileron angle, ξ and L_p is the rolling moment derivative due to roll rate, p and I_x is the moment of inertia of aircraft about the roll axis.

Referring to Eq. (5.16) in Chap. 5, viz.

$$\dot{\Phi} = p + q \sin \Phi \tan \theta + r \cos \Phi \tan \theta$$

If Φ and θ are small angles

$$\dot{\Phi} = p$$

That is,

$$\Phi = \int p \, dt \qquad (8.6)$$

These assumptions are used in the worked example which follows in order to simplify the aircraft dynamics. It is also assumed that there is no loss of height when the aircraft banks to correct heading, as appropriate changes to the pitch attitude and hence incidence are being effected by the height hold autopilot control, thus maintaining height.

8.2.3.1 Worked Example of Heading Control Autopilot

The aircraft in this example has a roll rate response time constant $T_R = 0.5$ seconds and $L_p/L_\xi = 20^\circ$/second per degree aileron deflection.

The response of the aileron servo actuators approximates to that of a simple first-order filter with a time constant of 0.08 seconds. The aircraft's forward velocity, $U = 250$ m/sec (approximately 500 knots) and is assumed to be constant.

Determine suitable values for the autopilot gains (or gearings) K_p, K_ϕ and K_ψ.

The overall loop for the heading control is as shown in Fig. 8.3. The block diagram for the bank angle demand loop is shown in Fig. 8.4.

It should be noted that the Laplace operator, s, has been substituted for operator $D(d/dt)$.

Determination of Roll Rate Error Gain, K_p

The open-loop transfer function of the innermost loop, namely, the roll rate feedback loop is

$$\frac{p}{p_E} = \frac{K_p 20}{(1 + 0.08s)(1 + 0.5s)} = K\,G(s) \qquad (8.7)$$

where $p_E = p_D - p$.

The closed-loop transfer function p/p_D is given by

$$\frac{p}{p_D} = \frac{K\,G(s)}{1 + KG(s)} \qquad (8.8)$$

The stability of the closed loop is determined by the roots of the characteristic equation

$$1 + K\,G(s) = 0 \qquad (8.9)$$

That is,

$$(1 + 0.08s)(1 + 0.5s) + 20K_p = 0$$
$$s^2 + 14.5s + 25(1 + 20K_p) = 0 \qquad (8.10)$$

Comparing with the standard form of second-order equation

$$2\zeta\omega_0 = 14.5$$
$$\omega_0^2 = 25(1 + 20K_p)$$

Choosing a damping ratio, $\zeta = 0.7$, requires the undamped natural frequency, $\omega_0 = 10.4$ rad/sec ($14.5/2 \times 0.7$). From which $K_p = 0.165°$ aileron angle per degree/second roll rate error.

$$K_p = 0.165 \text{ seconds}$$
$$\frac{p}{p_D} = \frac{0.767}{1 + \frac{2\zeta}{\omega_0}s + \frac{1}{\omega_0^2}s^2} \qquad (8.11)$$

where $\omega_0 = 10.4$ and $\zeta = 0.7$.

The roll rate response with and without roll rate feedback is shown in Fig. 8.5. (The step input to the system is adjusted to give the same steady-state rate of roll in each case.) The improved speed of response with roll rate feedback can be seen. The actuator lag limits the loop gain to 20×0.165, that is, 3.3, in order to achieve satisfactory damping without recourse to shaping (e.g. phase advance).

Determination of Bank Angle Error Gain, K_ϕ

Consider now the bank angle demand loop. The open-loop transfer function is

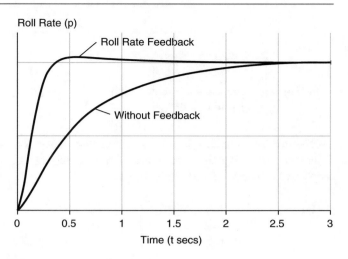

Fig. 8.5 Roll rate response

$$\frac{\Phi}{\Phi_E} = \frac{0.767K_\phi}{s\left(1 + \frac{2\zeta}{\omega_0}s + \frac{1}{\omega_0^2}s^2\right)} \qquad (8.12)$$

where $\omega_0 = 10.4$ and $\zeta = 0.7$.

The value of K_ϕ is determined using frequency response methods, as follows. The open-loop frequency response can be obtained by substituting $j\omega$ for s.

$$\frac{\Phi}{\Phi_E}(j\omega) = \frac{K}{j\omega\left(1 - \left(\frac{\omega}{\omega_0}\right)^2 + j2\zeta\frac{\omega}{\omega_0}\right)} \qquad (8.13)$$

$$= K F_1(j\omega) \cdot F_2(j\omega) \qquad (8.14)$$

where $K = 0.767K_\phi$.

$$F_1(j\omega) = \frac{1}{j\omega} \qquad (8.15)$$

$$F_2(j\omega) = \frac{1}{1 - \left(\frac{\omega}{\omega_0}\right)^2 + j2\zeta\frac{\omega}{\omega_0}} \qquad (8.16)$$

$F_1(j\omega)$ and $F_2(j\omega)$ can be expressed in terms of their respective moduli and phase angles (or arguments)

$$F_1(j\omega) = \left|\frac{1}{\omega}\right|, \angle -90° \qquad (8.17)$$

It is convenient to use a non-dimensional frequency, $u = \omega/\omega_0$, in $F_2(j\omega)$

$$F_2(j\omega) = \frac{1}{1 - u^2 + j2\zeta u} \qquad (8.18)$$

$$F_2(j\omega) = \left| \frac{1}{\sqrt{(1 - u^2)^2 + 4\zeta^2 u^2}} \right|, \quad \angle \tan^{-1} \frac{2\zeta u}{1 - u^2} \quad (8.19)$$

The evaluation of $KF_1(j\omega)\,F_2(j\omega)$ at specific values of ω is carried out using the logarithms of the moduli of the respective elements of the transfer function.

$$\begin{aligned} 20\,\log_{10}\, &| K F_1(j\omega) \cdot F_2(j\omega) \\ &| = 20\,\log_{10} K + 20\,\log_{10} | F_1(j\omega) \\ &| + 20\,\log_{10} | F_2(j\omega) | \end{aligned} \qquad (8.20)$$

The log modulus of the overall transfer function is thus the sum of the logs of the moduli of the individual elements.

The overall phase angle is similarly the sum of the individual phase angles of the respective elements

$$\angle F_1(j\omega) \cdot F_2(j\omega) = \angle F_1(j\omega) + \angle F_2(j\omega) \qquad (8.21)$$

The overall open-loop frequency response in terms of log gain and phase has been calculated from the above formulae and is set out in Table 8.1 for $K = 1$.

It should be noted that there are standard tables and graphs of the frequency responses of the common transfer function elements. There are also computer programs for evaluating the overall frequency response of a transfer function comprising a number of elements.

The open-loop frequency response is plotted on Fig. 8.6 which shows the log gain and phase at the specific values of ω chosen in the table. Such a graph is known as a 'Nichol's chart'. Lines of constant gain and phase loci for converting the open-loop frequency response into the closed-loop frequency response are normally superimposed on such a chart. The lines corresponding to -3 dB and $-90°$ are shown on the figure, but the rest have been omitted for clarity.

The Nichol's chart has the following advantages:

1. The logarithmic gain scale enables the response at all the frequencies of interest to be shown on the graph.
2. The effect of shaping the open-loop response can be readily seen. For example, increasing the loop gain, or introducing phase advance or one plus integral of error control, or the effect of additional lags in the system.
3. The closed-loop frequency response can be obtained using the constant gain and phase loci.
4. The gain and phase margins can be read off directly.

Table 8.1 Frequency response

Ω	$F_1(j\omega)$ 20 log M	Phase	$F_2(j\omega)$ 20 log M	Phase	$F_1(j\omega) \cdot F_2(j\omega)$ 20 log M	Phase
0.5	6 dB	−90°	0 dB	−3.9°	6 dB	−94°
1	0 dB	−90°	0 dB	−7.8°	0 dB	−98°
2	−6 dB	−90°	0 dB	−15.6°	−6 dB	−106°
3	−9.5 dB	−90°	0 dB	−23.8°	−9.5 dB	−114°
5	−14 dB	−90°	−0.2 dB	−41.2°	−14.2 dB	−131°
7	−16.9 dB	−90°	−0.7 dB	−59.9°	−17.6 dB	−150°
10.4	−20.3 dB	−90°	−2.9 dB	−90°	−23.2 dB	−180°
15	−23.5 dB	−90°	−7 dB	−118.1°	−30.5 dB	−208°
20	−26 dB	−90°	−11.6 dB	−135.1°	−37.6 dB	−225°
30	−29.5 dB	−90°	−18.4 dB	−151.1°	−47.9 dB	−241°

$M =$ Modulus

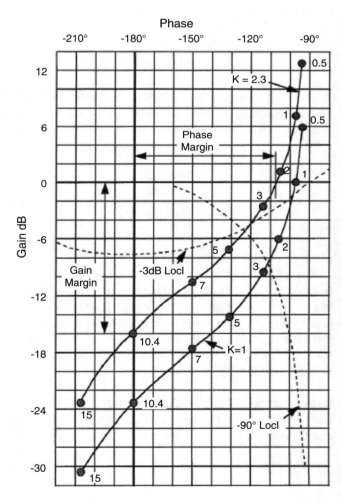

Fig. 8.6 Bank angle open-loop frequency response (Nichol's chart)

The value for the open-loop scalar gain, $K = 1$, gives a very stable loop, with a gain margin of 23.2 dB and a phase margin of 82°. The bandwidth, however, is relatively low. The closed-loop response is -3 dB down at $\omega = 1.2$ approximately (the intersection of the -3 dB loci with the open-loop response). The phase lag is 90° at $\omega = 2.8$ approximately (the intersection of the 90° loci with the open-loop response).

Increasing the open-loop scalar gain K to 2.3, that is, by 7.2 dB, improves the closed-loop bandwidth considerably. The closed-loop response is 3 dB down at $\omega = 3.8$ approximately, and there is 90° phase lag at $\omega = 4.2$ approximately. The gain and phase margins are reduced to 16 dB and 71°, respectively, but are still very adequate.

The open-loop transfer function is

$$\frac{\phi}{\phi_E} = \frac{2.3}{s(1 + 0.1346s + 0.0092456s^2)} \tag{8.22}$$

The closed-loop stability and damping can be confirmed by finding the roots of the characteristic equation

$$s(1 + 0.1346s + 0.0092456s^2) + 2.3 = 0$$

That is,

$$s^3 + 14.56s^2 + 108.16s + 246.77 = 0 \tag{8.23}$$

This factorises into

$$(s + 3.63)(s^2 + 10.93s + 68.5) = 0 \tag{8.24}$$

The $(s + 3.63)$ factor produces an exponential subsidence component in the response with a time constant of $1/3.63$ seconds, that is, 0.28 seconds. The quadratic factor $(s^2 + 10.93 s + 68.5)$ produces an exponentially damped sinusoidal component with an undamped natural frequency $\omega_0 = \sqrt{68.5}$, that is, 8.28 rad/s (1.32 Hz) and with a damping ratio $\zeta = 0.66$, confirming a reasonably well-damped closed-loop response. Hence

$$\frac{\Phi}{\Phi_D} = \frac{1}{(1 + 0.28s)(1 + 0.16s + 0.0146s^2)} \tag{8.25}$$

The required value of K_ϕ is obtained from $K = 0.767$ $K_\phi = 2.3$

$$K_\phi = 3 \tag{8.26}$$

Determination of Heading Error Gain, K_ψ

Referring to Fig. 8.3 it can be seen that the open-loop transfer function of the heading control loop is

$$\frac{\psi}{\psi_E} = \frac{K_\psi g/U}{s(1 + 0.28s)(1 + 0.16s + 0.0146s^2)} \tag{8.27}$$

Hence

$$\frac{\psi}{\psi_E}(j\omega) = \frac{K}{j\omega\left(1 + j\frac{\omega}{3.63}\right)\left[1 - \left(\frac{\omega}{8.28}\right)^2 + j0.16\omega\right]} \tag{8.28}$$

The factors in the denominator have been expressed in a form which enables the cut-off frequencies to be readily seen. The scalar gain $K = K_\psi\, g/U$. It can be seen from the above expression that at low frequencies (considerably lower than the first cut-off frequency of 3.63 rad/s), both the first- and second-order filter components of the transfer function approach unity. Hence at low frequencies

$$\frac{\psi}{\psi_E}(j\omega) \to \frac{K}{j\omega}$$

and

$$\left|\frac{\psi}{\psi_E}(j\omega)\right| \to \frac{K}{\omega}$$

The open-loop gain thus falls at -6 dB/octave at low frequencies.

The value of the frequency at which the open-loop gain is equal to unity (i.e. 0 dB) is known as the '0 dB cross-over frequency'. The value of K is chosen to give a 0 dB cross-over frequency such that the rate of change of open-loop gain with frequency is -6 dB/octave for well over an octave in frequency either side of the 0 dB cross-over frequency. This satisfies Bode's stability criterion.

The 0 dB cross-over frequency is chosen to be 0.7 rad/s. Hence

$$\frac{K}{0.7} = 1$$

That is,

$$K = 0.7 = K_\psi 9.81/250$$

$K_\psi = 18$ degrees bank angle per degree heading error.

From the open-loop transfer function, it can be seen that the gain starts to fall off at -12 dB/octave $(1/\omega^2)$ between $\omega = 3.63$ and $\omega = 8.28$ and then at -24 dB /octave $(1/\omega^4)$ above $\omega = 8.28$ rad/s. The phase changes very rapidly over this frequency range, for example $-115°$ at 1 rad/s, $-138°$ at 2 rad/s, $-180°$ at 4.1 rad/s, $-246°$ at 8.28 rad/s, $-310°$ at 15 rad/s and approaches $-360°$ as the frequency increases still higher.

The gain and phase margins with $K_\psi = 18$ are approximately 19 dB and 73°, respectively. The closed-loop bandwidth is about 1.4 rad/s when the phase lag is 90°. The 3 dB down frequency is just over 1 rad/s.

The loop gain clearly varies with forward velocity, U, and a generous gain margin has been allowed to cope with changes in speed. Gain scheduling with speed is required, however, to cover the whole of the aircraft speed envelope.

A bank angle limiter is also required to prevent too large a bank angle being demanded when there is a significant initial heading error on engaging the autopilot. Typical limits would be $\pm12°$ ($\pm0.2\ g$). As stated earlier, the aircraft dynamics have been simplified in this example to give an insight into the design of the inner and outer autopilot loops.

In practice, a full representation of the aircraft dynamics including the yaw/roll cross-coupling effects would be used to establish the autopilot design.

8.2.4   ILS/MLS Coupled Autopilot Control

8.2.4.1 Approach Guidance Systems

ILS is a radio-based approach guidance system installed at major airports and airfields where the runway length exceeds 1800 m which provides guidance in poor visibility conditions during the approach to the runway.

A small number of major airports are also now equipped with MLS – microwave landing system. MLS is a later and more accurate system which is superior in all aspects to ILS. ILS, however, is a very widely used system and it will be a long time before it is completely replaced. It will thus be supported and maintained for many years to come.

It should be noted that the advent of Satellite Based Augmentation Systems (SBAS) exploiting differential GPS techniques will be able to provide accurate and reliable approach path guidance in Cat. II visibility conditions (refer Chap. 6, Sect. 6.5.7). This will be increasingly used in the future.

The Ground Based Augmentation System, (GBAS), which is able to support Cat. III operation, is described later in Sect. 8.2.3.

The runway approach guidance signals from the ILS (or MLS) receivers in the aircraft can be coupled into the autopilot which then automatically steers the aircraft during the approach so that it is positioned along the centre line of the runway and on the descent path defined by the ILS (or MLS) beams. The autopilot control loops are basically the same for ILS or MLS coupling apart from some signal preconditioning.

Space does not permit a detailed description of either system and a very brief outline of the ILS system only is given as this is the most widely used system. The ILS system basically comprises a localiser transmitter and a glide slope transmitter located by the airport runway together with two or

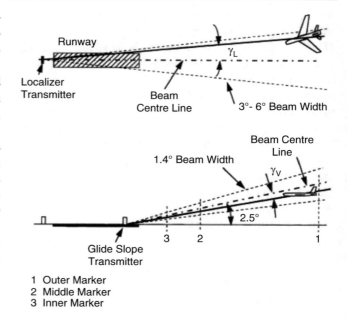

Fig. 8.7 ILS localiser and glide slope geometry

three radio marker beacons located at set distances along the approach to the runway. The airborne equipment in the aircraft comprises receivers and antennas for the localiser, glide slope and marker transmissions. The guidance geometry of the localiser and glide slope beams is shown in Fig. 8.7.

The localiser transmission, at VHF frequencies (108–122 MHz), provides information to the aircraft as to whether it is flying to the left or right of the centre line of the runway it is approaching. The localiser receiver output is proportional to the angular deviation γ_L, of the aircraft from the localiser beam centre line which in turn corresponds with the centre line of the runway.

The glide slope (or glide path) transmission is at UHF frequencies (329.3–335 MHz) and provides information to the aircraft as to whether it is flying above or below the defined descent path of nominally 2.5°, for the airport concerned. The glide slope receiver output is proportional to the angular deviation γ_V, of the aircraft from the centre of the glide slope beam which in turn corresponds with the preferred descent path. (The sign of the γ_L and γ_V signals is dependent on whether the aircraft is to the left or the right of the runway centre line, or above or below the defined glide slope.)

The marker beacon transmissions are at 75 MHz. The middle marker beacon is located at a distance of between 1000 and 2000 m from the runway threshold and the outer marker beacon is situated at a distance of between 4500 and 7500 m from the middle marker. The inner marker beacon is only installed with an airport ILS system which is certified to Category III landing information standards and is located at a distance of 305 m (1,000 ft) from the runway threshold.

It should be noted that ILS does not provide sufficiently accurate vertical guidance information down to touchdown.

The height limits and visibility conditions in which the autopilot can be used to carry out a glide slope coupled approach to the runway depend on the visibility category to which the autopilot system is certified for operation, the ILS ground installation standard, the runway lighting installation and the airport's runway traffic control capability.

Visibility conditions are divided into three categories, namely, Category I, Category II and Category III, depending on the vertical visibility ceiling and the runway visual range (RVR). These will be explained in more detail in Sect. 8.2.5, but briefly the visibility conditions deteriorate as the category number increases; 'Cat. III' includes zero visibility conditions. The required autopilot system capabilities in terms of safety and integrity for operation in these visibility categories are also discussed in Sect. 8.2.5.

An automatic glide slope coupled approach is permitted down to a height of 30 m (100 ft) above the ground, but only if the following conditions are met:

1. There is sufficient vertical visibility at a height of 100 ft with a runway visual range of at least 400 m for the pilot to carry out a safe landing under manual control (Category II visibility conditions). This minimum permitted ceiling for vertical visibility for the landing to proceed is known as the decision height (DH).
2. The autopilot system is certified for Cat. II operation. This will be explained in more detail in Sect. 8.2.5, but briefly a fail passive autopilot system is required. This is so that the pilot can take over smoothly in the event of a failure in the autopilot system.

Hence, when the decision height is reached the pilot carries out the landing under manual or automatic control. Alternatively, the pilot may execute a go-around manoeuvre to either attempt to land a second time or divert to an alternative airport/airfield.

A very high integrity autopilot system is required for fully automatic landing below a DH of 100 ft – Cat. III conditions. This is covered in Sect. 8.2.5.

8.2.4.2 Flight Path Kinematics

The mathematical relationships between the flight path velocity vector and the angular deviations of the aircraft from the guidance beam or 'beam errors' (γ_L and γ_V) are basically the same for the lateral and vertical planes. These relationships are derived below as they are fundamental to both the localiser and glide slope control loops. The particular parameters for each loop can be readily substituted in the general case which is shown in Fig. 8.8.

θ_F is the angle between the flight path velocity vector V_T and a chosen spatial reference axis. In the vertical guidance

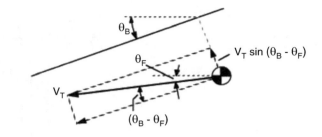

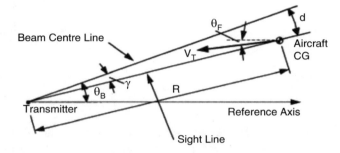

Fig. 8.8 Guidance geometry

case, the reference axis is the horizontal axis. In the lateral case, the reference axis is the runway centre line.

θ_B is the angle between the centre line of the guidance beam and the chosen reference axis. In the vertical case $\theta_B = 2.5°$ nominally and in the lateral case $\theta_B = 0$.

From Fig. 8.8, it can be seen that the component of the flight path velocity vector normal to the sight line is equal to $V_T \sin(\theta_B - \theta_F)$, that is, $U(\theta_B - \theta_F)$ as $V_T \approx U$ and $(\theta_B - \theta_F)$ is a small angle.

The rate of rotation of the sight line is thus equal to $U(\theta_B - \theta_F)/R$ where R is the slant range of the aircraft from the transmitter. θ_B is fixed, so that $\dot{\theta}_B = 0$. Hence

$$\dot{\gamma} = U(\theta_B - \theta_F)/R \qquad (8.29)$$

and

$$\gamma = \int \frac{U}{R}(\theta_B - \theta_F)\,dt \qquad (8.30)$$

The beam error $\gamma = d/R$, where d is the displacement of the aircraft's CG from the beam centre line. Hence, as R decreases, a given displacement produces an increasing beam error. For example, a displacement of 5 m at a range of 1500 m gives a beam error of 3.3 milliradians, that is, 0.2°. At a range of 300 m, the same offset produces a beam error of 1°. The guidance sensitivity thus increases as the range decreases, as can also be seen from Eq. (8.30).

8.2.4.3 ILS Localiser Coupling Loop

The ILS localiser coupling loop of the autopilot is shown in the block diagram in Fig. 8.9. It should be noted that the VOR

Fig. 8.9 Localiser coupling loop (ϕ is aircraft heading relative to runway centre line bearing)

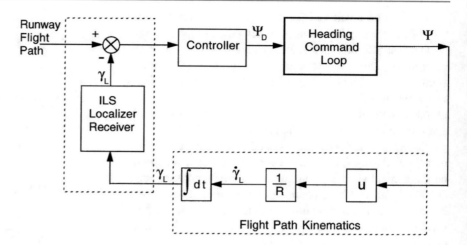

coupling loop is basically similar. The range of the ILS localiser is much lower, however, and is usually less than 24 km compared with up to 150 km for VOR. The heading control loop is the same as that described in Sect. 8.2.3, the heading angle ψ is referenced relative to the runway centre line bearing.

The flight path kinematics have been derived in the preceding section. Referring to Eq. (8.29) and then substituting into the equation the appropriate values of θ_B and θ_F, that is, $\theta_B = 0$ and $\theta_F = \psi$ gives

$$\dot{\gamma}_L = U\psi/R \qquad (8.31)$$

$$\gamma_L = \int \frac{U}{R}\psi dt \qquad (8.32)$$

The relationship is shown in the block diagram in Fig. 8.9.

It can be seen that the loop gain increases as R decreases and will reach a point where the loop becomes unstable. Gain scheduling with range is thus required.

The localiser controller in the autopilot provides a proportional plus integral of error control and generally a phase advance term. It should be noted that some filtering of the beam error signal, γ_L, is required to remove the 90 and 150 Hz modulation components inherent in the ILS system and specially to attenuate the noise present. This filtering inevitably introduces some lags.

8.2.4.4 ILS Glide Slope Coupling Loop

The ILS glide slope coupling loop is shown in the block diagram in Fig. 8.10. The pitch attitude command loop which controls the inclination of the flight path velocity vector is the same as that described in Sect. 8.2.2.

The flight path kinematics have been derived in Sect. 8.2.4.2. $\theta_F = (\theta\alpha)$ where θ is the aircraft pitch angle and α is the angle of incidence.

Substituting the appropriate values, $\theta_B = 0.044$ radians (2.5°) and $\theta_F = (\theta - \alpha)$ in Eq. (8.29) yields

$$\dot{\gamma}_V = U(0.044 - \theta + \alpha)/R \qquad (8.33)$$

Hence

$$\gamma_V = \int \frac{U}{R}(0.044 - \theta + \alpha)dt \qquad (8.34)$$

As in the localiser coupling loop, the loop gain increases as the range decreases and will ultimately cause instability. Gain scheduling with range is thus required.

The airspeed, U, is controlled by an auto-throttle system, as described later on in Sect. 8.2.6, and is progressively reduced during the approach according to a defined speed schedule.

The glide slope controller generally comprises a proportional plus integral and phase advance control terms with a transfer function of the form

$$K_C\left(1 + \frac{1}{T_1D}\right)\left(\frac{1 + T_2D}{1 + \frac{T_2}{n}D}\right)$$

where K_C is the controller scalar gain, T_1 is the integral term time constant, T_2 is the phase advance time constant and n is the phase advance gain.

As with the localiser coupling loop, the filtering of the beam error signal γ_V, introduces lags. The noise present also limits the phase advance gain, n.

The mixing of inertially derived position and velocity information from the INS (if fitted) with the ILS information in a suitable complementary filtering system using a Kalman filter can greatly improve the dynamic quality of the beam

Fig. 8.10 Glide path coupling loop

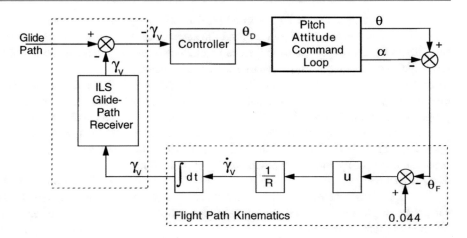

error control signals and virtually eliminate the noise. The loop performance and stability are thus improved. It can also smooth the spatial noise and some of the 'kinks' in the beam.

8.2.5 Automatic Landing

8.2.5.1 Introduction

The previous sections have described the automatic coupled approach phase of an automatic landing using the guidance signals from the ILS (or MLS) system in Cat. I or Cat. II visibility conditions. The pilot, however, takes over control from the autopilot when the decision height is reached and lands the aircraft under manual control.

Attempting to land an aircraft under manual control with decision heights of less than 100 ft, as in Cat. III conditions, is very demanding because of the lack of adequate visual cues and the resulting disorientation which can be experienced. There are only two alternatives for effecting a safe landing in such conditions:

(a) A fully automatic landing system with the autopilot controlling the landing to touchdown. A very high integrity autopilot system is required with failure survival capability provided by redundancy such that the probability of a catastrophic failure is less than $10^{-7}/$ hour. High integrity autopilot systems capable of carrying out a fully automatic landing in Cat. III conditions are now at a mature stage of development and large numbers of civil jet aircraft operated by major airlines worldwide are now equipped with such systems.

(b) The use of an enhanced vision system with a HUD as described in Chap. 2, using a millimetric wavelength radar sensor in the aircraft to derive a synthetic runway image. This is presented on the HUD together with the primary flight data, including the flight path velocity vector, and provides sufficient information for the pilot to land the aircraft safely under manual control.

The development of automatic landing systems has been a worldwide activity, with many companies and airlines, research establishments and regulatory authorities involved. It is appropriate, however, to mention the leading role of UK organisations and companies in the initial development of high integrity automatic blind landing systems. The UK had a distinct economic incentive to develop automatic landing systems because of the frequency of fogs and poor visibility conditions, especially in London, which necessitated frequent diversions. The first major contribution was made by the Blind Landing Experimental Unit, BLEU, of the former Royal Aerospace Establishment (now part of the UK Defence Research Agency, DRA). BLEU was formed in 1946 and by 1958 had developed the system which has subsequently formed the basis of all the automatic landing systems now in operation. The BLEU system is briefly described in Sect. 8.2.5.3. Another essential contribution was the formulation of the safety and certification requirements by the Air Registration Board around 1960 and the total system approach to safety and regulations including ground guidance equipment and airport facilities initiated by the UK Civil Aviation Authority (CAA). The next major contribution was made by the successful development and certification of two fully automatic landing systems by two British companies in the late 1960s. A monitored duplicate autopilot system for the VC10 airliner was developed by Elliott Brothers (London) Ltd., now part of BAE Systems Ltd., and a triplex autopilot system was developed for the Trident III airliner by Smiths Industries Ltd. The Trident III automatic landing system was in fact the first 'failure-survival fully automatic landing' system to be certified for Cat. III operation in the world, and is cleared to a decision height of 12 ft and a runway visual range of 100 m. The contributions made to the development of automatic landing systems during the 1960s by companies and organisations in the USA and France should also be acknowledged. At that time, efforts in both countries were directed towards the development of simpler fail passive automatic landing systems. The Sud Aviation SE

210 Caravelle airliner was in fact certified for Cat. III operation with a decision height of 50 ft around the same time as the Trident III. The experience and the design methodologies developed on these pioneering systems, such as redundancy management and techniques such as failure modes and effects analyses, have been of great value in subsequent programmes and have been disseminated worldwide.

It is also noteworthy that much of the technology developed for high integrity autopilot systems has provided the base for the subsequent development of fly-by-wire flight control systems.

It is appropriate at this point to describe the visibility categories in greater detail and the autopilot capabilities for operation in these categories.

8.2.5.2 Visibility Categories and Autopilot Requirements

As already explained, the two basic parameters used to define the visibility category are the decision height, that is, the minimum vertical visibility for the landing to proceed, and the runway visual range. Table 8.2 shows the various visibility categories.

The safety and integrity requirements for the autopilot system to be qualified for operation in the various visibility categories and the limits on its operation are shown in Table 8.3. The flight path guidance system must also meet the appropriate category standards and accuracy. For example, a Cat. II ILS system must provide accurate glide path guidance down to a height of 100 ft above the ground.

8.2.5.3 The BLEU Automatic Landing System

The BLEU automatic landing system is shown in Fig. 8.11 and is divided into four phases from the time the outer marker radio beacon is reached, about 8000 m from the threshold. These phases are briefly described below.

1. *Final approach.* This phase covers the approach from the outer marker beacon to the inner marker beacon. At the inner marker beacon, the aircraft flight path should be aligned with the defined glide path at a height of 100 ft above the ground and also aligned with the centre line of the runway. During this phase, the autopilot controls the aircraft flight path using the guidance signals from the ILS system. The aircraft height above the ground is measured by very accurate radio altimeters.

2. *Constant attitude.* The guidance signals from the ILS are disconnected from the autopilot when the aircraft reaches a height of 100 ft above the ground. The autopilot then controls the aircraft to maintain the pitch attitude and heading at the values set up during the approach until the height is reached at which the flare-out is initiated.

Table 8.2 Visibility categories

Category	Minimum visibility ceiling	Runway visual range
I	200 ft	800 m
II	100 ft	400 m
IIIa	12–35 ft	100–300 m
Depending on aircraft type and size		
IIIb	12 ft	<100 m
IIIc	0 ft	0 m

Table 8.3 Safety and integrity requirements

Category	Autopilot requirements and operational limits
I	Simplex autopilot system acceptable
	Pilot takes over the landing at a DH of 200 ft
II	Fail passive autopilot system required
	Pilot takes over the landing at a DH of 100 ft
IIIa	Full automatic landing system with automatic flare
	Failure survival autopilot system with a probability of catastrophic failure of less than 10^{-7} per hour required
	Pilot assumes control at touchdown
IIIb	Same as IIIa as regards autopilot system capability and safety and integrity requirements, but with automatic roll out control after touchdown incorporated. Runway guidance system required
	Pilot assumes control at some distance along the runway
IIIc	Same as IIIb as regards autopilot system capability and safety and integrity requirements, but with automatic taxi-ing control incorporated. Runway guidance required to taxi point. No system yet certified for Cat. IIIc operation

Fig. 8.11 BLEU automatic landing system

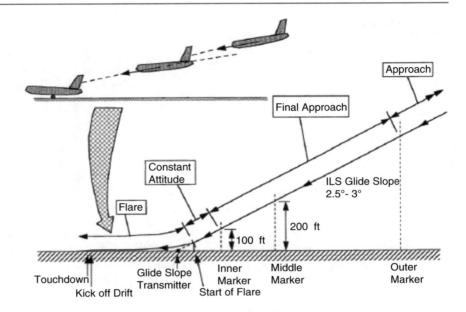

3. *Flare.* The aircraft pitch attitude is controlled by the feedback of the radio altimeter derived height to produce an exponential flare trajectory. The flare is initiated at a height of around 50 ft where the aircraft is over or very near the runway threshold. The aircraft is progressively rotated in pitch during the flare so that the flight path angle changes from the $-2.5°$ to $-3°$ value at the start of the flare to the positive value specified for touchdown. The vertical velocity is reduced from typically about 10 ft/s at the start of the flare manoeuvre to around 1 to 2 ft/s at touchdown.

4. *Kick off drift.* Just prior to touchdown a 'kick off drift' manoeuvre is initiated through the rudder control so that the aircraft is rotated about the yaw axis to align it with the runway. This ensures the undercarriage wheels are parallel to the runway centre line so that no sideways velocity is experienced by the wheels when they make contact with the runway.

8.2.5.4 Automatic Flare Control

The phases of the automatic landing have just been described. The automatic control loops for pitch attitude and heading control during the 'constant attitude' phase have been described earlier. The automatic flare control system, however, merits further explanation.

The mathematical law describing the exponential flare manoeuvre is

$$\dot{H} = -KH \qquad (8.35)$$

where H is the aircraft height above the ground and K is a constant.

The solution of Eq. (8.35) is

$$H = H_0 e^{-t/T} \qquad (8.36)$$

where H_0 is the aircraft height above the ground at the start of the flare manoeuvre, T is the time constant $= 1/K$. (Time, t, is measured from the start of the manoeuvre.) The horizontal velocity component of the aircraft is effectively equal to U as the flight path angle is a small angle. Assuming U is constant, the trajectory will thus be exponential.

The control law used for the autoflare is

$$H + T\dot{H} = H_{REF} \qquad (8.37)$$

H_{REF} is a small negative height, or bias, which ensures there is still a small downwards velocity at touchdown. This avoids the long exponential 'tail' to reach zero velocity and enables a reasonably precise touchdown to be achieved.

As mentioned earlier, the auto-flare is initiated at a height of around 50 ft where the aircraft is over or very near the runway threshold so that the radio altimeter is measuring the height of the aircraft above the runway. Low range radio altimeters are used to ensure accuracy. Safety and integrity considerations generally dictate a triplex or even quadruplex configuration of totally independent radio altimeters (rad. alt.).

The block diagram for the automatic flare control loop is shown in Fig. 8.12 which also indicates the redundancy necessary to meet the safety and integrity requirements in an automatic landing system.

The required control law response can be obtained by feeding back the rate of change of height suitably scaled by the required time constant, that is, $T\dot{H}$, together with the

Fig. 8.12 Automatic flare
control loop

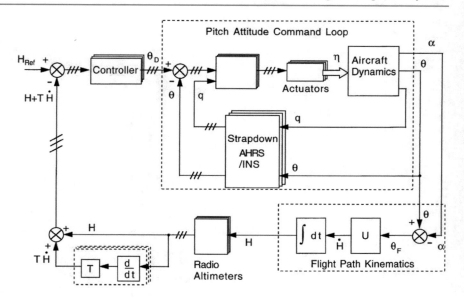

height measured by the rad.alt. This is because the response
of a closed-loop system approaches the inverse of the feedback path transfer function if the gain of the forward loop is
sufficiently high. The aircraft height response thus
approaches that of a simple first-order system with a transfer
function of $1/(1 + T D)$ at low frequencies where the forward
loop gain is high.

The $\dot{H}$ feedback term can be derived by differentiating the
suitably smoothed rad.alt. output. A filter is required to
smooth the noise present on the rad.alt. output and the differentiation process amplifies any high frequency noise
components which are present. The $\dot{H}$ signal so obtained
thus has lags present in its response because of the smoothing
filters required.

An alternative and superior source of $\dot{H}$ can be derived
from inertial mixing of the INS derived vertical velocity. This
is assuming there is adequate redundancy, for example, a
triplex INS installation. The inertial mixing enables an $\dot{H}$
output to be obtained with an excellent dynamic response
and low noise content as explained earlier in Chap. 6.

The auto-flare loop is a high-order system; apart from the
lags present in the filtered rad.alt. signals there are also the
lags present in the response of the pitch attitude command
loop. This loop controls $\dot{H}$ and its response is significantly
slower at the low speeds during the approach.

A proportional plus integral control term is used in the
auto-flare controller to ensure accuracy and some phase
advance is generally provided to compensate for the lags in
the loop and hence improve the loop stability and damping.

The approach speed is typically around 65 m/s (130 knots)
so that the vertical velocity at the start of the flare is around
$65 \sin 2.5°$, that is, 2.84 m/s or 9.3 ft/s. The flare time constant
is typically around 5 seconds so that the vertical velocity is
reduced exponentially from around 2.84 m/s at the start of the
flare to about 0.6 m/s at touchdown. The corresponding time

to touchdown is around 7.7 seconds. Hence, assuming the
approach speed stays constant, the touchdown point would be
around 500 m from the runway threshold.

8.2.6 Satellite Landing Guidance Systems

The navigation position accuracy of 1 m which can be
achieved with the differential GPS technique is being
exploited in the USA for landing guidance with a system
called the Ground Based Augmentation System, GBAS. The
Ground Based Augmentation System, when installed at an
airport, will be able to provide the high integrity and accurate
guidance necessary for landing in Cat. III visibility
conditions. The equipment is simpler and less expensive to
install and maintain than an Instrument Landing System
(ILS) or Microwave Landing System (MLS), so the GBAS
life-cycle operation costs are a fraction of the other systems.
It is therefore an attractive proposition for the many smaller
airports which are not equipped with ILS or MLS.

It is also a more flexible system. For example, the final
approach path need not be limited to straight line approaches,
but can be curved or stepped, horizontally or vertically.

The Ground Based Augmentation System is shown
schematically Fig. 8.13. It consists basically of several GPS
receivers connected to a base station in an equipment room.
The base station processes the measurements from the GPS
receivers, determines the differential corrections
(as explained in Chap. 6, Sect. 6.5.6), estimates their quality
and broadcasts this information to nearby aircraft. In addition, the co-ordinates of the final approach paths are transmitted to the aircraft.

The control laws exercised by the autopilot during the
automatic landing are basically similar whether the guidance
is provided by an ILS or MLS, or the GBAS. The flight path

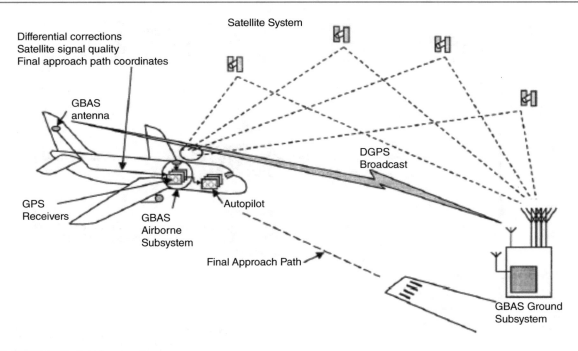

Fig. 8.13 GBAS landing guidance system

kinematics to align the aircraft's flight path with the commanded flight path are the same whether this is defined by a radio/microwave beam or the GBAS (refer Sect. 8.2.4.2).

The signals from the GBAS are tailored to provide an 'ILS look alike' guidance for the pilot's displays in terms of their sensitivity. This is to maximise pilot and operator acceptability.

An optional provision is made in the GBAS for additional ranging signals to be provided by ground-based transmitters called 'pseudolites' which can be installed to meet high availability requirements.

8.2.7 Speed Control and Auto-throttle Systems

Control of the aircraft speed is essential for many tasks related to the control of the aircraft flight path, for example, the position of the aircraft relative to some reference point.

The aircraft speed is controlled by changing the engine thrust by altering the quantity of fuel flowing to the engines by operating the engine throttles. Automatic control of the aircraft's airspeed can be achieved by a closed-loop control system whereby the measured airspeed error is used to control throttle servo actuators which operate the engine throttles. The engine thrust is thus automatically increased or decreased to bring the airspeed error to near zero and minimise the error excursions resulting from disturbances. A typical airspeed control system is shown in the block diagram in Fig. 8.14.

In any closed-loop system, the lags in the individual elements in the loop resulting from energy storage processes (e.g. accelerating inertias) exert a destabilising effect and limit the loop gain and hence the performance of the automatic control system. The dynamic behaviour of the engines over the range of flight conditions, the throttle actuator response and the aircraft dynamics must thus be taken into account in the design of the speed control system. The response of the jet engine thrust to throttle angle movement is not instantaneous and approximates to that of a simple first-order filter with a time constant which is typically in the range 0.3 to 1.5 seconds, depending on the thrust setting and flight condition. Clearly, the lag in the throttle servo actuator response should be small compared with the jet engine response. The aircraft dynamics introduces further lags as a change in thrust produces an acceleration (or deceleration) so that an integration is inherent in the process of changing the airspeed. The derivation of airspeed from the air data system can also involve a lag.

The rate of change of forward speed, $\dot{U}$, derived from a body mounted accelerometer with its input axis aligned with the aircraft's forward axis, can provide a suitable stabilising term for the control loop. (The $\dot{U}$ term could also be provided by a strap-down AHRS/INS.) A proportional plus integral of error control is usually provided to eliminate steady-state airspeed errors.

A duplicate configuration is generally used so that the systems fail passive. The throttle actuator is de-clutched in the event of a failure and the pilot then assumes control of the engine throttles.

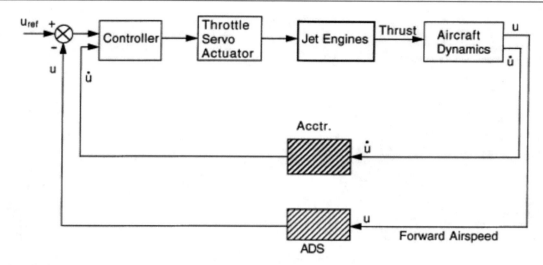

Fig. 8.14 Airspeed control system

8.3 Flight Management Systems

8.3.1 Introduction

The FMS has become one of the key avionics systems because of the major reduction in pilot work load which is achieved by its use. In the case of military aircraft, they have enabled single crew operation of advanced strike aircraft.

Flight management systems started to come into use in the mid-1980s and are now in very wide scale use, ranging from relatively basic systems in commuter type aircraft to 'all-singing, all-dancing' systems in long range wide body jet airliners. They have enabled two crew operation of the largest civil airliners and are generally a dual FMS installation because of their importance.

Figure 8.15 is a block diagram of a typical flight management system.

The benefits they confer are briefly set out below:

- *Quantifiable economic benefits* – provision of automatic navigation and flight path guidance to optimise the aircraft's performance and hence minimise flight costs.
- *Air traffic* – growth of air traffic density and consequently more stringent ATC requirements, particularly the importance of 4D navigation.
- *Accurate navigation sources* – availability of accurate navigation sources. For example, INS /IRS, GPS, VOR, DME and ILS / MLS.
- *Computing power* – availability of very powerful, reliable, affordable computers.
- *Data bus systems* – ability to interconnect the various sub-systems.

The FMS carries out the following tasks:

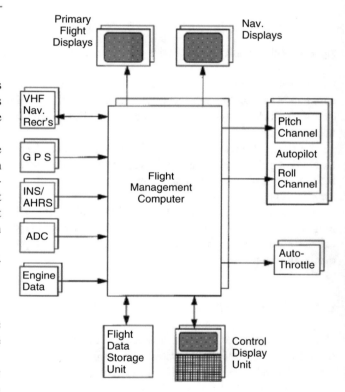

Fig. 8.15 Flight management system block diagram

1. Flight guidance and lateral and vertical control of the aircraft flight path.
2. Monitoring the aircraft flight envelope and computing the optimum speed for each phase of the flight and ensuring safe margins are maintained with respect to the minimum and maximum speeds over the flight envelope.
3. Automatic control of the engine thrust to control the aircraft speed.

Fig. 8.16 FMS architecture.
(By courtesy of Airbus)

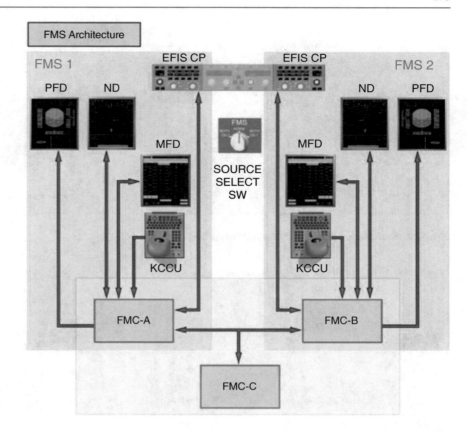

In addition, the FMS plays a major role in the flight planning task, provides a computerised flight planning aid to the pilot and enables major revisions to the flight plan to be made in flight, if necessary, to cope with changes in circumstances.

Figure 8.16 on the previous page shows the flight management system architecture of a modern airliner, in this case the Airbus A380.

Two independent Flight Management Systems FMS-1 on the Captain's side and FMS-2 on the First Officer's side carry out the flight management function.

The cockpit interfaces to the flight crew provided by each FMS comprise a Navigation Display (ND), a Primary Flight Display (PFD), a Multi-Function Display (MFD), a Keyboard and Cursor Control Unit (KCCU) and an Electronic Flight Instrument System (EFIS) Control Panel (EFIS CP).

The *Multi-Function Display* (MFD) displays textual data; over 50 FMS pages provide information on the flight plan, aircraft position and flight performance. The MFD is interactive; the flight crew can navigate through the pages and can consult, enter or modify the data via the Keyboard and Cursor Control Unit (KCCU).

The *Keyboard and Cursor Control Unit* (KCCU) enables the flight crew to navigate through the FMS pages on the MFD and enter and modify data on the MFD, as mentioned above, and can also perform some flight plan revisions on the lateral Navigation Display (ND).

The *EFIS Control Panel* (EFIS CP) provides the means for the flight crew to control the graphical and textual FMS data that appear on the ND and PFD.

There are three *Flight Management Computers*; FMC-A, FMC-B and FMC-C to carry out the functional computations, which can be reconfigured to maintain the system operation in the event of failures.

There are three different FMS operating modes: Dual Mode, Independent Mode and Single Mode dependant on the system status.

8.3.1.1 Dual Mode

Both flight management systems, FMS-1 and FMS-2, are healthy. Figure 8.17 shows the configuration in normal operation in the left side illustration and the configuration after a single flight management computer failure in the right side illustration.

In normal operation, FMC-A provides data to FMS-1, FMC-B provides data to FMC-2 and FMC-C is the standby computer.

Of the two active computers, one FMC is the 'master' and the other is the 'slave', depending on which autopilot is active and the selected position of the FMS Source Select Switch.

The two active FMCs independently calculate data, and exchange, compare and synchronise these data. The standby computer does not perform any calculations, but is regularly updated by the master FMC.

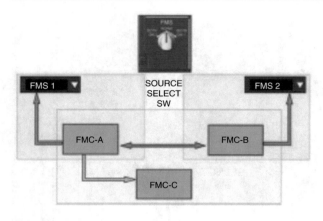
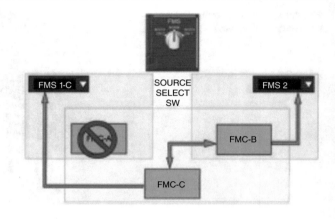

Fig. 8.17   Dual Mode operation. (By courtesy of Airbus)

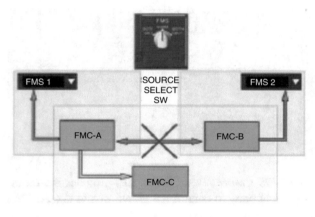
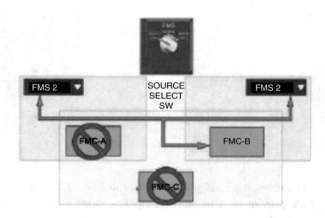

Fig. 8.18   Independent Mode and Single Mode configurations. (By courtesy of Airbus)

In the case of a single FMC failure, for example, FMC-A, FMC-C provides data to FMS-1. As shown in the right side illustration in Fig. 8.17.

8.3.1.2 Independent Mode

In the Independent Mode, FMS-1 and FMS-2 are both operative, but there is no data exchange between them because they disagree on one or more items such as aircraft position and gross weight. This case is shown in the left side illustration in Fig. 8.18.

8.3.1.3 Single Mode

The loss of two FMC's causes the loss of either FMS-1 or FMS-2.

The data from the operative FMS is displayed to the flight crew by operating the Source Select Switch. This case is shown in the right side illustration in Fig. 8.18.

8.3.2   Radio Navigation Tuning

The FMS automatically tunes the radio navigation aids, (NAVAIDs), used for the radio position computation, the

NAVAIDS for display on the Navigation Displays and the landing system NAVAIDS.

In 'Dual' and 'Independent' modes, each FMS tunes its onside NAVAIDS.

These comprise, in the case of the A380, one VOR, four DMEs, one ILS (MLS/GLS optional), one ADF (optional).

In 'Single' FMS mode, or in the case of a communications failure between an FMS and its onside Radio Management Panel (RMP), the available FMS will tune the NAVAIDS on both sides.

The tuning of the onside NAVAIDS passes through the onside RMP, to synchronise the NAVAIDS tuning between the FMS and the RMP.

The A380 FMS radio navigation tuning system is shown in Fig. 8.19 together with the 'POSITION/ NAVAIDS Page' display on the Multi-Function Display.

The NAVAIDS displayed on the Navigation Displays and the landing system NAVAIDS can also be tuned manually on the 'POSITION/NAVAIDS' Page of the MFD, or on the Radio Management Panel.

Manual tuning always has priority over automatic tuning.

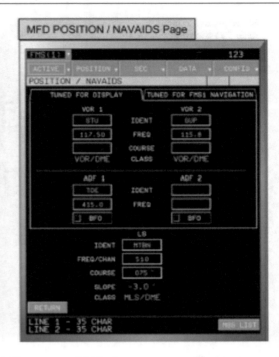

Fig. 8.19 NAVAIDS tuning and POSITION/NAVAIDS Page displays on MFD. (By courtesy of Airbus)

8.3.3 Navigation

The FMS combines the data from the navigational sources, comprising the inertial systems, GPS and the radio navigation systems, in a Kalman filter to derive the best estimate of the aircraft position. The accuracy of this estimate is also evaluated.

Figure 8.20 is a block diagram of the Kalman filtering of the navigational sources (refer Chap. 6, Sect. 6.3).

Each FMS computes the aircraft position and the position accuracy. The FMS computed position is an optimum combination of the inertial position and the GPS or radio position, depending on which equipment provides the most accurate data. This results in four navigation modes in decreasing order of priority:

- Inertial (IRS) – GPS
- Inertial (IRS) – DME/DME
- Inertial (IRS) – VOR/DME
- Inertial (IRS) only

The FMS aircraft position always uses the inertial position. This computation is not possible if the inertial position is not valid, and in this case all the FMS navigation and flight planning functions are no longer available.

The FMS continually computes the Estimated Position Uncertainty (EPU), and the EPU is used, together with the Required Navigation Performance (RNP,) to define the aircraft navigation accuracy.

The FMS continuously compares the actual EPU with the current RNP, and defines the navigation class as follows:

- HIGH, if the EPU is less than, or equal to the RNP.
- LOW, if the EPU is greater than the RNP.

The navigation class has to satisfy the Airworthiness Authorities Accuracy Requirements (AAAR).

The FMS computes ground speed, track, wind direction and velocity. (It should be noted that the air data system provides the height information for vertical navigation.) As stated earlier, the FMS provides both lateral and vertical guidance signals to the autopilot to control the aircraft flight path.

In the lateral case, the FMS computes the aircraft position relative to the flight plan and the lateral guidance signals to capture and track the flight path specified by the flight plan.

Three-dimensional vertical guidance is provided to control the vertical flight profile including the time dimension as will be explained in more detail later. This is of particular benefit during the descent and approach.

Fig. 8.20 Kalman filtering of navigational data sources

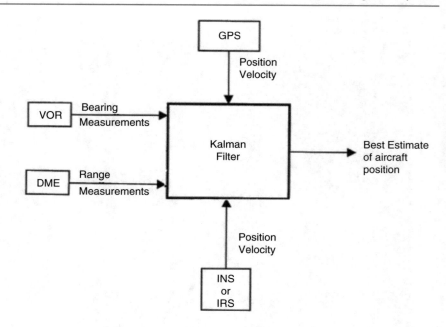

8.3.4   Flight Planning

As explained earlier, a major function of an FMS is to help the flight crew with flight planning and it contains a database of the following:

- *Radio NAVAIDS* – VOR, DME, VORTAC, TACAN, NDB, comprising identification, latitude/longitude, altitude, frequency, magnetic variation, class, air-line figure of merit.
- *Waypoints* – usually beacons.
- *Airways* – identifier, sequence number, waypoints, magnetic course.
- *Airports* – identifier, latitude, longitude, elevation, alternative airport.
- *Runways* – length, heading, elevation, latitude, longitude.
- *Airport procedures* – ICAO code, type, SID, STAR, ILS, profile descent.
- *Company routes* – original airport, destination airport, route number, type, cruise altitudes, cost index.

The navigation data base is updated every 28 days, according to the ICAO AiRAC cycle, and is held in non-volatile memory. It is clearly essential to maintain the recency and quality of the data base and the operator is responsible for the detail contents of the data base which is to ARINC 424 format.

Figure 8.21 illustrates a typical airline route section and shows the information content.

The flight crew can enter the flight plan in the FMS including all the necessary data for the intended lateral and vertical trajectory.

When all the necessary data is entered, the FMS computes and displays the speed, altitude, time and fuel predictions that are associated with the flight plan.

The flight crew can change the flight plan at any time; a change to the lateral plan is called a 'lateral revision' and a change to the vertical plan a 'vertical revision'.

The FMS can simultaneously memorise four flight plans:

- One *active* flight plan for lateral and vertical long-term guidance and for radio navigation auto-tuning.
- Three *secondary* flight plans with drafts to compare predictions, to anticipate a diversion or to store company, ATC and Onboard Information System flight plans.

The lateral flight plan includes the departure, cruise and arrival and is composed of waypoints that are linked with flight plan legs and transitions between legs.

Figure 8.22 shows the displays on the MFD of the 'Active Initialisation' Page (ACTIVE INIT) and the 'Active Fuel' Page (ACTIVE FUEL).

The ease and visibility of the data entry process can be appreciated.

A flight plan can be created in three ways:

1. By inserting an origin/destination pair and then manually selecting the departure, waypoints, airways and arrival.
2. By inserting a company route stored in the database.
3. By sending a company request to the ground for an active Flight Plan (F-PLN) uplink.

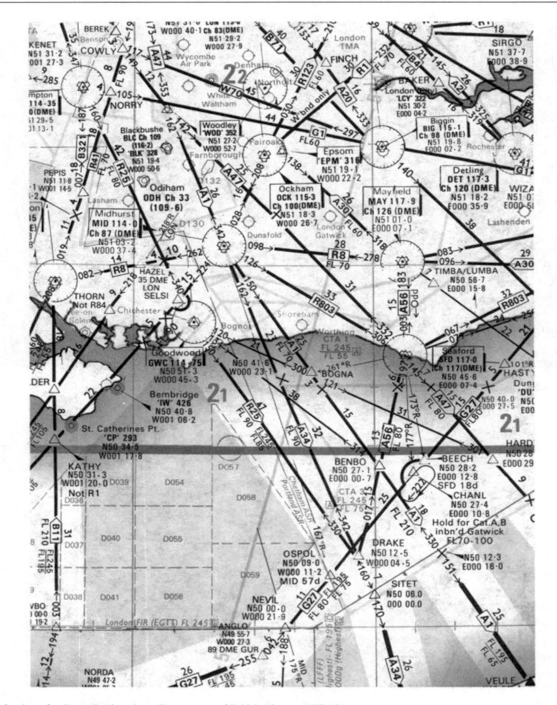

Fig. 8.21 Section of radio navigation chart, (By courtesy of British Airways AERAD)

The flight crew can perform the following lateral revisions:

- Delete and insert waypoints.
- Departure procedures: Take-off runway, Standard Instrument Departure (SID) and transition.
- Arrival procedures: Runway, type of approach, Standard Terminal Arrival Route (STAR), via, transition.

- Airways segments.

The flight crew can also perform the following vertical revisions:

- Time constraints.
- Speed constraints.
- Constant Mach segments.

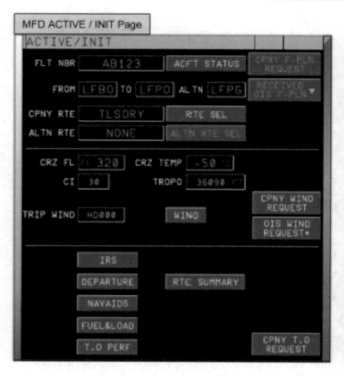

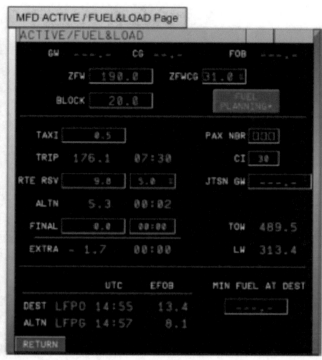

Fig. 8.22 'Active Initialisation' Page and 'Active Fuel' Page displays on MFD. (By courtesy of Airbus)

- Altitude constraints.
- Step altitudes.
- Wind.

Figure 8.23 illustrates the 'Active Flight Plan' page display on the MFD for carrying out flight plan revisions.

8.3.5 Performance Prediction and Flight Path Optimisation

The FMS is able to optimise specific aspects of the flight plan from a knowledge of the aircraft type, weight, engines and performance characteristics, information on the wind and air temperature and the aircraft state – airspeed, Mach number, height, etc.

The FMS continually monitors the aircraft envelope and ensures that the speed envelope restrictions are not breached. It also computes the optimum speeds for the various phases of the flight profile. This is carried out taking into account factors such as:

- Aircraft weight – computed from a knowledge of the take-off weight and the fuel consumed (measured by the engine flow meters). It should be noted that fuel can account for over 50% of the aircraft weight at take-off.

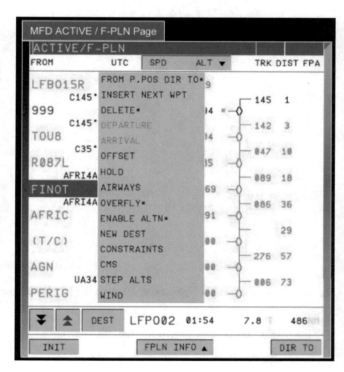

Fig. 8.23 'Active Flight Plan' Page display on MFD. (By courtesy of Airbus)

- CG position – computed from known aircraft loading and fuel consumed.
- Flight level and flight plan constraints.
- Wind and temperature models.
- Company route cost index.

The recommended cruise altitude and the maximum altitude are also computed from the above information.

The flight crew enter the following data to enable the performance computations and flight plan predictions to be made:

- Zero Fuel Weight (ZFW) and Zero Fuel Centre of Gravity (ZFCG)
- Block fuel
- Airline Cost Index (CI)
- Flight conditions (Cruise Flight Level (CRZ FL), temperature, wind)

The FMS computes the following predictions from the flight plan and the flight crew data entries:

- Wind and temperature
- Speed changes
- Pseudo waypoint computation: T/C, T/D, LVL OFF
- For each waypoint or pseudo waypoint:
  - Distance
  - Estimated Time of Arrival (ETA)
  - Speed
  - Altitude
  - Estimated Fuel on Board (EFOB)
  - Wind for each waypoint or pseudo waypoint
- For primary and alternate destination
  - ETA
  - Distance to destination
  - EFOB at destination

These predictions are continually updated depending on the following:

- Revisions to the lateral and vertical flight plans
- Current winds and temperature
- Actual position versus lateral and vertical flight plans
- Current guidance modes

The predictions and the lateral flight plan combine to form a vertical profile that has six flight phases.

The Multi Function Display Page display used to carry out the performance calculation and optimisation process is shown in Fig. 8.24.

Flight envelope protection is achieved by computing maximum and minimum selectable speed, stall warning, low

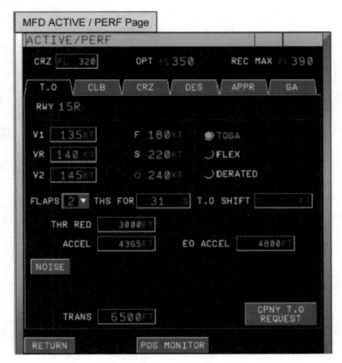

Fig. 8.24 'Active Performance' Page display on MFD. (By courtesy of Airbus)

energy threshold, alpha floor signal and reactive wind shear detection. The FMS also computes manoeuvring speed and flap and slat retraction speeds.

Figure 8.25 shows these limits displayed on the Primary Flight Displays.

8.3.6 Control of the Vertical Flight Path Profile

The FMS selects the speeds, altitudes and engine power settings during climbs, cruises and descents taking into account the flight plan, the prevailing conditions and the optimisation of the operation of the aircraft.

The vertical definition of a typical flight plan is shown in Fig. 8.26.

The tasks which can be carried out and the facilities provided by the FMS during the various phases of the flight are briefly summarised below:

- TAKE-OFF – The critical speeds V1, VR and V2 are inserted by the crew and displayed on the primary flight displays.
- CLIMB – The FMS uses the manually input speed, the ATC constraint speed or the economical speed. It determines the start of the climb during take-off and predicts the end of the climb and the optimum cruising flight level.

Fig. 8.25 Speed scale showing safe limiting values on PFD. (By courtesy of Airbus)

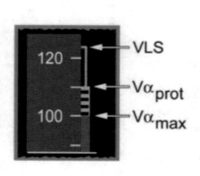

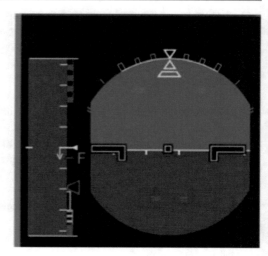

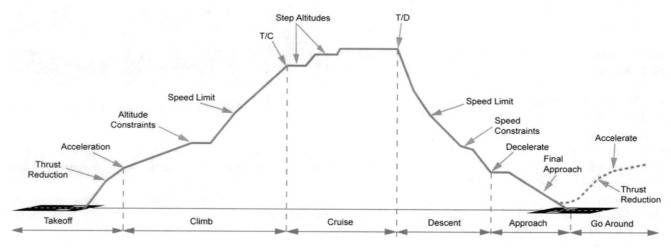

Fig. 8.26 Flight plan – vertical definition. (By courtesy of Airbus)

- CRUISE – Five flight levels can be defined manually in the FMS. Two flight levels can be stored for every route in the navigation data base. During the cruise, ATC or the crew may change the cruise altitude and the FMS can perform a 'step' climb at economical speed or a 'step' descent at 1000 ft/min at an economical speed. These events are also displayed symbolically on the navigation display.

- DESCENT – The FMS uses the manually input speed, the ATC constraint speed or the economic speed. The altitude and speed during the descent are computed as a function of the distance to the destination and a geometric profile is formed. The flight path is then computed backwards to satisfy the constraints.

- APPROACH – The FMS can be coupled to the autopilot or alternatively provide guidance information to the pilot for manual control of the aircraft. Speed is critical during

this phase and the approach speed is computed with respect to V_{REF} and the landing configuration (flaps, slats, etc.) and the wind at the destination.

The approach mode is entered at the end of the descent and the approach ends either with landing or go around. Lateral guidance is provided by the FMS from the computed aircraft position and vertical guidance from barometric altitude when an RNAV approach has been selected. The FMS also provides speed control.

At the end of an RNAV approach the crew takes control to carry out the landing using visual references. When an ILS approach has been selected, the FMS tunes the ILS frequency and selects the runway heading as required for the runway selected by the crew. The approach and landing guidance is carried out by the autopilot using the ILS localiser references for horizontal guidance and the ILS glide slope references for the vertical guidance until the

Fig. 8.27 Aircraft trajectory in vertical plane displayed on lower section of Navigation Display. (By courtesy of Airbus)

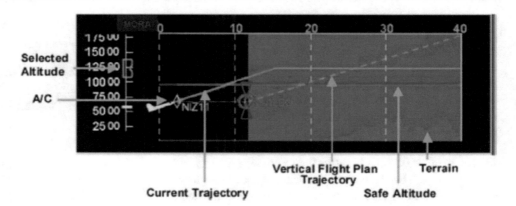

Selected Altitude

A/C

Current Trajectory

Vertical Flight Plan Trajectory

Safe Altitude

Terrain

glide extension and flare phases, unless the crew elect to carry out an automatic go around or elect to take over control.

- GO AROUND – This is always assumed. The FMS manages the climb to the accelerating altitude or a selected altitude and provides track guidance from the outbound track defined in the flight plan.

The lower section of the Navigation Display is known as the 'Vertical Display Zone' and is used to display the vertical flight profile. (Refer to Fig. 2.37 in Chap. 2, which shows a typical Navigation Display.)

Figure 8.27 is an annotated display of the aircraft's flight path in the vertical plane where the terrain height information is combined with the vertical flight profile to provide a synthetic view of the aircraft's vertical situation.

8.3.7 Operational Modes

The FMS provides a number of very useful operational modes which are shown in the little sketches in Fig. 8.28a–d.

- *Tangential go direct to mode* – This is shown in Fig. 8.28a and provides navigation from the current position to any waypoint in the flight plan or entered during the flight.
- *Turn anticipation* – This is shown in Fig. 8.28b and avoids overshooting waypoints. It reduces both the distance flown and off-track manoeuvring.
- *Parallel offset tracking* – This is illustrated in Fig. 8.28c. The lateral offset allows ATC to increase traffic flows in certain cases.
- *Holding pattern* – This is illustrated in Fig. 8.28d. The FMS produces a precision holding pattern based on published ICAO entry procedure to reduce the pilot work load.

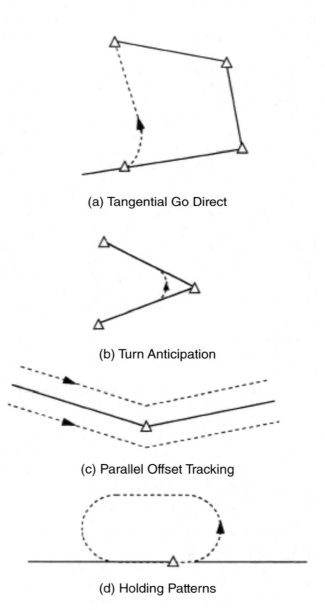

(a) Tangential Go Direct

(b) Turn Anticipation

(c) Parallel Offset Tracking

(d) Holding Patterns

Fig. 8.28 Operational modes

8.3.8 4D Flight Management

4D navigation has already been briefly referred to a number of times; in fact '4D flight management' is a better description of the process. 4D flight management covers the optimisation of the aircraft's flight along the most fuel conservative 3D path through climb, cruise and descent within the constraints of the air traffic control environment. Most importantly, the arrival time of the aircraft is also controlled so that it fits into the air traffic flow without incurring or causing delays. This is achieved by the automatic closed-loop control exercised by the FMS through the autopilot and auto-throttle systems. These control the aircraft's flight path so that its 3D position at any time corresponds closely with the optimum time referenced flight path generated by the FMS computer.

When instructed to resume the descent, the FMS then re-engages and guides the aircraft down to the ATC specified position at the pre-established time.

The process of automatic 4D descent control is very briefly described below. An ideal 4D trajectory, which is defined by a table of time, range and altitude, is pre-computed as briefly explained earlier.

Flexibility is essential for 4D flight management and there must be several minutes of arrival time variation possible so that the arrival time of the aircraft will coincide with the arrival time required by the air traffic controller. Some flexibility is also needed to compensate for differences between the actual and predicted wind and for performance modelling discrepancies with the actual aircraft in the FMS computations, although these are typically quite small.

Figure 8.29 shows the altitude and range errors from the ideal time referenced trajectory.

The air traffic control system also specifies the arrival time of the aircraft to the nearest integral minute, so the aircraft must have at least 30 seconds of arrival time flexibility to fit. Delays caused by traffic or weather also require some flexibility. Besides arrival time flexibility, there must also be some path flexibility. For example, air traffic control may require the aircraft to vector off course because of traffic. The 4D guidance and control exercised by the FMS then automatically re-engage the descent and compute the new course to the position specified by the ground ATC and make good the original or newly specified arrival time. Occasionally aircraft are required to level-off in their descent, for example, when the high-altitude air traffic controller hands the aircraft off to the low altitude controller.

For the current aircraft position (current range), the required altitude is computed from the ideal trajectory, and a combination of pitch attitude and direct drag or engine thrust is used to control both the altitude error and the speed.

Pitch attitude is controlled through the pitch channel of the autopilot by the FMS to maintain the aircraft on the ideal

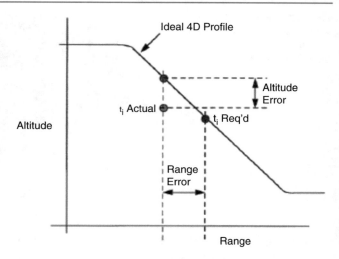

Fig. 8.29 4D descent control variables

trajectory or to return to it, while drag or thrust is used to control the speed as necessary.

For the current time, the required range is computed and speed changes are effected to control the range error and hence the required time.

The FMS exercises speed control for the 'too low' case by engine thrust control through the auto-throttle system. The 'too high/too fast' case is controlled by direct drag control by the spoilers/speed brakes. For safety/comfort reasons, the speed brake extension is generally controlled by the crew in response to a PFD message and not directly by the FMS/autopilot.

The arrival time uncertainty with automatic closed-loop 4D guidance and control is typically less than 4–8 seconds for 95% of all arrivals, assuming ATC related errors are zero. The effectiveness of the system can be seen.

The corresponding arrival time uncertainty for performing an open-loop high profile descent without time control is estimated to be of the order of 40 seconds.

The major reduction in arrival time uncertainty together with the ability to provide arrival time flexibility to accommodate delays caused by traffic or weather, etc., coupled with flight path flexibility to meet ATC requirements make 4D flight management a very important advance.

Further Reading

Combs, S.R., Sanchez-Chew, A.P., Tauke, G.J.: Flight Management System Integration on the F-117A. American Institute of Aeronautics and Astronautics AIAA-92-1077 (1991)

Lee, H.P., Leffler, M.F.: Development of the L1011 four-dimensional flight management system, National Aeronautics and Space Administration Report No. NASA CR 3700, 1 February 1984

McLean, D.: Automatic Flight Control Systems. Prentice Hall, Hemel Hempstead, UK (1990)

Avionics Systems Integration

9

9.1 Introduction and Background

Major avionic systems generally comprise a number of smaller sub-systems which are combined to form an overall system. The combination, interconnection and control of the individual sub-systems so that the overall system can carry out its tasks effectively is referred to as 'systems integration'. The number of sub-systems which need to be integrated to form a major system can be appreciated from the previous chapter on flight management systems.

It is instructive to review the development of avionic systems and their integration into overall systems in the light of the technology available and the circumstances prevailing at the time. The object is to put the development of today's advanced systems and the even more advanced systems currently under development in perspective. In many cases, the current concepts and philosophy are not new – often the originators of particular system developments in the past were far sighted in their concepts, but, as always, were limited by the technology available at the time.

The Second World War (WWII) resulted in a major growth in the electronic equipment installed in aircraft and the birth of avionics, with the very rapid development of airborne radar systems and associated displays, radar warning systems and ECM and more advanced autopilot systems exploiting electronics. Installation of the electronic equipment (or 'black boxes'), however, was very much on an ad hoc basis due to the very rapid developments and time scale pressures in war time. Some very limited degree of integration between sub-systems was introduced, for example, coupling the bomb sight to the autopilot – as readers who have seen the film 'Memphis Belle' may have noted. In general, however, the systems were 'stand alone' systems and their integration into an overall system was carried out by specific crew members such as the navigator, bomb aimer or radar operator.

The 1950s period saw the emergence of a number of avionic sub-systems (some of which were initiated during WWII) which have since undergone continual development and now form part of the avionic equipment suite of most civil and military jet aircraft and helicopters. For example, auto-stabilisers (or stability augmentation systems), ILS, VOR/DME, TACAN, Doppler, air data computers, attitude heading reference systems and inertial navigation systems.

The first major step towards integrating avionic systems was taken in the mid-1950s with the establishment of the 'weapon system' concept. These concepts were incorporated in the 1960s generation of aircraft, some of which are still in service. The concept requires a total system approach to the task of carrying out the mission effectively with a high probability of success. The aircraft, weapons and the avionic systems required by the crew to carry out the mission effectively must thus be considered as an integrated combination. It should be appreciated, of course, that the total system approach is applicable to any project, military or civil. As with many methodologies, however, military applications provided the spur and the initial funding. The very widely used 'programme evaluation review technique', or 'PERT' networks, and 'critical path analysis', for example, were originally developed on the POLARIS missile program.

As an example of the overall system approach, consider the requirements for a naval strike aircraft. The aircraft must be able to operate from an aircraft carrier in all weathers and be able to find the target and attack it with a suitable weapon (or weapons) with a high probability of success. Operational analysis shows that to minimise the probability of detection and alerting the enemy's defences, the aircraft needs to approach the target at high subsonic speeds (550–600 knots) at very low level at a height of 100 ft. or so above the sea so as to stay below the radar horizon of the target as long as possible. The avionic sub-system specifications can then be determined from the overall system requirements with an aircraft crew comprising pilot and observer/navigator. Hence, in the above example, the avionic equipment fit would comprise the following:

© The Author(s), under exclusive license to Springer Nature Switzerland AG 2023
R. P. G. Collinson, *Introduction to Avionics Systems*, https://doi.org/10.1007/978-3-031-29215-6_9

- Radar – target acquisition in all weather conditions.
- Doppler – accurate (4 knots) velocity sensor for DR navigation. (Note: IN systems capable of accurate initial alignment at sea on a moving carrier were still under development in the early 1960s.)
- Attitude heading reference system (or master reference gyro system – UK terminology) – attitude and heading information for pilot's displays, navigation computer, weapon aiming computer, autopilot.
- Air data computer – height, calibrated airspeed, true airspeed, Mach number information for pilot's displays, weapon aiming, reversionary DR navigation, autopilot.
- Radio altimeter – very low level flight profile during attack phase and all weather operation.
- Navigation computer – essential for mission.
- Autopilot – essential for reduction of pilot work load.
- Weapon aiming computer – essential for mission.
- HUD – all the advantages of the HUD plus weapon aiming for low level attack, for example, 'toss' bombing.
- Stores management system – control and release of the weapons.
- Electronic warfare (EW) systems – radar warning receivers, radar jamming equipment. Essential for survivability in hostile environment.
- Identification system (identification friend or foe – 'IFF') – essential to avoid attack by friendly forces.
- Radio navigation aids – location of parent ship on return from mission.
- Communications radio suite – essential for communicating to parent ship, cooperating aircraft, etc.

A significant degree of integration was required between the avionic sub-systems. For example, the weapon aiming system required the integration of the HUD, weapon aiming computer, AHRS, air data computer and the radar system.

The basic avionic systems specified for a naval strike aircraft for service introduction in the 1960s, for example, the Royal Navy Buccaneer aircraft would also be required in a 2010–2020 naval strike aircraft. The systems would need to be of much higher performance, however, and there would also be additional systems in a 2010–2020 aircraft which did not exist in the early 1960s. For example, laser gyro INS, helmet mounted displays, GPS, FLIR, 'smart weapons', etc., and a much greater level of systems integration would be required. The aircraft would also be a single crew aircraft. The point being made is the essential continuity of the role of avionic systems and their development, although their implementation changes as new technologies become available.

It should be stressed that although the above example is a military one, the same principles apply to civil avionic systems in terms of their role in enabling the mission to be carried out safely and effectively with the minimum flight crew. Most civil avionic sub-systems have also been directly developed from military avionic systems (e.g. air data computer, INS and GPS) and, until fairly recently, military aircraft generally required a larger number of avionic sub-systems with a higher degree of integration.

A major step towards facilitating the integration of avionic sub-systems in civil aircraft was taken in the early 1950s with the adoption of ARINC specifications for avionic systems and equipment. ARINC is a non-profit-making organisation in the USA which is run by the civil airlines with industry and establishment representation, which defines systems and equipment specifications in terms of functional requirements, performance and accuracy, input and output interfaces, environmental requirements and physical dimensions and electrical interfaces. For example, it defines systems and equipment specifications for air data computers, attitude heading reference systems, INS, communication radio equipment and data bus systems. Equipment made to an ARINC specification by one manufacturer should thus be completely interchangeable with equipment made by another manufacturer to the same ARINC specification. The electronic implementation of the two systems can be totally different provided they conform to the ARINC specification in terms of form–fit–function. For example, the LRUs (line replaceable units) must conform to the ARINC specifications dimensionally with the specified rack fixing arrangements, connectors and the pin allocations in the connectors. The systems must also meet the ARINC performance and accuracy specifications and the environmental requirements in terms of temperature range, acceleration, shock and vibration and EMC. The use of avionic equipment qualified to ARINC specifications thus ensures a competitive situation enabling procurement to be made from manufacturers on a worldwide basis.

The early 1960s saw the development of the first real-time airborne digital computers, and these were progressively introduced in military aircraft from the late 1960s for tasks such as navigation, mission management, weapon aiming, radar processing and displays processing.

These first-generation airborne digital computers were very expensive and it is interesting to note that a number of proposals were made during the 1960s to promote the concept of carrying out as many of the avionic sub-system computing tasks as possible with a powerful central digital computer. These proposals were not taken up for a number of reasons, such as:

1. Vulnerability – 'all the eggs in one basket', whereby a failure in the central digital computer affected all the sub-systems sharing its computing facilities.
2. Inflexibility – changes in an individual sub-system could involve changes in the main computer software with possible ramifications and 'knock on' effects on all the other sub-systems sharing the computing facility. This was particularly relevant with the computing speeds achievable at

Fig. 9.1 Interconnection of avionic sub-systems before introduction of multiplexed data buses

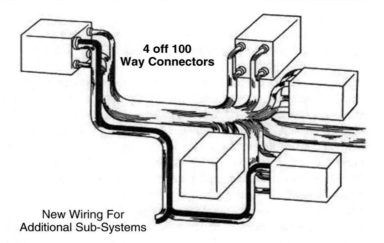

4 off 100
Way Connectors

New Wiring For
Additional Sub-Systems

the time and the high cost and limited capacity of the non-volatile memory technology which depended on magnetic core stores. (Typical store sizes were 8 K to 16 K at that time.)

3. Cost and weight of redundant central computer configurations was unacceptable.

The availability today, however, of affordable, very powerful processors with large memory capacity and high-speed data buses has radically changed the situation. The concept of sharing avionic sub-system processing tasks between a number of processors with spare capacity to take over particular tasks in the event of a processor failure is now economically attractive. The integrated modular avionics architectures which have been developed for the new generation of both military and civil aircraft which entered service from around 2005 onwards, in fact, exploit these concepts as will be discussed later.

As stated earlier, many of today's concepts are not new. It is the technology available at the time which limits their economic exploitation.

By the latter half of the 1960s, the development of affordable 'task orientated processors', that is processors which are sized to carry out one (or possibly two) avionic sub-system computing tasks, had become viable with the development of integrated circuit technology. A number of avionic sub-systems had thus been developed and were entering service in both military and civil aircraft by the end of the 1960s which contained their own internal digital computers, for example, digital waveform generators in HUDs, air data computers and inertial navigation systems.

Many of the aircraft sub-systems up to the early 1970s, however, were still largely analogue in their implementation with synchro and potentiometer outputs/inputs requiring point to point wiring to interconnect them. The interface units carrying out the necessary analogue to digital (A-D) and digital to analogue (D-A) conversions to enable the

sub-systems to communicate with each other were inevitably bulky and could exceed the size of the digital computer. The complexity of the inter-unit cabling looms and wiring harnesses can be seen in Fig. 9.1 which is typical of a late 1960s/early 1970s aircraft installation. It should be noted that some of these aircraft are still being operated. The weight of the cable looms could be very significant and exceed several hundred kilograms. The large number of multi-way plugs and connectors also inevitably degraded the overall system reliability and introduced intermittent faults.

The first development to reduce the number of interconnecting wires was the use of time division multiplexing (TDM) of the signals to be transmitted between two units.

TDM enables information from several signal channels to be transmitted through one communication system by staggering the different channel samples in time to form a composite pulse train. Each signal channel is transmitted as a serial digital pulse train at a given time slot in a clock cycle. Thus, given the clock time information and the address of the signal (suitably pulse coded), the receiver can then decode and distribute the individual signals. Hence, if there are, say, 30 different signals to be transmitted between two units, only two wires are required in principle when the data is multiplexed compared with 60 wires. This type of single-source–single-sink communications link between two units (say A and B) is frequently referred to as an 'A to B' link. Early systems, in fact, used three screened twisted pairs of wires to transmit the signal, address and clock data. This was then reduced to two screened twisted pairs of wires. Later single- and multi-source–multi-sink systems, however, operating at higher clock rates encode the signal data and address with the clock data using Manchester II bi-phase encoding (covered in the next section). A single screened twisted pair of wires only is required with such an encoding system to transmit information data rates of up to 1 to 2 Mbits/s. Higher data rates of up to 100 Mbits/s can be transmitted using a suitably screened coaxial electrical cable and information at even higher data rates can be transmitted as coded light pulses using a fibre optic cable. This will be covered in Sect. 9.2.

By the mid-1970s, it became possible to implement many more avionic sub-systems digitally, exploiting task orientated processors and the newly developed micro-processors and so eliminate the analogue computing elements and the analogue input/output components such as synchros and potentiometers. It thus became possible to interconnect the individual avionic sub-systems by means of a digital data bus system. This enabled the systems to communicate with each other and transfer serial digital information using time division multiplexing in conjunction with a suitable protocol system to control the data transmission to and from each individual sub-system. This will be covered in the next section but the dramatic reduction in inter-unit wiring can be seen in Fig. 9.2 which illustrates the interconnection of the sub-systems using a MIL STD 1553 data bus system.

An essential parallel development in the late 1970s and early 1980s was the implementation of the complex circuitry required by the MIL STD 1553 data bus system terminals in LSI (and later by VLSI). This enabled the system complexity to be encapsulated in highly reliable IC chips with a 'chip set' costing less than some multi-way connectors.

The avionic system architectures which had become established by the mid-1970s are still the standard architectures of current aircraft in service and are referred to as 'federated architectures'. Federated architectures essentially consist of a number of interconnected but functionally independent sub-systems with some degree of central computer control of overall operating modes. The advance of digital electronics through the late 1970s and 1980s enabled the development of very sophisticated sub-systems which,

although still addressing self-contained functional areas, shared information via data links such as the 1553 bus.

Federated architectures, built up from individual proprietary designs and available from a wide and experienced supplier base, became firmly established as the low-risk avionics system design approach. Also, because of the well-defined system boundaries and the clear division of areas of responsibility, federated systems were universally accepted as the low-risk approach commercially and contractually as well as technically.

In a federated architecture, however, there is inevitably some duplication of functions in each of the individual sub-systems which makes the overall system heavier and more costly than necessary. LRUs are each designed and optimised separately resulting in many different implementations of similar functions. Items such as processors, memory, software, interfaces and power supplies are individually optimised for each sub-system, and the design and support of these unique proprietary parts drastically increase the overall system development and support costs. The sub-system design approach also limits the introduction of fault tolerance because the proprietary sub-systems cannot easily share resources and the cost of building redundancy into every sub-system is unacceptable. The use of proprietary sub-systems also creates additional problems whenever an optimised proprietary sub-system interface must be accessed as part of an upgrade.

The maintenance of federated systems is constrained by the proprietary boundaries of the architecture and requires the complete removal of sub-system black boxes from the aircraft for repair in second-, third- and fourth-line maintenance facilities. Flexible deployment of such systems in military applications was seriously hampered by the need to deploy sophisticated second line, or 'avionics intermediate shop' support equipment. With the reducing numbers of new military designs, the need for mid-life upgrades focused attention on the high cost of upgrading federated architectures. Essentially, whole systems have to be removed and replaced and a small improvement in system functionality can result in very large costs.

Since the late 1980s, avionics R&D programmes in the USA and Europe have been directed towards finding a better way to build avionics. Over this period the concept of modular systems built from a small number of standard module types, housed in standard racking and communicating over standard data networks has been developed and has become known as *modular avionics*. However, to achieve the full benefits from modular systems, it is necessary to dissolve the sub-system boundaries which exist in conventional avionics so that a common pool of spare resources, shared between sub-systems, can be used to improve system availability. Hence, integrated modular avionic architectures offer

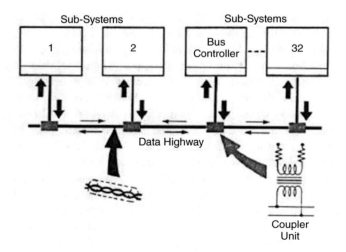

Fig. 9.2 Interconnections of avionic sub-systems by multiplexed data bus

additional benefits of fault tolerance and application flexibility through reconfiguration.

Integrated modular avionic architectures have been accepted as the way forward for both civil and military avionics and systems based on these concepts are now in service, or will be shortly entering service.

The topic of integrated avionic architectures is covered at an overview level in Sect. 9.3.

9.2 Data Bus Systems

Data bus systems are the essential enabling technologies of avionic systems integration in both federated and integrated modular avionics architectures.

They can be broadly divided into electrical data bus systems where the data are transmitted as electrical pulses by wires, and optical data bus systems where the data are transmitted as light pulses by optical fibres. These are discussed and the basic principles outlined in Sects. 9.2.1 and 9.2.2, respectively.

Serial digital data buses are used for interconnecting sub-units and sub-systems.

Parallel data buses are used within a unit or rack for interconnecting the individual modules and are briefly described in Sect. 9.2.3.

9.2.1 Electrical Data Bus Systems

There are several electrical serial digital data bus systems in use in avionics systems. These systems can be broadly divided into two categories in terms of their data rate transmission capabilities, namely, data bus systems operating with a maximum throughput of 1 to 2 Mbits/s and high-speed data bus systems with a throughput of 50 Mbits/s to 100 Mbits/s.

Space constraints have restricted the coverage of the lower speed systems to the MIL STD 1553B data bus system. This system is very widely used in military aircraft worldwide, although it originated in the USA. It has become the established and dominant standard data bus system since its introduction in 1975.

It transmits and receives data at 1 Mbit/s and is also a relatively sophisticated data bus system. An understanding of its operation reads across to the other systems in many areas, such as ARINC 429 which is a point-to-point system of lower capabilities (10 Kbits/s data rate) used in civil avionic systems and the more recent ARINC 629 data bus system.

The ARINC 629 data bus system has many similarities with the MIL STD 1553 B system; the main difference is that it is an autonomous system, whereas the '1553' system is a 'command response' system operated through a Bus Controller, It also operates at 2 Mbits/s as opposed to 1 Mbits/s for

'1553' The ARINC 629 data bus system is installed in the Boeing 777 airliner which entered airline service in 1995.

There are two standard high-speed data buses which have been developed in the USA for military applications. These are the 'Linear Token Passing Bus', LTPB, which operates at 50 Mbits/s and the 'High Speed Ring Bus', HSRB, which operates at 100 Mbits/s.

The high-speed data bus system, however, which is becoming widely adopted, particularly in new civil aircraft (e.g. the Airbus A380 airliner), is a system based on the 'Ethernet' data bus. The Ethernet data bus system is very widely used in commercial computing system applications. It has a data rate transmission capability of 100 Mbits/s and is mainly used for data file transfer.

The version which has been adapted for airborne applications is known as the 'Avionics Full Duplex Switched Ethernet', which has been shortened to 'AFDX Ethernet' network. It meets the civil aircraft avionic system requirements in all aspects and its commercially sourced components make it a very competitive system.

Its adoption in military aircraft avionic systems would appear to be a likely future development because of its cost advantages.

9.2.1.1 MIL STD 1553 Bus System

MIL STD 1553B is a US military standard which defines a TDM multiple-source– multiple-sink data bus system which is in very wide scale use in military aircraft in many countries. It is also used in naval surface ships, submarines, and land vehicles such as main battlefield tanks. The system is a half duplex system, that is operation of a data transfer can take place in either direction over a single line, but not in both directions on that line simultaneously.

The system was initially developed at Wright Patterson Air Force Base in the early 1970s and, as MIL STD 1553A, was first introduced in service on the F15 fighter programme in 1975. The standard was upgraded to MIL STD 1553B in 1978 to incorporate additional modes and facilities. This specification has since been progressively refined and amendments incorporated in the light of user experience by the SAE 2 K Committee under the auspices of the Society of Automotive Engineers (SAE) in the USA. The standard has also been adopted as a NATO standard and has been given the NATO codification STANAG 3838.

The basic bus configuration is shown in Fig. 9.3; the system is a command-response system with all data transmissions being carried out under the control of the bus controller. Each sub-system is connected to the bus through a unit called a remote terminal (RT). Data can only be transmitted from one RT and received by another RT (or RTs as there may be more than one sub-system requiring the same data) following a command from the bus controller (BC) to each RT. (The operation of the data bus system such that

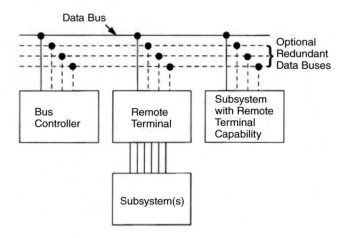

Fig. 9.3 Typical multiplex data bus system architecture

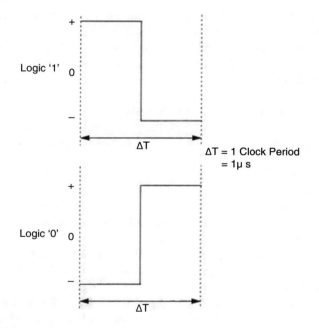

Fig. 9.4 Logic '1' and logic '0'

ΔT = 1 Clock Period
= 1μ s

Fig. 9.5 Data encoding

information transmitted by the bus controller or a remote terminal is addressed to more than one of the terminals connected to the data bus is known as the 'broadcast' mode.)

The protocol exercised by the bus controller hence ensures that there are no data clashes on the bus as only one RT is transmitting at any time. The bus controller thus initiates all data transfers and monitors the status of all transfers. It is generally incorporated in one of the sub-systems – usually the one generating the most traffic. The bus is formed as a single twisted cable pair with one layer of shielding and jacketing and with a maximum length of 100 m (328 ft). Although direct coupling to the bus is allowed, this is generally not used in order to avoid the risk of one terminal shorting out the bus. The bus connection is typically via a transformer coupled stub so that shorting of the stub is isolated from the bus. The maximum stub length allowed is 6 m (20 ft).

The data are transmitted at 1 Mbit/s. The data word size is 20 bits so that the maximum data transmission rate is 50,000 words/s.

A maximum of 31 terminals can be connected to the bus. The bus operation is asynchronous, each terminal having an independent clock source for transmission. Decoding is achieved in receiving terminals using clock information derived from the messages.

The technique adopted for data encoding is known as 'Manchester bi-phase' encoding where there must be an active transition for every bit, that is, for '0' and '1' signals. This is shown in Fig. 9.4. Apart from the SYNC bits all data bits must conform to these requirements. This eliminates 'stuck high' or 'stuck low' faults as there must be a transition during one clock period.

Figure 9.5 shows the data encoded waveform.

The standard requires the transmission rate to be 1 Mbit/s with a combined accuracy and long-term stability of 0.01% (i.e. 100 Hz). The short-term stability (i.e. stability over a 1.0 second period) is required to be at least 0.001% (i.e. 10 Hz). The word size is 16 bits plus the SYNC waveform and the parity bit for a total of 20 bit times. There are three types of words transferred; command words, status

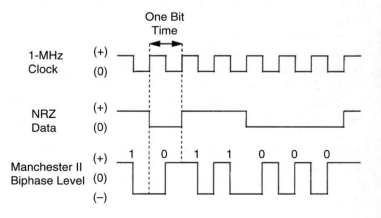

Fig. 9.6 Word formats

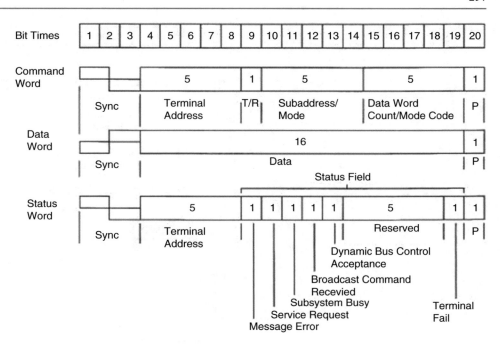

words and data words. The formats for these words are illustrated in Fig. 9.6.

A command word comprises six separate fields. These are briefly explained as follows:

- The SYNC signal field is an invalid Manchester waveform so that it cannot be 'confused' with any data bits.
- The RT address field occupies 5 bits, each RT being assigned a unique 5 bit address. Decimal address 31(11111) is not assigned as a unique address and is a broadcast address.
- The T/R bit is 0 if the RT is to receive, and 1 if the RT is to transmit.
- The sub-address/mode field, comprising 5 bits, is used for either an RT sub-address or mode control. The sub-address is used to route data to and from a location in the RT. A code of all zeros (00000) in the sub-address/mode field indicates that the contents of the word counts/mode field are to be decoded as a five bit mode command.
- The data word count/mode code field, comprising 5 bits, is generally used for data transfers. The word count field indicates the number of data words to be transferred in any one message block, the maximum number being 32 (indicated by all zeros).
- The parity bit is 1 if there is an odd number of bits in fields 1–19.

A status word is the first word of a response by an RT to a BC command. It provides the following:

(a) A summary of the status/health of the RT.

(b) The word count of the data words to be transmitted in response to a command.

The fields are briefly described as follows:

- The SYNC signal field is the same as with a command word.
- The RT address field (5 bits) confirms the correct RT is responding.
- The status field comprises 11 bits. The message error bit is set if the previous command was not correctly understood. The instrumentation bit =0 to distinguish the word from a command word.
- The parity bit is set by the RT in the same sense as a command word.

Data words contain the actual data transmitted between stations. The data field is 16 bits. For commands which imply a data content, data words are transmitted corresponding to the word count in the command or status word.

The SYNC signal is the inverse of the command and status word syncs. The most significant bit of the data is transmitted after the SYNC bits.

There are ten possible transfer formats, but the three most commonly used formats are as follows:

- BC to RT.
- RT to BC.
- RT to RT.

These are shown in Fig. 9.7. An intermission gap time of at least 2 μs is provided by the bus controller between

Fig. 9.7 Transfer formats

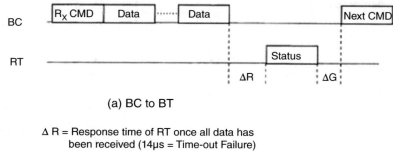

(a) BC to BT

Δ R = Response time of RT once all data has been received (14μs = Time-out Failure)

Δ G = Inter-message Gap (Typically 2 – 5μs)

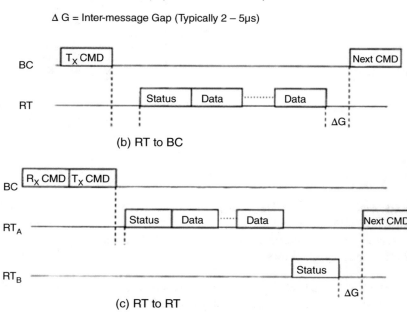

(b) RT to BC

(c) RT to RT

messages. A status word gap time of at least 2 μs but not more than 10 μs is provided by the RT before transmitting a status word.

A high degree of data checking and monitoring is built into the MIL STD 1553B system. For example:

- *Message data validation* – the terminal is designed to detect improperly coded signals, data drop-outs or excessively noisy signals.
- *Word validation* – the terminal checks that each word conforms to the following minimum criteria:
  - Word begins with a valid SYNC field.
  - Bits are in valid Manchester II code.
  - Information field has 16 bit plus parity.
  - Word parity is odd.

When a word fails to conform to the above criteria the word is considered invalid.

- *Transmission continuity* – the terminal checks the message is contiguous as shown in the formats in Fig. 9.7. Improperly timed data SYNCs are considered a continuity error.

- *Excessive transmission* – the terminal includes a signal time-out which precludes a signal transmission greater than 1 ms plus or minus 0.34 ms.

The data word is deemed valid when the data meet the above criteria and are received in contiguous fashion. The RT responds with a status word when a valid command word and the proper number of contiguous valid data words are detected as specified in the word count field of the command word, or as specified for the mode code contained in the command word.

A fault tolerant configuration of two (or more) identical MIL STD 1553B buses can be connected to the sub-system equipment.

Space constraints limit further discussion of the MIL STD 1553B data bus, in particular the bus operation and organisation to meet the data transfers between the sub-systems connected to the bus in terms of the required words/s. It is hoped, however, that the above coverage will enable the reader to understand and appreciate the basic system and know the questions to ask or look up in the reference literature.

The '1553' data bus system is a mature system which has now built up a very large operational usage and experience over a period of over 30 years since its introduction. It is by far and away the most widely used avionic data bus system and its performance can be summed up in one word – excellent.

9.2.2 Optical Data Bus Systems

Most readers are probably familiar to some extent with the use of optical fibres to transmit light signals. A brief explanation is set out below for those readers who need to refresh themselves on the subject and also to make clear the difference between multi-mode and single mode optical fibres and their respective applications.

The transmission of light signals along any optical fibre depends on the optical property of total internal reflection. This property is illustrated in Fig. 9.8, which shows four rays of light travelling through a medium of refractive index n_1 to a medium of lower refractive index n_2. Ray 1 is refracted in passing through the second medium, the relationship between the angle the incident ray makes with the normal, i, and the angle the refracted ray makes with the normal, r, being given by Snell's law:

$$\frac{\sin i}{\sin r} = \frac{n_2}{n_1}$$

At the critical incidence angle, i_{crit}, ray 2 is refracted through an angle of $90°$ and does not pass through the second medium ($i_{crit} \sin^{-1} n_2/n_1$). All rays with incident angles greater than i_{crit} such as rays 3 and 4 are thus reflected back into the first medium. This condition is known as total internal reflection and is effectively a loss free process.

An optical fibre basically comprises a central core of a suitable glass material (e.g. pure silica) with a very low optical transmission loss and with an outer cladding of a material with a slightly lower refractive index than the core. In multimode fibres, as shown in Fig. 9.9, the diameter of the core is large compared with the wavelength of the light being transmitted. For example, a typical core diameter is around 100 μm and the operating wavelength around 1 μm. A ray entering the fibre at an incident angle θ to the axis of the fibre less than the critical angle, $\theta_{crit} = \cos^{-1} n_2/n_1$, will undergo total internal reflection at the core/cladding interface. This ray will then undergo total internal reflection at the lower interface and will thus be guided through the core by repeated internal reflections as shown in Fig. 9.9. There are, however, a large number of different ways or modes by which light can be guided along the fibre depending on the incident angle θ. Hence, the term 'multimode fibre' is used to describe optical fibre that carries multiple light rays or modes at the same time, each at a slightly different angle inside the core. The time taken by a ray to travel along a fibre of length L is thus a function of θ and is equal to $n_1L/(c \cos \theta)$, where c is the velocity of light. The rays thus travel with different velocities so that a series of light pulses each of width t_1 at the input end of the fibre emerge after transmission through the fibre as a series of pulses of width t_2 (see Fig. 9.10). If the broadening of the pulses due to this time dispersion is large, then ultimately adjacent pulses will overlap at the output and cannot be resolved. This pulse broadening effect is generally tolerable for the current data bus rates of 50 Mbits/s and the relatively short lengths involved in aircraft installations which are generally less than 100 metres.

This pulse dispersion, however, is totally unacceptable for telecommunications applications which require very high data rates and long distances between repeaters to minimise the number of repeaters. This has resulted in the development of highly efficient single mode optical fibres. The major difference between single mode fibre and multi-mode fibre

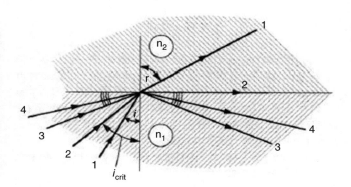

Fig. 9.8 Total internal reflection

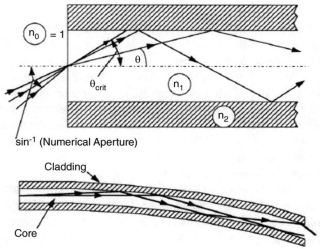

Fig. 9.9 Multi-mode optical fibre

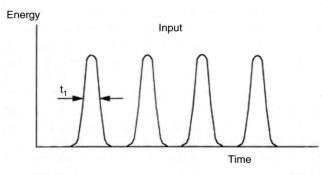

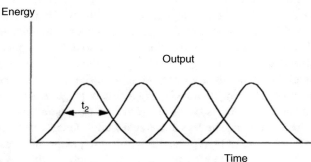

Fig. 9.10 Pulse broadening

is that the core diameter of single mode fibre is of the same order of magnitude as the wavelength of the light source (laser diode).

It can be shown that as the core diameter is decreased and the refractive index difference between the core and the cladding reduced, the number of possible guided modes for transmitting light along the fibre decreases. There is a normalised parameter known as the *waveguide parameter* which is equal to

$$\frac{2\pi a n_1 (2\Delta)^{1/2}}{\lambda_0}$$

where λ_0 is the operating wavelength, a is the core radius, n_1 is the refractive index of core, n_2 is the refractive index of cladding and $\Delta = (n_1 n_2)/n_1$.

When the waveguide parameter is less than a certain critical value (2.4048 for step index fibre) then only one guided mode is possible for transmitting light along the fibre, and the fibre is known as a single mode fibre. Practical single mode fibres have Δ varying from 0.002 to 0.005 and typical core diameters in the range 5–10 μm. Typical operating wavelength is around 1.5 μm.

Single mode means there is only value of θ and hence only one ray path for the light to travel along the fibre by multiple reflections so that there is only one velocity of propagation for a light pulse. The pulse broadening resulting from the differences in velocity for the different ray paths in multi-mode fibre is thus eliminated. It should be noted that there are

other smaller sources of dispersion, namely, material dispersion and waveguide dispersion which must be minimised in order to achieve very high data rates and very long transmission distances. Material dispersion is the dispersion resulting from the dependence of the refractive index of the fibre material on wavelength. Waveguide dispersion is the dispersion resulting from the spectral width of the source; the different wavelength components experience different refractive indices. These effects can be minimised by techniques such as grading the refractive index profile across the core, for example, varying the refractive index linearly from the centre of the core to the cladding. Dispersions of a few pico-seconds per kilometre are attained in modern single mode fibres compared with 50 nano-seconds per kilometre for a typical step index multi-mode fibre. Fourth-generation fibre optic communication systems are now coming into service with an information carrying capacity in excess of 10 Gbits/s and repeater spacing of over 100 km; the transmission loss in the fibre is less than 0.3 dB/kilometre.

As stated earlier, multi-mode optical fibre can be used in avionic system applications because of the relatively short lengths involved and the current data rate requirements of 50 Mbits/s. The reason for its use is primarily due to the need for demountable connectors in avionic equipment for ease of servicing and replacement of a failed unit. While demountable connectors for single mode fibres are feasible, they present a number of mechanical alignment problems and have not progressed beyond the laboratory stage. (The telecommunications industry uses fusion splicing techniques instead of connectors and allows sufficient extra fibre in a small coil at the end of the fibre to re-splice when necessary.)

The task of designing suitable low loss, robust connectors for multi-mode fibres is eased by the following factors:

1. Multi-mode fibres have a considerably larger numerical aperture than single mode fibres. The numerical aperture, NA, defines the semi-angle of the cone within which the fibre will accept light and is a measure of the light gathering power of the fibre.

$$\mathrm{NA} = n_1 (2\Delta)^{1/2}$$

Typical values for a multi-mode fibre are n_1 1.46 and $\Delta = 0.01$ giving a NA of 0.2. The fibre will thus accept light incident over a cone with a semi-angle of $\sin^{-1} 0.2$, that is 11.5° about the axis. Typical NAs for single mode fibres result in acceptance semi-angles in the region of 4° to 8° as Δ is in the region of 0.002 to 0.005.

The larger NA eases the alignment tolerances of the two halves of the connector.

Table 9.1 Features of transmission systems

Parameter	STANAG 3910 data bus	LTPB (Linear Token Passing Bus)	HSRB (High Speed Ring Bus)
Data rate	20 Mbits/s	50 Mbits/s	50–100 Mbits/s
Encoding technique	Manchester bi-phase	Manchester bi-phase	4B/5B data encoding
Topology	Bus structure	Bus structure	Point to point linked ring
Max message transfer	4096 data words	4096 data words	4096 data words
Number of stations	31	128	128
Bus control philosophy	Central control	Distributed control	Distributed control
Controlling mechanism	1553 bus control	Token passing control	Token passing control
Bus length	Dependent on 1553 network	1000 m	1500 m
Interconnect media	Fibre optics	Fibre optics or electrical	Fibre optics or electrical
Standard	STANAG	SAE	SAE
Country of origin	Europe	USA	USA

2. LEDs can be used for the modulated light source. These approach a Lambertian source with a hemispherical power profile which together with the reasonable NA of multi-mode fibre enables a simple and efficient optical coupling arrangement to couple the light source to the fibre to be implemented.

3. The larger core diameter eases the mechanical tolerancing problems in aligning the two halves of the connector.

It is in fact possible to adapt existing electrical connectors to incorporate multi-mode optical fibres.

The application of single mode optical fibres in avionic systems is thus currently confined to optical sensors such as the fibre optic gyro. Future requirements for very high data rates, however, could well lead to their adoption. The very high reliability of future avionic systems makes the need for demountable connectors questionable and the use of fusion splicing techniques feasible on the rare occasions when such equipment needs to be removed and replaced.

To summarise, the use of optical fibres to transmit data offers major advantages:

• High data rate capability (>10 Gbit/s using single mode fibre).
• Insensitivity to electro-magnetic interference.
• Electrical isolation.
• No line capacitance or mutual coupling.
• Low cross-talk.
• Lower power dissipation.
• Reduced weight and volume requirements.

The integration of the various avionic sub-systems to increase mission effectiveness requires an inter-connection network system capable of two-way communication of serial digital data at high speed. Present electrical systems, such as the widely used MIL STD 1553B data bus system, have the following limiting restrictions:

• Relatively slow transmission rate limited by medium (1 Mbit/s).
• Restricted number of terminals for communication (max 31).
• Restricted number of words transferred per message (max 32).
• Central control unit managing all data transfer.

To overcome these problems, a number of higher speed transmission systems have been developed both in Europe and the USA. Table 9.1 summarises their salient features. The systems are described briefly in the following sections.

9.2.2.1 STANAG 3910 Data Bus System

STANAG 3910 is a European data bus with a 20 Mbit/s data rate which has been adopted for the Eurofighter Typhoon, now in squadron service in the UK, Germany, Italy and Spain.

The bus provides an evolutionary increase in capability by using MIL STD 1553B (STANAG 3838) as the controlling protocol for high speed (20 Mbit/s) message transfer over a fibre optic network as shown in Fig. 9.11.

The optical star coupler is a passive optical coupler which enables light signals from each fibre stub to be coupled into the other fibre stubs and thence to the sub-systems.

9.2.2.2 Linear Token Passing High-Speed Data Bus

The linear token passing high-speed data bus (HSDB) has been developed in the USA for the new-generation modular avionic systems as discussed later in Sect. 9.3. The basic bus configuration is shown in Fig. 9.12. The system uses distributed control by means of a token passing protocol and operates at 50 Mbits/s.

9.2.2.3 Avionics Full Duplex Switched Ethernet, AFDX, Communication Network

As explained briefly in the introduction, the new generation of civil airliners (e.g. Airbus A380), exploit modular avionic architectures and use a communication network adapted from

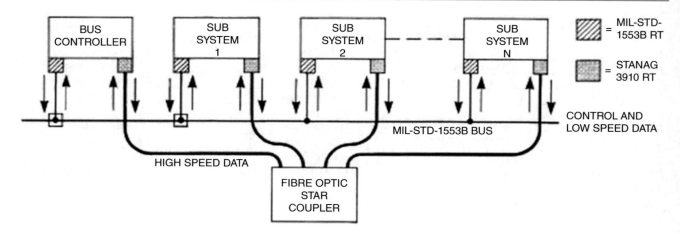

Fig. 9.11 STANAG 3910 data bus system

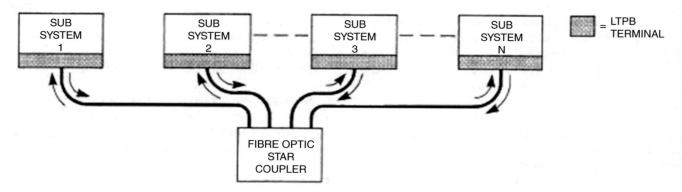

Fig. 9.12 Linear token passing high-speed bus

the widely commercially used Full Duplex Switched Ethernet (FDX). Additional features have been incorporated to meet avionic system requirements and it is referred to as the 'Avionics Full Duplex Switched Ethernet', or AFDX network.

The network provides 100 Mbits/s full duplex (two way) communication and provides flexibility to manage any change in the data communication between the connected systems without wiring modifications.

An example of the use of the AFDX network to interconnect the Air Data Inertial Reference Units (ADIRUs), GPS and the radio NAVAIDs to the Flight Management Computers (FMCs) to carry out 'Position' computation is shown in Fig. 9.13. (Refer to Sect. 8.3.3 and Fig. 8.13.)

Note. 'MMR' stands for 'Multi-Mode Receiver', and 'IOM' for 'Input Output Module' which provides interfaces between AFDX and other signal types such as ARINC 429, discrete and analogue.

Communication through the AFDX network is managed by the switches which route the data transmitted by each network subscriber to one or more other subscribers. The topology of the network is defined according to safety,

availability and communication constraints with the objective of connecting subscribers which exchange large amounts of data with each other to the same switches.

The basic principle of the switched Ethernet is illustrated in Fig. 9.14.

Referring to Fig. 9.14, the central equipment forwards the frames through to the destination port (and only to this one); the routing decisions are based on the Ethernet addresses. The system is collision free. In case of simultaneous transmissions, the messages are buffered in the switch before treatment.

Referring to the addressing example:

* *A* sends a frame to *D*.
* The switch analyses the frame and looks at the destination address.
* According to the 'routing table', the switch sends the frame to the destination port (port 4 in the example) and not to the others.

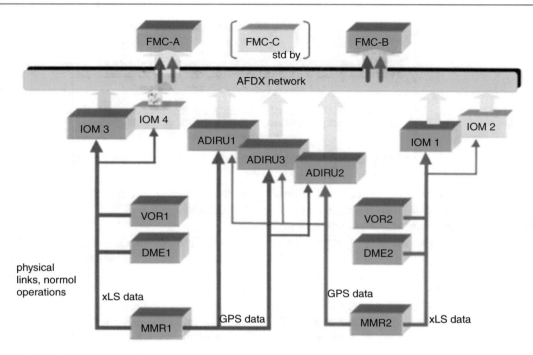

Fig. 9.13 Interconnection of navigation system units by the AFDX network (by courtesy of Airbus)

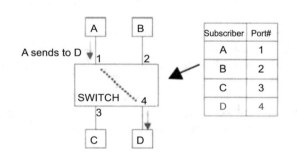

Addressing example:

| Destination: @D | Source: @A | | DATA |

64 to 1518 bytes

(simplified frame)

- A sends a frame to D

Fig. 9.14 Switched Ethernet principle

9.2.3   Parallel Data Buses

Parallel data buses are used within the units or racks of the avionic systems. These are almost invariably electrical at the present time and use a variety of standards.

Before the availability of complex microprocessor devices, a processor could require a complete circuit module, with the memory devices being located on separate modules, the bus merely being an extension of the processor signals, suitably buffered.

As microprocessor and high-density memory devices became available, and the functionality implemented in the systems grew, bus technology was developed to provide inter-processor communication with common interfaces. Buses either use commercial standards, or have been developed to meet the special needs of the system. Examples of the former are PC and VME buses; the latter includes the PIBUS widely used in the F-22 aircraft. (PIBUS was originated by JIAWG, Joint Industry Avionics Working Group.)

In future, with a greater level of integration and the use of commercial off-the-shelf (COTS) modules, there will be increased standardisation in the parallel buses.

The use of an optical backplane for interconnecting modules is a possible future development.

9.3   Integrated Modular Avionics Architectures

The background to integrated modular avionics architectures has been explained in Sect. 9.1. The importance of finding a better way of implementing avionic systems can be appreciated when it is realised that avionics currently account for some 30% of the total cost of a new aircraft. Reducing these costs must thus play a major role in containing overall system costs and halting the cost spiral inherent in federated architectures as increasing performance and capability are sought. The IMA architectures provide higher levels of performance and system capability, increased equipment availability and reduced levels of maintenance, so lowering costs

Fig. 9.15 The 'three layer stack' modular software concept

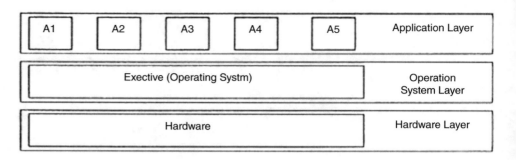

right across the system life cycle. Space constraints limit the treatment of this topic to that of an overview explaining the basic philosophy, aims and objectives of the architectures and the implications of their implementation using standardised electronic modules.

The term 'avionics architecture' is a deceptively simple description for a very complex and multi-faceted subject. Essentially, an avionic architecture is the total set of design choices which make up the avionic system and result in it performing as a recognisable whole. In effect, the architecture is the total avionics system design.

The system complexity means that there are very many parts of an avionics architecture and the architecture is best viewed as a hierarchy of levels comprising the following:

1. Functional allocation level. The arrangement of the major system components and the allocation of system functions to those components.
2. Communications level. The arrangement of internal and external data pathways and data rates, transmission formats, protocols and latencies.
3. Data processing level. Central or distributed processing, processor types, software languages, documentation and CASE (computer aided software engineering) design tools.
4. Sensor level. Sensor types, location of sensor processing, extent to which combining of sensor outputs is performed.
5. Physical level. Racking, box or module outline dimensions, cooling provisions, power supplies.

This is not an exhaustive list and there are many other important aspects of the avionics, for example, control and displays and maintenance philosophy, which are certainly a part of the avionics architecture.

The influence of the architecture also continues down to lower levels of implementation and technology detail. It is the higher levels, however, which are most often referred to as the 'architecture' and it is at these levels that integrated modular avionic concepts are most able to influence overall system costs.

The software concept is also modular, comprising a number of application programs running under the control of an executive operating system. The basic system requirements are as follows:

- Suitable stable specifications for the interconnection between modules, both hardware and software.
- Hardware that is independent of the application in which it will be used.
- Executive and application software that is independent of the hardware on which it will run.
- Standard interface between the executive and the hardware for input/output.

This is often referred to as a three-layer stack and is shown schematically in Fig. 9.15.

The modular avionics concept relies on the use of a limited range of standard modules which are packaged in a standardised modular format and installed in a small number of common racks.

The concept of modular equipment is not new, however, and avionics manufacturers have frequently used modular packaging to seek a competitive cost advantage within their own equipment. What is different with the integrated modular avionic approach is as follows:

(i) The use across a range of aircraft platforms (including helicopters) of standard 'F^{3}I' form, that is 'form, fit, function and interface', interchangeable modules procured from an 'avionics supermarket'.

(ii) The integration of data and signal processing across traditionally separate aircraft sub-systems enabling wide scale use of reconfiguration to improve availability.

(iii) The use of sufficient built in test to enable faulty modules to be correctly diagnosed and replaced at first line without additional test equipment.

The use of a small number of standard modules potentially reduces the initial development and procurement cost of equipment through competition and eventually through economics of scale in the production of the modules. It also reduces the maintenance costs by reducing spares holding.

Similarly, the integration of functions across traditional equipment boundaries promises to reduce the proliferation of different module designs across the aircraft and it allows reconfiguration strategies to be adopted economically – it is cheaper to carry a spare set of modules that can be configured to act as different systems than it is to carry complete spare systems. The requirement for a second line avionics intermediate shop is eliminated since the high levels of fault detection and isolation (typically >98%) using built in test circuitry allow maintenance to consist of the replacement of a faulty module on the aircraft with the faulty module being returned to the original manufacturer for repair.

However, with these advantages come many implications for the way equipment is currently contracted for and built. The integrated system design increases enormously both the system complexity and the potential for interactions between sub-systems. At the same time, it blurs the traditional lines of responsibility that exist in the industry and it therefore requires a very careful and systematic design approach if integrated systems are to be put together successfully. To implement such highly integrated systems, very close collaboration between systems engineers from different avionic, airframe and software suppliers is required.

The above is a summary of the aims and objectives of modular avionics architectures and the broad issues that are being addressed.

The problems of component obsolescence in a standard module can be seen by looking at the very rapid advances in technology that are being made. Provision must therefore be made in a standard module design, particularly in the software, to allow for component obsolescence and updating with later technology components.

The alternative is living with older technology and procuring sufficient devices in the initial purchase to provide replacement spares for the service life of the equipment.

To date, modular avionics is generally limited to the digital processing and communication areas of the mission systems, and the power supplies necessary to run them. Here, the complex functionality is implemented in software, and can be developed largely independent of the actual platform.

Both the Lockheed F-22 'Raptor' fighter in service with the USAF, and the Lockheed Martin F-35 'Lightning 2' Joint Strike Fighter, currently under development, exploit modular avionics in their mission avionics systems.

An overall avionic systems architecture for a military aircraft and designed for implementation using standard avionic modules is shown in Fig. 9.16. The essential intercommunication system provided by the high-speed multiplex data buses can be seen. The grouping together of the systems which carry out flight critical functions, such as flight control, propulsion control, electrical power supply control, sensors and actuators, into the 'aircraft management system' is a noteworthy feature.

9.3.1 Civil Integrated Modular Avionic Systems

As in military systems, the use of new hardware, software and communication technologies has enabled the design of new system architectures based on resource sharing between different systems.

Current microprocessors are able to provide computing capabilities that exceed the needs of single avionics functions. Specific hardware resources, coupled with the use of Operating Systems with a standardised Application Programming Interface provide the means to host independent applications on the same computing resource in a segregated environment.

The AFDX Communication Network provides high data throughput coupled with low latencies to multiple end users across the bus network. The basic concepts are illustrated in Fig. 9.17.

The basic Line Replaceable Unit, LRU, becomes an avionics application which is hosted on one, or more, Integrated Avionic Modules (IAMs), providing shared computing resources (processing and memory and I/O).

External components like displays, sensors, actuators and effectors can be connected to standard or specific interfaces in the module or to *Remote Data Concentrators* (RDCs), normally located close to the sensors and actuators. The RDCs are connected to the IMA modules through data buses (ARINC 429 or CAN).

Note. CAN is a data bus system developed by the automobile industry, and is now being used in certain application areas in avionic systems.

The application software for a function will execute on one or more *Core Processing and Input Output Modules* (CPIOMs) in a partitioned environment providing segregation from other functions.

Several modules may be used for a single function to provide high integrity operation through cross-checking and/or increased availability.

The CPIOM provides a standard Application Programming Interface, API, to the applications and segregated computing resources (processing time, memory and I/O) to each application partition.

Input Output Modules (IOMs) provide interfaces between AFDX and other signal types (ARINC 429, CAN, discrete and analogue), but do not host applications.

Both IOMs and Core Processing and CPIOMs are configurable through loadable configuration tables and also provide standard services such as data loading and Resource BITE (Built In Test Equipment).

Figure 9.18 shows the integrated modular avionic systems on the Airbus A380.

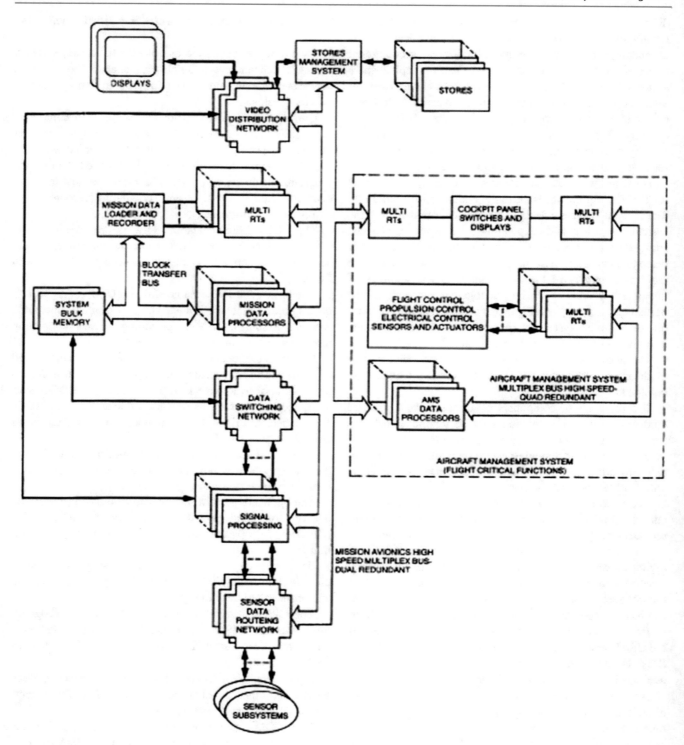

Fig. 9.16 Integrated avionic systems architecture

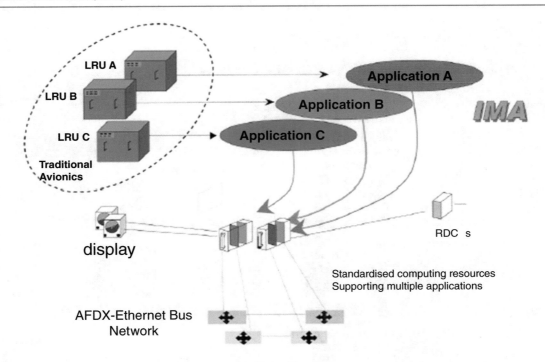

Fig. 9.17   Principles of Integrated Modular Avionics (by courtesy of Airbus)

9.4   Commercial Off-the-Shelf (COTS)

The term 'Commercial off-the-shelf' (COTS) refers to the use of commercially available electronic hardware and/or software for the implementation of avionics systems. These hardware and software are designed for the general electronics marketplace, especially in the industrial control and personal and industrial computing sectors.

Until the mid-1990s, the majority of avionic systems were specifically designed for the application, although these used commercial components where suitable parts were available. The technical development of semi-conductor devices was largely driven by the needs of the military avionics sector until the mid-1980s, but since then the commercial and industrial sectors have taken over, with the development and continual advancement of items such as personal computers, computer games and cell phones. The use of complex electronic systems in the automotive sector is a growth area that mirrors some of the environmental constraints required by avionic systems, although to date such systems have not employed COTS technology.

COTS systems and equipment are mainly being used in commercial and military transport aircraft where the operating environment is relatively benign. This is initially found in areas where there is minimal risk from the failure of the systems, for example, cabin entertainment systems, communications and long-term navigation, and includes both hardware and software. The use of COTS displays surfaces such as matrix addressable LCDs has already been discussed briefly in Chap. 2.

The application of COTS equipment to military fighter and strike aircraft is limited due to all aspects of the operating environment; mechanical (vibration, shock), climatic (temperature and pressure) and electromagnetic (including lightning and radiation effects). Special 'ruggedised' hardware is available at a significant cost premium; whether this should be referred to as COTS is questionable, since its application to other types of system would be very limited.

The use of COTS hardware and software equipment for applications that are safety-critical, for example, flight and propulsion control sub-systems, raises issues associated with the certification of such systems. Certification imposes demonstration of fitness for purpose, and assessments that all reasonable actions have been taken to ensure that the risk of failure is at an acceptably low level. Experience has shown that there are significant and possibly unacceptable risks with the use of COTS, both hardware and software, because of the following:

- Lack of design quality, documentation, guarantees and warranty.
- Lack of stable standards and specifications.
- Short lifetime dictated by commercial pressures.
- Lack of guaranteed forward and backward compatibility.

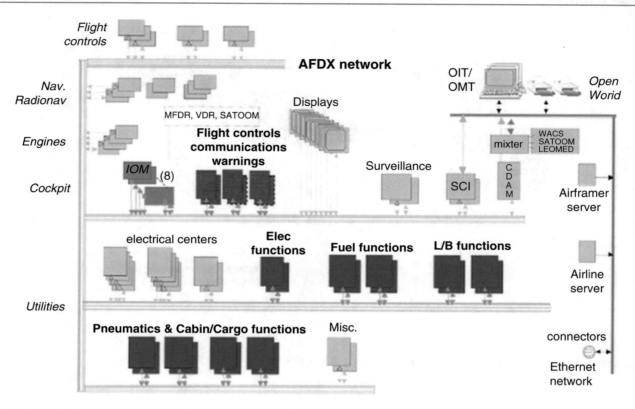

Fig. 9.18 Integrated modular avionic systems on the A380 (by courtesy of Airbus)

Problems of software integrity would appear to preclude the use of most COTS software in safety critical applications. The use of COTS components is, however, likely to increase in military applications because of their intrinsic cost/performance benefits. Provision in the system software to accommodate new COTS components to replace obsolescent COTS components is clearly essential, as mentioned earlier. The degree of environmental isolation, which can be provided by a suitably designed electronic rack/cabinet in terms of vibration isolation and cooling provision (forced air or liquid cooling or possibly both), could extend the use of COTS components in military applications.

The increasing use of COTS components in unmanned aircraft is also very likely.

Further Reading

Ghatak, A.K., Thyagarajan: Optical Electronics. Cambridge University Press

Unmanned Air Vehicles

10

10.1 Importance of Unmanned Air Vehicles

The growing importance of unmanned air vehicles, UAVs, for carrying out an increasing number of military roles has become very clear over the last decade. They also have civil applications, particularly in surveillance and monitoring tasks, although safety and legislative issues need to be resolved.

UAVs depend totally on avionic systems in order to function and carry out their mission, and this chapter provides a brief overview of the current situation. Space constraints limit the coverage to a few pages only for a subject which can occupy several volumes.

It should be noted that a number of alternative names have been used for unmanned air vehicles such as unmanned aircraft, UMAs, unmanned air vehicles, UAVs, remote piloted vehicles, RPVs, uninhabited combat air vehicles, UCAVs. Unmanned air vehicles. UAVs are the terminology adopted in this book for the generic classification for both civil and military systems. UCAV is used for UAVs with an offensive combat role as opposed to passive surveillance.

The fundamental advantages of UAVs are their ability to perform dangerous, sensitive or dull tasks in a cost-effective manner. They are now widely used in a range of surveillance tasks from battlefield surveillance and target acquisition at fairly low altitudes of around 3000 to 15,000 ft. or more above the terrain to long range surveillance/reconnaissance missions carried out at very high altitude.

UCAVs are being developed for offensive roles such as suppression of enemy defences.

Small UCAVs which can be launched from a parent fighter aircraft are also being investigated for interception tasks.

Among the advantages of removing the pilot is that combat '*g*' can be increased by a factor of two or more over the 9 *g* limit imposed by a pilot, so that a UCAV is more manoeuvrable and the survivability increased. The air vehicle can be made smaller and more 'stealthy'. More fuel, payload or weapons can be carried in place of the pilot and all the pilot's support equipment, including ejector seat, enabling a smaller airframe to have greater capabilities.

UCAVs are likely to be a third of the cost of manned aircraft and to be 75% cheaper to operate and maintain.

UAVs/UCAVs are certain to become one of the dominating elements in any future conflicts over this century as technology advances and they incorporate an increasing degree of artificial intelligence. The era of 'robot wars' appears to be approaching and what is now science fiction ('sci fi') may well become a reality in the not too distant future.

UAVs are also able to fulfil a number of civilian roles, carrying out surveillance and monitoring tasks.

Such tasks include border patrol for drug smuggling or illegal immigrants, coastguard patrol, inspecting electrical power lines, monitoring gas or oil pipe lines, forest and agricultural/environmental monitoring, traffic monitoring and urban surveillance. They have also been used for crop spraying in Japan.

The main obstacles to their civilian application are, as mentioned earlier, resolving the safety and legislative issues.

For instance, the basic visual flight rules (VFR) airspace concept is one of 'see and avoid' and implies a pilot on board. The regulatory authorities will need to define a similar concept of 'sense and avoid' using a variety of sensing methods if UAVs are to share the same airspace as civil airliners. Such sensing methods could include automatic traffic monitoring (ATM), radar, electro-optical or other sensors that would perform the same function. The introduction of automatic satellite-based, traffic sensing anti-collision systems on board the majority of aircraft may, in time, solve this problem.

R. P. G. Collinson, *Introduction to Avionics Systems*, https://doi.org/10.1007/978-3-031-29215-6_10

Small portable surveillance UAVs, comparable in size and weight to radio-controlled model aircraft, can be operated within the same restrictions as model aircraft, namely, the operator must be in visual line of sight contact with the air vehicle at all times. Small UAVs are increasingly being used by police and traffic control authorities for traffic surveillance tasks.

10.2   UAV Avionics

The avionic systems covered in this book are, in most cases, equally applicable to both manned and unmanned aircraft. The specific aspects of UAV application are commented on briefly below.

10.2.1   Displays and Man-Machine Interaction

HUDs are obviously not required in a UAV system, but head down displays are essential in the ground control stations, or parent aircraft, for the UAV operators.

Displays, for example, are required of mission information, UAV state (height, velocity, altitude, heading, position, etc.), UAV systems status, colour moving map showing navigational situation, target, etc., video and IR sensor imagery, radar (SAR), threat situation. 'Friend or foe' identification by remote viewing, or other means, will be necessary if so-called 'friendly fire' incidents are to be avoided in a ground attack or airborne interception by a UCAV.

Large display surfaces can be used in a ground control station as these are free of the severe space constraints of an aircraft cockpit.

Mobile ground control stations operate in a more rugged environment and space constraints will generally restrict the size of the display surfaces to that of a commercial work station. In fact, 'hardened' commercial work stations are being used in some systems.

HMDs can be a cost-effective solution for providing displays as they avoid having to install fixed head down displays. Binocular HMDs mounted on an AFV (Armoured Fighting Vehicle) helmet could be used in small, highly mobile ground control stations to display all the UAV situation and control information currently displayed on ground station displays. Direct voice input control could be used to switch the display information required.

A binocular HMD presenting left eye and right eye stereo pair images can also provide a 3D presentation to the UAV operator and give a better situational awareness; in effect, a virtual cockpit.

10.2.2   Aerodynamics and Flight Control

The material in Chap. 3 is entirely relevant to UAVs.

The mathematical transfer functions which determine the dynamic response of the basic aircraft to the control surface deflections are the same whether the aircraft is manned or unmanned. The UAV may well be aerodynamically unstable, however, because of the performance advantages which can be achieved with aerodynamically unstable configurations.

10.2.3   Fly-by-Wire Flight Control

Again, the material in Chap. 4 is entirely relevant; all UAVs are inherently fly-by-wire. The new generation of UCAVs currently under development exploits stealthy configurations and will generally be aerodynamically unstable for maximum performance, reduced weight and improved stealth characteristics.

A redundant, failure survival FCS will generally be required, but a lower failure probability could probably be accepted in view of the absence of the pilot or crew. For example, a figure of 10^{-5}/h could be acceptable, depending on the UCAV mission.

10.2.4   Inertial Sensors and Attitude Derivation

The inertial sensors (gyros and accelerometers) coverage and attitude derivation from strap down sensors are equally applicable to manned aircraft and UAV sensor systems. Fibre optic gyros and skewed axis sensor configurations offer many advantages in UAV applications (as well as manned aircraft in terms of cost, reliability, performance, weight, ruggedness and availability).

10.2.5   Navigation Systems

The navigation system coverage in Chap. 6 is again equally relevant to manned and unmanned aircraft. The vital role played by satellite navigation systems (GPS) and satellite communication links in current and future UAVs should be noted.

Small UAVs use GPS combined with inertial mixing of velocity and position data derived from a low cost, strap down inertial sensor unit using a Kalman filter, basically complementary filtering. The inherent latency in the GPS processing and the resulting short term errors during manoeuvres can be virtually eliminated. The inertial 'fill in' is over very short periods of time, generally less than 60 s because of the inherent limitations in the accelerometer and rate gyro accuracy.

The triad of low cost, solid state rate gyros and accelerometers in the inertial sensor unit exploits Micro Electro-Mechanical Systems (MEMS) fabrication technology (refer Chap. 5, Sect. 5.2.2). These MEMS gyros have orders of magnitude lower accuracy than even sub-standard IN gyros. Typical drift rate is of the order of 2 degrees/minute, or more, and also changes with temperature. The rate gyros in any modern digital strap down system act as integrating rate gyros, either inherently as in the case of the RLG or IFOG, or by suitable processing, and output angular increments. It can be seen that the heading error alone, with this level of gyro accuracy, would rapidly accumulate and produce unacceptable cross-track errors.

However, the error sources in the accelerometers and gyros and the resulting inertial derived velocity and position errors can be modelled in a multi-state Kalman filter. These errors can then be corrected during the periods of good GPS accuracy. Sensor errors are thus only required to remain stable over relatively short periods of time (providing the periods of manoeuvring are not excessively prolonged).

The low cost, high speed microprocessors now available can readily execute the strap down attitude algorithms and a multi-state Kalman filter, enabling a small, light weight, high accuracy and low cost navigation solution to be achieved.

Satellite navigation systems, however, are vulnerable to enemies able to jam or disrupt them.

Terrain reference navigation, TRN, systems are used in cruise missiles and aircraft. They have the major advantage in UAV/UCAV applications of being completely self-contained and unjammable as well as of high accuracy. They are completely independent of any external systems such as satellites or ground stations.

10.2.6 Air Data Systems

The coverage in Chap. 7 is equally applicable to UAVs/UCAVs – an air data system is essential for the control and guidance of the air vehicle.

10.2.7 Autopilots and Flight Management Systems

Much of the material in this chapter is applicable to UAVs/UCAVs, particularly the basic autopilot modes. A flight management system is essential in all but the simplest of UAVs. In the smaller UAVs, the FMS function is generally carried out by the Ground Station Computer.

10.2.8 Integrated Avionics Systems

UAVs are prime candidates for integrated avionic systems and the exploitation of COTS hardware and also COTS software in some areas, for example, in ground stations. The overall costs, including operating and maintaining costs, will clearly benefit from an integrated avionics system architecture approach.

10.3 Brief Overview of Some Current UAVs/UCAVs

Some typical UAVs/UCAVs are briefly reviewed in this section to show the wide spectrum of UAV capabilities.

10.3.1 'Watchkeeper' Battlefield Surveillance System

Figure 10.1 is a photo of the UAV element of the British Army's 'Watchkeeper' battlefield surveillance system which will be entering service in the very near future. The system is a development by Thales UK of the Israeli 'Hermes 450' UAV surveillance system.

The Watchkeeper system replaces the British Army's earlier 'Phoenix' battlefield surveillance system. It has a much greater range of sensors and navigation capability and can remain airborne for up to 16 h, compared with 4 h with Phoenix.

The Watchkeeper UAV has a sophisticated suite of enhanced electro-optical sensors including a thermal imaging infrared sensor system, low light and high-definition daylight video cameras together with a laser target designator. The electro-optical systems are mounted in two turrets underneath the fuselage.

It is also fitted with an advanced synthetic aperture radar (SAR)/ground moving target indicator radar. Watchkeeper is able to operate from semi-prepared airstrips and includes a de-icing system to expand its ability to operate in an all-weather operational environment. It incorporates a high degree of automation with automatic take-off and landing. The I-STAR (Intelligence-Surveillance-Target Acquisition-Reconnaissance) images and information from the Watchkeeper UAV are provided to a network of highly mobile ground control stations and remote terminals where British military operators will control the whole mission and interface within a network-enabled environment.

Fig. 10.1 'Watchkeeper' UAV
(by courtesy of Thales UK)

10.3.2 MQ-9 'Reaper' UCAV System

Figure 10.2 shows the General Atomics MQ-0 'Reaper' unmanned air combat vehicle (UCAV) which is in service with the USAF. A small number of Reapers have also been procured by the UK and are being operated by the RAF. This is truly a twenty-first century weapon system which has demonstrated the ability to hunt for and destroy targets with laser guided air-to-ground missiles under the control of an operator several thousand miles away, based at a USAF Air Base near Las Vegas, USA. It is reported that an operator's command takes 1.2 s to reach the UCAV via a satellite link.

The USAF operators controlling the Reapers are all qualified pilots because of the current USAF mode of operation where they are fully integrated into air operations and often flying missions alongside manned aircraft. A short runway is required for take-off and landing.

The typical MQ-9 system consists of multiple aircraft, ground control station, communications equipment and links, maintenance spares and military personnel. The crew controlling the UCAV consists of a pilot and sensor operator.

Sensor payloads comprise a multi-spectral targeting sensor suite which includes a colour/monochrome daylight TV, infrared and image intensified TV with laser rangefinder/target designator. The synthetic aperture radar enables guided bomb targeting and is capable of very fine resolution in both spotlight and strip modes.

It also has ground moving target indicator capability. The MQ-9 Reaper has a wingspan of 66 ft. (20 m) and a maximum take-off weight of 10,500 lb. (4760 kg) and is powered by a 950 HP (712 kW) turboprop engine. Operational altitude is 25,000 ft. and service ceiling 50,000 ft.; endurance is 14–28 h (14 h fully loaded) and the range is 3200 nm. Maximum speed is 260 knots (300 mph) and cruising speed is 150–170 knots.

Maximum payload is 3800 lb. (1700 kg) comprising an internal load of 800 lb. (360 kg) and an external load of 3000 lb. (1400 kg). The external load is carried on hard points under the wings and can comprise a wide range of weapons. For example, 14 'Hellfire' laser guided air-to-ground missiles, or four Hellfire missiles and two 500 lb. (230 kg) laser guided bombs.

A civil, unarmed version of the MQ-9 Reaper system has also been developed for border patrol and surveillance duties. These versions carry the full suite of electro-optical/infrared sensors and synthetic aperture radar together with comprehensive communications equipment, including a satellite command and control link.

10.3.3 'Taranis' UCAV Demonstrator

'Taranis' is the name given to an unmanned combat air vehicle demonstrator programme funded by the UK MOD and British industry to bring together a number of technologies, capabilities and systems to produce a technology demonstrator based around a fully autonomous intelligent system. Figure 10.3 illustrates its configuration. (The name 'Taranis' comes from the Celtic British god of thunder.)

Fig. 10.2 MQ-9 'Reaper' UCAV (US Air Force photo by Lt. Col. Leslie Pratt)

Fig. 10.3 'Taranis' UCAV demonstrator (by courtesy of BAE Systems)

The programme was awarded to a team led by BAE Systems in December 2006. The industrial team led by BAE Systems includes Rolls Royce, Qinetiq and GE Aviation (formerly Smiths Aerospace).

The programme is being directed towards designing and flying an unmanned aircraft and gathering the evidence needed to inform decisions about a future long range offensive aircraft, and evaluating UAVs contribution to the RAF's future mix of aircraft.

The configuration and shape of the Taranis air vehicle could be described as the shape of future UCAVs in general, as a number of UCAV programmes in the USA and Europe exploit similar 'stealthy', aerodynamically efficient, blended fuselage/delta wing configurations – in effect a 'flying wing'.

The Taranis configuration is highly unstable aerodynamically in yaw (no vertical surfaces) and possibly unstable in pitch and requires a high integrity FBW flight control system to provide automatic stability and good control characteristics.

An essential requirement, for future UCAVs, is a very high level of autonomous operation and this appears to be a key element in the Taranis systems architecture and design from the limited press releases to date.

A comprehensive 'C4 ISTAR' system is also being developed for the Taranis programme. (C4 ISTAR is an acronym for Computers-Command-Control-Communications-Intelligence-Surveillance-Target Acquisition-Reconnaissance.)

The Taranis air vehicle will be one of the world's largest unmanned air vehicles and has been described in press releases as being approximately the same size as the BAE Systems Hawk advanced jet trainer, which has a wingspan of 9.9 m, a length of 11.35 m and a height of 3.98 m. On the same basis, the weight of Taranis is expected to be approximately 8 tonnes. The engine appears to be a development of the Rolls-Royce Adour (which is installed in the Hawk) and thrust could be in the region of 6500 lb. A retractable tricycle undercarriage is fitted, and take-off and landing appear to be from a paved runway.

Fig. 10.4 'Draganflyer' X-6 portable helicopter UAV (by courtesy of www.draganfly.com)

10.3.4 Draganflyer X-6 Portable Surveillance Helicopter UAV

Figure 10.4 is a photo of a small portable surveillance helicopter UAV, the *Draganflyer X-6* produced by Draganfly Innovations Inc., Canada. This small UAV helicopter can be used in a variety of civil surveillance applications. Its small size (just under 1 m diameter), very low weight (1 kg plus maximum payload of 0.5 kg), very low noise levels and the ability to hover are particular advantages. The helicopter features an innovative six rotor configuration, with the six rotors arranged as three contra-rotating offset pairs mounted at the ends of the three arms. Differential thrust from these three equally spaced points enables it to manoeuvre quickly and precisely. The offset layout doubles the thrust without increasing the size of the footprint, and naturally eliminates loss of efficiency due to torque compensation.

The helicopter is powered by a Lithium Polymer battery; the rotors being driven directly by brushless DC motors. The body frame and rotor blades are of carbon fibre construction for low weight and high strength. Its basic mechanical simplicity provides inherent reliability; no maintenance is required.

Unassisted visual reference is required for its operation.

Approximate maximum speed is 50 km/h (30 mph), maximum altitude is 8000 ft.

Maximum flight time is quoted as 20 min duration without payload. Duration with various payloads is not quoted, and obviously depends on the payload weight.

A wide variety of cameras can be carried, for example, high-definition digital video camera, low light camera, thermal infrared camera.

RF communications comprise a 2.4 GHz data link and a 5.8 GHz video link. It also carries a miniature Flight Data Recorder. The on-board sensors comprise a triad of solid-state MEMS gyros and accelerometers, three magnetometers, barometric pressure sensor, GPS receiver.

The software in the on-board microprocessor uses the data from the sensors to control the rotor drive motors differentially and hence the magnitude and direction of the resultant thrust vector. Suitable control law algorithms are executed to provide automatic stability and manoeuvre command FBW flight control. The task of controlling the helicopter flight path by the ground operator is thus greatly eased and does not require continuous control to keep it on its flight path.

Further Reading

British International Conference Proceedings on Unmanned Air vehicle Systems, Department of Aerospace Engineering, Queens Building, University walk, Bristol, BS8 1TR. Note: Conferences were held annually at Bristol University, UK

Abbreviations

A to D	Analogue to Digital Converter
AAAR	Airworthiness Authorities Accuracy Requirements
ACE	Actuator Control Electronics
ADC	Air Data Computer
ADI	Attitude and Direction Indicator
ADIRS	Air Data Inertial Reference System
AFCS	Automatic Flight Control System
AFDX	Avionics Full Duplex Switched Ethernet
AHRS	Attitude Heading Reference System
AiRAC	Aeronautical Information Regulation and Control
AMLCD	Active Matrix Liquid Crystal Display
ARINC	Aeronautical Radio Incorporated
ASI	Airspeed Indicator
ASIC	Application-Specific Integrated Circuit
ATC	Air Traffic Control
ATM	Automatic Traffic Management
BC	Bus Controller
BCM	Backup Control Module
BHMD	Binocular Helmet-Mounted Display
BIT	Built-In Test
BLUE	Blind Landing Experimental Unit
BPS	Backup Power Supply
BVR	Beyond Visual Range
C/A	Coarse Acquisition Code
CAA	Civil Aviation Authority
CAN	Controller Area Network
CASE	Computer-Assisted Software Engineering
CCD	Charge-Coupled Device
CCIP	Continuously Computed Impact Point
CG	Centre of Gravity
CGH	Computer-Generated Hologram
CI	Cost Index
COTS	Commercial Off The Shelf
CP	Control Panel
CPIOM	Core Processing and Input Output Modules
CRT	Cathode Ray Tube
CRZ FL	Cruise Flight Level
D to A	Digital to Analogue Converter
DCM	Direction Cosine Matrix
DGPS	Differential Global Positioning System

R. P. G. Collinson, *Introduction to Avionics Systems*, https://doi.org/10.1007/978-3-031-29215-6

DLE	Digital Light Engine
DME	Distance Measuring Equipment
DR	Dead Reckoning
DTG	Dynamically Tuned Gyro
DVI	Direct Voice Input
ECM	Electronic Counter Measures
EFIS	Electronic Flight Instrument System
EFOB	Estimate Fuel On Board
EHA	Electro-Hydrostatic Actuator
EM	Electro-Magnetic
EMC	Electro-Magnetic Compatibility
EMI	Electro-Magnetic Interference
EMP	Electro-Magnetic Pulse
EPU	Estimated Position Uncertainty
ETA	Estimate Time of Arrival
EW	Electronic Warfare
F3I	Form Fit Function Interface
FAA	Federal Aviation Authority
FADEC	Full Authority Digital Engine Control
FATP	Functional Area Test Plan
FBW	Fly-by-Wire
FCS	Flight Control System
FE	Flight Envelope
FG	Flight Guidance
FLIR	Forward Looking Infrared
FMC	Flight Management Computer
FMS	Flight Management System
FOG	Fibre Optic Gyro
FOV	Field of View
GBAS	Ground-Based Augmentation System (GPS)
GCA	Ground Control Approach
GDOP	Geometric Dilution of Precision
GLONASS	Global Navigation Satellite System
GPS	Global Positioning System
GPWS	Ground Proximity Warning System
HDD	Head-Down Display
HF	High Frequency
HMD	Helmet-Mounted Display
HMS	Helmet-Mounted Sight
HMS	His Majesty's Ship
HOL	High-Order Language
HSDB	High-Speed Data Bus
HSI	Horizontal Situation Indicator
HSRB	High-Speed Ring Bus
HUD	Head-Up Display
IAM	Integrated Avionic Module
IAS	Indicated Air Speed
IC	Integrated Circuit
ICAO	International Civil Aviation Organization
IFF	Identification Friend or Foe
IFOG	Interferometric Fibre Optic Gyro
IFOV	Instantaneous Field of View
IIT	Image Intensifier Tube

IKBS	Intelligent Knowledge-Based System
ILS	Instrument Landing System
IMA	Integrated Modular Avionics
IMU	Inertial Monitoring Unit
IN	Inertial Navigation
INS	Inertial Navigation System
IOM	Input Output Module
IR	Infra-red
I-STAR	Intelligence-Surveillance-Target Acquisition Reconnaissance
JIAWG	Joint Industry Avionics Working Group
KCCU	Keyboard and Cursor Control Unit
LCD	Liquid Crystal Display
LED	Light-Emitting Diode
LLTV	Low-Light TV
LOS	Line of Sight
LRU	Line Replaceable Unit
LSI	Large-Scale Integration
LTPB	Linear Token Passing Bus
MCDU	Multi-purpose Control Display Unit
MEMS	Micro-Electro-Mechanical Systems
MF	Medium Frequency
MFD	Multi-Function Display
MIMO	Multi-Input Multi-Output
MLS	Microwave Landing System
MMR	Multi-Mode Receiver
MTBF	Mean Time Between Failures
NA	Numerical Aperture
NATO	North Atlantic Treaty Organisation
NAVAID	Navigation Aids
ND	Navigation Display
NDB	Non-directional Beacon
NED	North East Down
nm	Nautical Mile
NVG	Night-Viewing Goggles
OLED	Organic Light Emitting Diode
OOD	Object-Oriented Design
PCU	Power Control Unit
PERT	Project Evaluation Review Technique
PFC	Primary Flight Computer
PFD	Primary Flight Display
PIO	Pilot-Induced Oscillation
PRIM	Primary
QA	Quality Assurance
QFE	Ground pressure
QNH	Mean Sea level pressure
RDC	Remote Data Concentrator
RFOG	Resonator Fibre Optic Gyro
RLG	Ring Laser Gyro
RMP	Radio Management Panel
RMS	Root Mean Square
RNAV	Area Navigation
RNP	Required Navigation Performance
RT	Remote Terminal

RVR	Runway Visual Range
SAE	Society of Automotive Engineers
SAM	Surface-to-Air Missile
SAR	Search and Rescue
SAR	Synthetic Aperture Radar
SAS	Stability Augmentation System
SATCOM	Satellite Communications
SBAS	Satellite-Based Augmentation System
SCD	Source Control Documentation
SEC	Secondary
SID	Standard Instrument Departure
SIMO	Single Input Multi-Output
SINS	Ships Inertial Navigation System
SISO	Single Input Single Output
SMAC	Scene Matching Area Correlation
SMTP	Software Module Test Plan
SRD	Software Requirements Document
SSEC	Static Source Error Correction
STANAG	Standardisation Agreement
STAR	Standard Terminal Arrival Route
SWDD	Software Design Document
T/C	Traffic Collision
T/F	Terrain Following
TACAN	Tactical Air Navigation
TCAS	Traffic Collision Avoidance System
TCM	Terrain Characteristic Matching
TCN	Terrain Contour Navigation
TDM	Time Division Multiplexing
TERCOM	Terrain Contour Matching
TERPROM	Terrain Profile Matching
TFOV	Total Field of View
THS	Trimmable Horizontal Stabiliser
TLSRD	Top Level System Requirements Document
TRN	Terrain Reference Navigation
TV	Television
UAV	Unmanned Air Vehicle
UCAV	Uninhabited Combat Air Vehicle
UHF	Ultra High Frequency
UMA	Unmanned Aircraft
UTC	Universal Time Coordinate
VFR	Visual Flight Rules
VHF	Very High Frequency
VLSI	Very Large-Scale Integration
VOR	VHF Omni-directional Range
VRD	Virtual Retinal Display
VSI	Vertical Speed Indication
WAAS	Wide Area Augmentation System (GPS)
WDM	Wavelength Division Multiplexing
WW2	World War 2
ZFCG	Zero-Fuel Centre of Gravity
ZFW	Zero-Fuel Weight

Symbols

Because of the number of different disciplines covered by the blanket name 'Avionics', it is inevitable that there are occasions when the same symbol is used to denote different quantities. Where this is the case, the appropriate chapters are indicated in brackets. In any case, the context should make clear which quantity is being denoted.

A	Local speed of sound (Chap. 2)
A	Area enclosed by closed path of optical gyroscope (Chap. 5)
A	Diameter of HUD crt (Chap. 2)
$\mathbf{A}$	State coefficient (or plant) matrix
A_0	Speed of sound at standard sea level conditions ($A_0 = 340.3$ m/sec)
$A_{1'}$	Constant coefficient in q/η transfer function ($A_1 = M_q/I_y + Z_\alpha/mU$)
A_2	Constant coefficient in q/η transfer function ($A_2 = M_\alpha/I_y + M_q Z_\alpha/I_y mU$)
a	Aircraft acceleration (Chaps. 5 and 6)
a	Core radius of optical fibre (Chap. 9)
a_D	Downwards (vertical) axis accelerometer output
a_E	East axis accelerometer output
a_N	North axis accelerometer output
a_X	X-axis accelerometer output
a_Y	Y-axis accelerometer output
a_Z	Z-axis accelerometer output
a_0	GPS time correction coefficient representing satellite clock phase offset (Chap. 6)
a_1	GPS time correction coefficient representing satellite clock frequency offset (Chap. 6)
a_1	Magnetic fluxgate error coefficient (Chap. 6)
a_2	GPS time correction coefficient representing satellite clock ageing term (Chap. 6)
a_2	Magnetic fluxgate error coefficient (Chap. 6)
a_3	Magnetic fluxgate error coefficient (Chap. 6)
B	Accelerometer bias error
B_E	East accelerometer bias error
B_N	North accelerometer bias error
$\mathbf{B}$	Driving matrix
b_1	Magnetic fluxgate error coefficient
b_2	Magnetic fluxgate error coefficient
b_3	Magnetic fluxgate error coefficient
C_D	Drag coefficient ($C_D = D_W / \frac{1}{2}\rho V_T^2 S$)
C_L	Lift coefficient ($C_L = L_W / \frac{1}{2}\rho V_T^2 S$)
C_{Lt}	Tailplane lift coefficient
$C_{L\,max}$	Maximum value of lift coefficient
C_M	Pitching moment coefficient ($C_M = M / \frac{1}{2}\rho V_T^2 S c$)
C_{M0}	Pitching moment coefficient at zero lift
C_u	Range equivalent of GPS user clock offset
c	Speed of light ($c = 3 \times 10^8$ m/sec)

© The Editor(s) (if applicable) and The Author(s), under exclusive license to Springer Nature Switzerland AG 2023
R. P. G. Collinson, *Introduction to Avionics Systems*, https://doi.org/10.1007/978-3-031-29215-6

c	Aerodynamic mean chord = wing area/wing span (Chap. 3)
c_p	Specific heat at constant pressure
c_v	Specific heat at constant volume
c_1	Magnetic fluxgate error coefficient
c_2	Magnetic fluxgate error coefficient
c_3	Magnetic fluxgate error coefficient
D	Operator d/dt
D	Diameter of collimating lens (Chap. 2)
D_W	Wing drag force ($D_W = \frac{1}{2}\rho V_T^2 S C_D$)
d_1	Magnetic fluxgate error coefficient
d_2	Magnetic fluxgate error coefficient
d_3	Magnetic fluxgate error coefficient
e_0	Euler parameter ($e_0 = \cos \mu/2$)
e_1	Euler parameter ($e_1 = \alpha \sin \mu/2$)
e_2	Euler parameter ($e_2 = \beta \sin \mu/2$)
e_3	Euler parameter ($e_3 = \gamma \sin \mu/2$)
F	Effective focal length of HUD collimating lens
f	Frequency
Δf	Frequency difference
G	Stick to actuator gearing in power control unit
G_q	Pitch rate gearing (or gain)
g	Local value of gravitational acceleration $g = \dfrac{R_0^2}{(R_0+H)^2} \cdot g_0$
g_{equ}	Value of gravitational acceleration at equator at sea level ($g_{equ} = 9.780327714$ m/s)
g_x	Gravitational acceleration component along aircraft OX (forward) axis
g_y	Gravitational acceleration component along aircraft OY (side slip) axis
g_z	Gravitational acceleration component along aircraft OZ (vertical) axis
g_0	Local value of gravitational acceleration at sea level $$g_0 = \frac{\left(1+0.00193185 \sin^2 \lambda\right)}{\left(1 - 0.00669438 \sin^2 \lambda\right)^{1/2}} \cdot g_{equ}$$
H	Angular momentum of gyro ($H = J\omega_R$) (Chap. 5)
H	True altitude of aircraft above sea level (Chap. 6)
H_G	Geometric altitude – altitude above sea level assuming standard ICAO law and allowing for variation of g with altitude
H_H	Horizontal component of Earth's magnetic field $\left(H_H = \sqrt{H_1^2 + H_2^2}\right)$
H_I	Barometric/inertial height
H_P	Pressure altitude – geo-potential altitude above sea level assuming standard ICAO pressure/altitude law and $g = g_0$
H_S	Altitude of stratopause (20,000 m)
H_T	Altitude of tropopause (11,000 m)
H_X	Earth's magnetic field component along aircraft OX (forward) axis
H_Y	Earth's magnetic field component along aircraft OY (side-slip) axis
H_Z	Earth's magnetic field component along aircraft OZ (vertical) axis
H_1	Horizontal component of Earth's magnetic field along aircraft heading axis
H_2	Horizontal component of Earth's magnetic field at right angles to aircraft heading axis
h_{ix}	Partial derivative of range to ith satellite with respect to GPS users X co-ordinate $h_{ix} = \partial R_i/\partial X$)
h_{iy}	Partial derivative of range to ith satellite with respect to GPS users Y co-ordinate ($h_{iy} = \partial R_i/\partial Y$)
h_{iz}	Partial derivative of range to ith satellite with respect to GPS users Z co-ordinate ($h_{iz} = \partial R_i/\partial Z$)
	h_{ix}, h_{iy}, h_{iz} are the direction cosines of the angles between the range vector to the ith satellite and the X, Y, Z co-ordinates

I_x	Moment of inertia of aircraft about OX (roll) axis
I_y	Moment of inertia of aircraft about OY (pitch) axis
I_z	Moment of inertia of aircraft about OZ (yaw) axis
J	Moment of inertia of gyro rotor about spin axis (Chap. 5)
J	Mechanical√ equivalent of heat (Joule's constant) (Chap. 7)
j	$\sqrt{-1}$
K	Scalar gain of transfer function KG(D)
K	q/η transfer function gain ($K = M_\eta/I_y T_2$)
K	Kalman gain matrix
K_p	Roll rate error gain
K_v	PCU actuator velocity constant
K_Φ	Bank angle error gain
K_Ψ	Heading error gain
K_0	Gyro output scale factor (Chap. 5)
K_1	Feedback gain
K_2	Feedback gain
K_3	Integral term feedback gain
k_t	Tailplane efficiency factor $\left(k_t = \frac{\text{Dynamic pressure at tailplane}}{\text{Freestream dynamic pressure}}\right)$
L	Lapse rate – rate of change of static air temperature with altitude (Chap. 7)
	(Troposphere lapse rate $= 6.5 \times 10^{-3}$ °C/m)
	(Chemosphere rise rate $= 1.0 \times 10^{-3}$ °C/m)
L	Perimeter of closed light path in an optical gyro (Chap. 5)
L	Length of optical fibre (Chap. 9)
L	Distance of observer's eyes from collimating lens of HUD (Chap. 2)
L	Resultant aerodynamic rolling moment (Chap. 3)
L_f	Distance of aerodynamic centre of fin from aircraft CG
L_p	Rolling moment derivative due to rate of roll
L_r	Rolling moment derivative due to rate of yaw
L_v	Rolling moment derivative due to side-slip velocity
L_W	Wing lift force ($L_W = \frac{1}{2}\rho V_T^2 S C_L$)
L_ζ	Rolling moment derivative due to rudder deflection
L_ξ	Rolling moment derivative due to aileron deflection
l_t	Distance of aerodynamic centre of tailplane from aircraft CG
M	Mach number ($M = V_T/A$)
M	Resultant aerodynamic pitching moment
M_q	Pitching moment derivative due to pitch rate
M_u	Pitching moment derivative due to forward velocity increment
M_w	Pitching moment derivative due to vertical velocity increment
$M_{\dot{w}}$	Pitching moment derivative due to rate of change of vertical velocity
M_α	Pitching moment derivative due to incidence change
M_η	Pitching moment derivative due to tailplane/elevator deflection
m	Aircraft mass
m_a	Pendulous mass of accelerometer
N	Resultant aerodynamic yawing moment
N_p	Yawing moment derivative due to rate of roll
N_r	Yawing moment derivative due to rate of yaw
N_v	Yawing moment derivative due to side-slip velocity
N_ζ	Yawing moment derivative due to rudder deflection
N_ξ	Yawing moment derivative due to aileron deflection
n	Phase advance gain (Chaps. 4, 5, and 8)
n	Refractive index (Chaps. 5 and 9)
n_1	Refractive index of optical fibre core (Chap. 9)

n_2	Refractive index of optical fibre cladding (Chap. 9)
$\mathbf{P}$	Covariance matrix
P_s	Hydraulic supply pressure
P_S	Static pressure
P_{S0}	Static pressure at standard sea level conditions ($P_{S0} = 101.325$ kN/m^2 = 1013.25 mbar)
P_{SS}	Static pressure at stratopause altitude (20,000 m) ($P_{SS} = 54.75$ mbar)
P_{ST}	Static pressure at tropopause altitude (11,000 m) ($P_{ST} = 226.32$ mbar)
P_T	Total pressure ($P_T = Q_C + P_S$)
ΔP	INS position error (Chap. 6)
ΔP	Angular increment in roll ($\Delta P = p\Delta t$) (Chap. 5)
p	Roll rate – aircraft angular velocity in roll
p	Pressure (Chap. 7)
p_D	Roll rate demand
p_E	Roll rate error ($p_E = p_D - p$)
Q	Dynamic pressure ($Q = \frac{1}{2}\rho V_T^2 S$)
Q_C	Impact pressure
ΔQ	Angular increment in pitch ($\Delta Q = q\Delta t$)
q	Pitch rate – aircraft angular velocity in pitch
R	Distance of aircraft from Earth's centre ($R = R_0 + H$) (Chap. 6)
R	Slant range of aircraft from ILS localiser or glide slope transmitters (Chap. 8)
R_a	Gas constant for unit mass of dry air ($R_a = 287.0529$ Joules/°K/kg)
R_i	Range from ith GPS satellite ($i = 1, 2, 3,..., 24$)
R_{iu}	Estimated range from ith satellite computed from estimates of GPS users co-ordinates
R_0	Radius of earth
R_1	Range of GPS satellite 1 from user
R_2	Range of GPS satellite 2 from user
R_3	Range of GPS satellite 3 from user
R_4	Range of GPS satellite 4 from user
R_{1p}	Pseudo-range of GPS satellite 1 from user
R_{2p}	Pseudo-range of GPS satellite 2 from user
R_{3p}	Pseudo-range of GPS satellite 3 from user
R_{4p}	Pseudo-range of GPS satellite 4 from user
ΔR	Angular increment in yaw ($\Delta R = r\Delta t$)
ΔR_1	Range error GPS satellite 1
ΔR_2	Range error GPS satellite 2
ΔR_3	Range error GPS satellite 3
ΔR_4	Range error GPS satellite 4
ΔR_i	Range error ith GPS satellite ($i = 1,2,3,4,...,24$)
r	Yaw rate – aircraft angular velocity in yaw
r	Recovery ratio – constant dependent on the temperature probe installation (Chap. 7)
S	Surface area of wing
S_t	Surface area of tailplane
s	Laplace operator
T	Temperature (Chap. 7)
T	Iteration period (Chap. 6)
T	Time constant (Chaps. 3, 4, 5, 6, and 8)
T	Range equivalent of GPS user's clock offset ($T - c\Delta T$)
ΔT	GPS user time offset (or bias)
δT	Correction to estimate of range equivalent of GPS user time offset ($T = T_u + \delta T$)
T_{Act}	PCU actuator time constant
T_m	Measured (or indicated) air temperature

T_R	Roll time constant ($T_R = I_x/L_p$)
T_S	Static air temperature – temperature of free air-stream
T_{sp}	Time constant of spiral divergence
T_T^*	Static air temperature at tropopause altitude ($T_T^* = 216.65\,°\text{K}$)
T_T	Total air temperature
T_0	Static air temperature at standard sea-level conditions ($T_0 = 288.15\,°\text{K}$)
T_1	Time constant ($T_1 = I_y/M_q$)
T_2	Time constant ($T_2 = mU/Z_\alpha$)
t	Time
Δt	Time increment
Δt_r	GPS relativistic correction term
$t_{s/c}$	Effective GPS satellite time at signal transmission
$\Delta t_{s/c}$	Time offset between GPS satellite and GPS master time
Δt_1	Time for signal from GPS satellite 1 to reach user
Δt_2	Time for signal from GPS satellite 2 to reach user
Δt_3	Time for signal from GPS satellite 3 to reach user
Δt_4	Time for signal from GPS satellite 4 to reach user
$t_{0/c}$	GPS epoch time at which polynomial coefficients a_0, a_1, a_2 are referenced
U	Forward velocity – velocity of aircraft CG along OX (forward) axis in disturbed flight ($U = U_0 + u$)
$\mathbf{U}$	Control input vector
U_A	Forward velocity derived from air data system
U_0	Velocity of aircraft CG along OX (forward) axis in steady flight
u	Forward velocity increment in disturbed flight
V	Air velocity (Chap. 7)
V	Side-slip velocity – velocity of aircraft CG along OY (side-slip) axis in disturbed flight
ΔV	INS velocity error
V_A	Side-slip velocity derived from air data system
V_C	Calibrated airspeed
V_D	Downwards (vertical) velocity
V_E	Velocity along east axis of local north, east, down axis frame
ΔV_E	East axis velocity error
V_F	Horizontal component of aircraft forward velocity
V_G	Ground speed ($V_G = \sqrt{V_F^2 + V_S^2}$)
V_H	Horizontal component of true air-speed ($V_H = V_T \cos\theta$)
V_N	Velocity along north axis of local north, east, down axis frame
ΔV_N	North axis velocity error
V_S	Horizontal component of aircraft side-slip velocity
V_T	True air-speed
V_W	Wind velocity
v	Side-slip velocity increment in disturbed flight
W	Gyro drift rate (Chap. 6)
W	Vertical velocity – velocity of aircraft CG along OZ (vertical) axis in disturbed flight ($W = W_0 + w$) (Chaps. 3 and 6)
W_A	Vertical velocity derived from air data system (along aircraft OZ axis)
W_E	Gyro drift rate about East axis
W_N	Gyro drift rate about North axis
W_0	Velocity of aircraft CG along OZ (vertical) axis in steady flight
w	Vertical velocity increment in disturbed flight
$\mathbf{X}$	System state vector
X	X-axis co-ordinate of GPS user

ΔX	Correction to estimate of X-axis coordinate of GPS user ($X = X_\mathrm{u} + \Delta X$)
X_a	Incremental change in aerodynamic force along OX (forward) axis following a disturbance
X_i	X-axis co-ordinate of ith GPS satellite
X_u	Forward force derivative due to forward velocity increment (Chap. 3)
X_u	Best estimate of X-axis coordinate of GPS user (Chap. 6)
X_w	Forward force derivative due to vertical velocity increment
X_1	X-axis coordinate of GPS satellite 1
X_2	X-axis coordinate of GPS satellite 2
X_3	X-axis coordinate of GPS satellite 3
X_4	X-axis coordinate of GPS satellite 4
x_ε	PCU control valve displacement
x_m	Forward axis magnetic fluxgate output
x_i	Pilot's input movement to PCU
x_0	PCU actuator displacement
Y	Y-axis coordinate of GPS user
ΔY	Correction to estimate of Y-axis coordinate of GPS user ($Y = Y_\mathrm{u} + \Delta Y$)
Y_a	Incremental change in aerodynamic force along OY (side-slip) axis following a disturbance
Y_i	Y-axis co-ordinate of ith GPS satellite
Y_u	Best estimate of Y-axis coordinate of GPS user
Y_v	Side-force derivative due to side slip velocity increment
Y_ζ	Side-force derivative due to rudder deflection
Y_1	Y-axis coordinate of GPS satellite 1
Y_2	Y-axis coordinate of GPS satellite 2
Y_3	Y-axis coordinate of GPS satellite 3
Y_4	Y-axis coordinate of GPS satellite 4
y_m	Side-slip axis magnetic fluxgate output
Z	Z-axis coordinate of GPS user
ΔZ	Correction to estimate of Z axis coordinate of GPS user ($Z = Z_\mathrm{u} + \Delta Z$)
Z_a	Incremental change in aerodynamic force along OZ (vertical) axis following a disturbance
Z_i	Z-axis co-ordinate of ith GPS satellite
Z_m	Vertical axis magnetic fluxgate output
Z_q	Vertical force derivative due to pitch rate
Z_u	Vertical force derivative due to forward velocity increment (Chap. 3)
Z_u	Best estimate of Z-axis coordinate of GPS user
Z_w	Vertical force derivative due to vertical velocity increment
$Z_{\dot{w}}$	Vertical force derivative due to rate of change of vertical velocity
Z_α	Vertical force derivative due to incidence change
Z_η	Vertical force derivative due to tailplane/elevator deflection
Z_1	Z-axis coordinate of GPS satellite 1
Z_2	Z-axis coordinate of GPS satellite 2
Z_3	Z-axis coordinate of GPS satellite 3
Z_4	Z-axis coordinate of GPS satellite 4
α	Angle of incidence (angle of attack – US) (Chaps. 3, 4, 7, and 8)
α	Direction cosine used to specify aircraft attitude and derive Euler parameters e_0, e_1, e_2, e_3 (Chap. 5)
α_max	Maximum value of angle of incidence
α_T	Angle of incidence for trimmed flight
β	Side-slip incidence (Chaps. 3, 4, 5, and 7)
β	Direction cosine used to specify aircraft attitude and derive Euler parameters e_0, e_1, e_2, e_3 (Chap. 5)

γ	$\dfrac{\text{Specific heat at constant pressure}}{\text{Specific heat at constant volume}} \left(\gamma = \dfrac{c_p}{c_v} \right) = 1.4$ for air (Chap. 7)
γ	Dihedral angle (Chap. 3)
γ	Direction cosine used to specify aircraft attitude and derive Euler parameters e_0, e_1, e_2, e_3 (Chap. 5)
γ_F	Flight path angle – angle aircraft velocity vector makes with chosen reference axis (usually horizontal axis)
γ_L	Angular displacement of aircraft from ILS localiser beam centre line
γ_V	Angular displacement of aircraft from ILS glide slope beam centre line
Δ	Refractive index difference ratio of optical fibre $(\Delta = (n_1 - n_2/n_1))$
δ	Drift angle – angle between aircraft track and heading
δ_i	Pilot's stick input
ε	Error signal
ζ	Angular deflection of rudder from trimmed position (Chaps. 3 and 4)
ζ	Damping ratio of second-order system (standard form: $D^2 + 2\zeta\omega_0 D + \omega_0^2$)
η	Tailplane/elevator angular deflection from trimmed position
η_D	Demanded tailplane/elevator angle
θ	Pitch angle (Chaps. 3, 4, 5, and 8)
θ	Incidence angle of light ray (Chaps. 2 and 9)
θ_B	Angle between ILS guidance beam centre line and reference axis
θ_{crit}	Critical incidence angle of light ray at which total internal reflection occurs
θ_D	Demanded pitch angle
θ_i	Input quantity
θ_o	Output quantity
θ_p	Angular rotation of stable platform with respect to inertial axis frame
θ_1	Quantity 1
θ_2	Quantity 2
$\Delta\theta$	Tilt angle error
$\Delta\theta_X$	Tilt angle error about X axis
$\Delta\theta_Y$	Tilt angle error about Y axis
$\Delta\theta_E$	Tilt angle error about East axis
$\Delta\theta_N$	Tilt angle error about North axis
λ	Latitude angle (Chaps 5 and 6)
λ	Wave length (Chaps 5 and 9)
λ	Root of characteristic equation (Chap. 3)
λ_0	Latitude of initial position
μ	Longitude angle (Chap. 6)
μ	Single angular rotation used to specify aircraft attitude and Euler parameters e_0, e_1, e_2, e_3 (Chap. 5)
μ_0	Longitude of initial position
ξ	Angular deflection of ailerons from trimmed position
ρ	Air density
ρ_0	Air density at standard sea level conditions $(\rho_0 = 1.225 \text{ kg/m}^3)$
τ_1	GPS satellite 1 clock correction
τ_2	GPS satellite 2 clock correction
τ_3	GPS satellite 3 clock correction
τ_4	GPS satellite 4 clock correction
Φ	Bank angle (Chaps. 3, 4, 5, and 8)
Φ	Phase shift (Chaps. 5 and 8)
Φ_D	Demanded bank angle
Φ_E	Bank angle error $(\Phi_E = (\Phi_D - \Phi))$
Φ_n	State transition matrix
Φ_s	Sagnac phase shift

ψ	Heading (or yaw) angle
ψ_D	Demanded heading angle
ψ_E	Heading error ($\psi_E = (\psi_D - \psi)$)
$\psi_{G/M}$	Combined gyro/magnetic heading angle
ψ_M	Magnetically derived heading angle
ψ_T	Track angle
ψ_W	Wind direction with respect to true north
$\Delta\psi$	INS heading error
Ω	Earth's angular velocity about polar axis
ω	Angular frequency
ω_R	Angular velocity of gyro rotor about spin axis
ω_0	Undamped natural frequency of second$_2$) order system (standard form: $D^2 + 2\zeta\omega_0 D + \omega_0^2$)

Glossary

Adiabatic process A process where no heat enters or leaves the system.

Aerodynamic centre The point about which the pitching moment does not change with the angle of incidence (providing the velocity is constant).

Aerodynamic derivative The partial derivative of the aerodynamic force or moment with respect to a particular variable. For small changes in that variable, the resulting incremental force or moment is equal to:(Derivative) × (Incremental change in the variable).

Aliasing The effects from sampling data at a sampling frequency below the frequency of the noise components present in the signal so that spurious low-frequency signals are introduced from the sampled noise.

Air density The mass per unit volume of air.

Air density ratio The ratio of the air density to the value of the air density at standard sea level conditions. See 'Standard sea level conditions'.

Altitude The height of the aircraft above the ground.

Attitude The angular orientation of the aircraft with respect to a set of Earth-referenced axes. This is defined by the three Euler angles – yaw (or heading) angle, pitch angle, bank angle. See 'Euler angles'.

Bank angle The angle through which the aircraft must be rotated about the roll axis to bring it to its present orientation from the wings level position, following the pitch and yaw rotations. See 'Euler Angles'.

Bus controller The unit which controls the transmission of data between the units connected to an MIL STD 1553B multiplexed data bus system.

Calibrated airspeed The speed which under standard sea level conditions would give the same impact pressure as that measured on the aircraft. See 'Standard sea level conditions'.

Category I, II, III landing conditions These three categories define the landing visibility conditions in terms of the vertical visibility ceiling and the runway visual ranges; the visibility decreasing with increasing category number.

Chemosphere The region above the stratopause altitude of 20,000 m (65,617 ft) up to 32,004 m (105,000 ft) where the temperature is assumed to rise linearly with increasing height at 1.0×10^{-3} °C/m.

Collective pitch control The angle of incidence of the helicopter rotor blades to the airstream, i.e. pitch angle, is changed individually by the *same amount* to increase or decrease the rotor lift to control the helicopter vertical flight profile.

Collimation An optically collimated display is one where the rays of light from any particular point on the display are all parallel after exiting the collimating system.

Combiner The optical element of the head-up display through which the pilot views the outside world and which combines the collimated display image with the outside world scene.

Complementary filtering The combination of data from different sources through appropriate filters which select the best features of each source so that their dissimilar characteristics can be combined to complement each other.

Coning The motion resulting from a body experiencing two angular vibrations of the same frequency and 90° out of phase about two orthogonal axes of the body.

Consolidation The process of deriving a single value for a quantity from the values obtained for that quantity from several independent sources; for example, selecting the median value.

Coriolis acceleration The acceleration introduced when the motion of a vehicle is measured with respect to a rotating frame of reference axes.

Cyclic pitch control (Refer to collective pitch control.) The pitch angle of each individual rotor blade is varied over the cycle of a complete revolution to change the direction of the rotor lift vector, through the rotor and helicopter dynamics, and hence enable control of the forward and lateral motion of the helicopter.

Decision height This is the minimum vertical visibility ceiling for a landing to be safely carried out.

R. P. G. Collinson, *Introduction to Avionics Systems*, https://doi.org/10.1007/978-3-031-29215-6

Derivative See 'Aerodynamic derivative'.

Differential GPS The positioning of a mobile station in real time by corrected (and possibly Doppler or phase smoothed) GPS pseudo-ranges. The corrections are determined at a static 'reference station' and transmitted to the mobile station. A monitor station may be part of the system, as a quality check on the reference station transmissions.

Drag coefficient A non-dimensional coefficient which is a function of the angle of incidence and which is used to express the drag generation characteristics of an aerofoil.

Drift angle The angle between the horizontal projection of the aircraft's forward axis and the horizontal component of the aircraft's velocity vector.

Dynamic pressure The pressure exerted to bring a moving stream of air to rest assuming the air is incompressible.

Ephemeris parameters These comprise 16 parameters which define the GPS satellite orbital position data with respect to Earth reference axes.

Epoch GPS receivers provide a stream of position estimates and associated parameters. The time mark given to each observation is known as an epoch.

Euler angles The aircraft attitude is defined by a set of three ordered rotations, known as the Euler angles, from a fixed reference axis frame; the aircraft is assumed to be initially aligned with the reference axes:1. A clockwise rotation in the horizontal plane through the *yaw angle* about the yaw (or vertical) axis.2. A clockwise rotation through the *pitch angle* about the pitch (or side-slip) axis.3. A clockwise rotation through the *bank angle* about the roll (or forward) axis.

Euler symmetrical parameters These four parameters are used to derive the vehicle attitude in a strap-down system and are functions of the three direction cosines of the axis about which a single rotation will bring the vehicle from initial alignment with a reference axis frame to its present orientation, and the single rotation angle.The four parameters are equal to:cosine (half rotation angle)(direction cosine 1) × sine (half rotation angle)(direction cosine 2) × sine (half rotation angle)(direction cosine 3) × sine (half rotation angle)

Exit pupil diameter The diameter of the circle within which the observer's eyes are able to see the whole of the display; the centre of the exit pupil is located at the design eye position.

Fluxgate A magnetic field sensor which provides an electrical output signal proportional to the magnetic field.

Fly-by-wire control system A flight control system where all the command and control signals are transmitted electrically and the aerodynamic control surfaces are operated through computers which are supplied with the pilot's command signals and the aircraft state from appropriate motion sensors.

Free azimuth axes A local level set of axes where the horizontal axes are not rotated in space about the local vertical axis.

Gain margin The amount the loop gain can be increased in a closed-loop system before instability results because the open-loop gain at the frequency where there is 180° phase lag has reached 0 dB (unity).

Gearing A term used in flight control systems to specify a feedback gain in terms of the control surface angular movement per unit angular change, or, unit angular velocity change in the controlled quantity, e.g. 1° tailplane angle/1° per second pitch rate.

Geometric dilution of precision The loss of accuracy in the position computation by the GPS receiver due to the geometry of the user and the satellites in view.

Great circle A circle on the surface of a sphere whose plane goes through the centre of the sphere.

Gyro compassing A method for determining the direction of true north by using inertial quality gyros to measure the components of the Earth's angular velocity.

Heading angle The angle between the horizontal projection of the aircraft's forward axis and the direction of true north.

Impact pressure The pressure exerted to bring the moving airstream to rest at that point.

Incidence angle The angle between the direction of the relative wind vector to the aerofoil chord line (a datum line through the aerofoil section). Also known as the 'Angle of attack' in the USA.

Indicated airspeed The speed under standard sea level conditions which would give the same impact pressure as that measured by the air-speed indicator (ASI). It is basically the same quantity as the calibrated airspeed but includes instrument errors and static source pressure errors.

Indicated air temperature See 'Measured air temperature'.

Instantaneous field of view The angular coverage of the imagery which can be seen by the observer at any specific instant.

Kalman filter A recursive data processing algorithm which processes sensor measurements to derive an optimal estimate of the quantities of interest (states) of the system using a knowledge of the system and measurement device dynamics, uncertainties, noises, measurement errors and initial condition information.

Knot A measure of vehicle speed; one knot being equal to one nautical mile per hour.

Lapse rate The rate at which the temperature is assumed to decrease with increasing altitude.

Latency The time delay between sampling a signal and processing it so that the processed signal output lags the real signal in time. The resulting phase lag can exert a destabilising effect in a closed-loop control system.

Latitude The angle subtended at the Earth's centre by the arc along the meridian passing through the point and measured from the equator to the point.

Lift coefficient A non-dimensional coefficient which is a function of the angle of incidence and which is used to express the effectiveness of an aerofoil in generating lift.

Longitude The angle subtended at the Earth's centre by the arc along the equator measured east or west of the prime meridian to the meridian passing through the point.

Mach number The ratio of the true airspeed of the aircraft to the local speed of sound.

Magnetic deviation The error introduced by the distortion of the Earth's magnetic field in the vicinity of the magnetic sensor by the presence of magnetic materials.

Magnetic dip angle The angle between the Earth's magnetic field vector and the horizontal.

Magnetic variation The angular difference between the direction of true north and magnetic north.

Mean aerodynamic chord This is equal to the wing area divided by the wing span.

Measured air temperature The temperature measured by a sensing probe where the air may not be brought wholly to rest.

Meridian A circle round the Earth passing through the North and South poles.

Multi-mode fibre An optical fibre whose dimensions are such that there are a large number of ways or modes by which light can be guided along the fibre depending on the incidence angle.

Nautical mile One nautical mile is equal to the length of the arc on the Earth's surface subtended by an angle of one minute of arc measured at the Earth's centre.

Neutral point The position of the aircraft's CG where the rate of change of pitching moment coefficient with incidence is zero.

Newton The force required to accelerate a mass of one kilogram at one metre per second per second.

Nichol's chart A chart of logarithmic gain versus phase which is used to plot the open-loop frequency response of a closed-loop automatic control system. The chart shows the open-loop logarithmic gain and phase lag at a particular frequency and enables the phase margin and gain margin to be read off directly. Loci of constant gain and phase are superimposed on the chart which enable the closed-loop response to be easily mapped.

Notch filter A filter designed to provide a very high attenuation over a narrow band of frequencies centred at a specific frequency.

Numerical aperture This defines the semi-angle of the cone within which an optical fibre will accept light and is a measure of the light-gathering power of the fibre.

Pascal The pressure exerted by a force of one Newton acting on an area of one square metre.

Phase margin The additional phase lag within a closed-loop system that will cause instability by producing 180° phase lag at the frequency where the open-loop gain is 0 dB (unity gain).

Phugoid A very lightly damped long period oscillation in height and airspeed in the longitudinal plane; the angle of incidence remaining virtually unchanged.

Pitch angle The angle between the aircraft's forward axis and the horizontal, being the angle the aircraft must be rotated about the pitch axis following the rotation in yaw to arrive at its present orientation with respect to a fixed reference frame. See 'Euler angles'.

Pitching moment coefficient A non-dimensional coefficient which is equal to the pitching moment about the aircraft's CG divided by the product of the dynamic pressure, wing surface area and the mean aerodynamic chord.

Pressure The force per unit area.

Pressure altitude The height above sea level calculated from the measured static pressure assuming a standard atmosphere.

Precession The behaviour of the gimbal suspended spinning rotor of a gyroscope which causes it to turn about an axis which is mutually perpendicular to the axis of the applied torque and the spin axis; the angular rate of precession being proportional to the applied torque.

Prime meridian The meridian passing through Greenwich, England.

Quaternion A quantity comprising a scalar component and a vector with orthogonal components. The Euler symmetrical parameters are quaternions.

Recovery ratio A correction factor to allow for the air not being brought wholly to rest at the temperature probe.

Relative wind Velocity of the airstream relative to the aircraft (equal and opposite to the true airspeed vector).

Remote terminal The interface unit which enables a sub-system to communicate with other systems by means of a time division multiplexed data bus system; in particular, the MIL STD 1553B data bus system.

Sagnac effect Two counter propagating coherent light waves experience a relative phase difference on a complete trip around a rotating closed path; this phase difference is proportional to the input rotation rate. The effect is known as the Sagnac effect and the phase difference the Sagnac phase shift.

Schuler period The period of oscillation of a Schuler tuned IN system which is equal to that of a simple pendulum with a length equal to the radius of the Earth, that is 84.4 minutes approximately.

Schuler tuning The feedback of the inertially derived vehicle rates of rotation about the local level axes of an IN system so that the system tracks the local vertical as the vehicle moves over the spherical surface of the Earth.

Single mode fibre An optical fibre which is specifically designed so that only one guided mode is possible for transmitting light along the fibre. This is achieved by suitable choice of the core radius (of the same order as the transmitted wavelength) and the refractive index difference between the core and cladding.

Spiral divergence A slow build up of yaw and roll motion resulting in a spiral dive which is due to the rolling moment created by the rate of yaw.

Standard sea level conditions These are assumed in the 'standard atmosphere' to be a sea level pressure of 101.325 kPa (1013.25 mb) and a temperature of 288.15 °K (15 °C).

Static air temperature The temperature which would be measured moving freely in the air stream.

Static margin The distance of the centre of gravity of the aircraft from the *neutral point* divided by the mean aerodynamic chord.

Static pressure The pressure of the free airstream due to the random motion of the air molecules.

Static source error The error in the measured static pressure due to the effects of Mach number and incidence on the static source.

Strapdown system A system where the gyros and accelerometers are mounted in a block which is fixed rigidly to the airframe (as opposed to being mounted in a gimbal system).

Stratopause altitude The altitude in the 'standard atmosphere' where the temperature is assumed to start rising linearly with increasing altitude and the region of the chemosphere begins. It is equal to 20,000 m (65,617 ft).

Stratosphere The region between the tropopause altitude of 11,000 m (36,089.24 ft) and the stratopause altitude of 20,000 m (65,6127 ft) where the temperature is assumed in the 'standard atmosphere' to be constant at 216.65 °K (−56.5 °C).

Tilt errors The gravitational acceleration component errors introduced by the tilt angles from the horizontal of the input axes of the nominally horizontal accelerometers of an IN system (or derived horizontal acceleration components in the case of a strap-down system).

Total air temperature The temperature that would exist if the moving airstream where brought wholly to rest at that point. It is thus the free airstream temperature plus the temperature rise due to the kinetic heating of the air because of the air being brought to rest.

Total field of view The total angular coverage of the display imagery which can be seen by moving the observer's eye position around.

Total pressure The pressure that would exist if the moving airstream were brought to rest at that point. It is equal to the impact pressure plus the static pressure.

Track angle The direction of the ground speed vector relative to true North.

Tropopause altitude The altitude where the region of constant temperature known as the stratosphere is assumed to start in the 'standard atmosphere'. It is equal to 11,000 m (36,089.24 ft).

Troposphere The region from sea level up to the tropopause altitude of 11,000 m (36,089.24 ft) where the temperature is assumed to decrease linearly with increasing altitude.

Yaw angle The angle measured in the horizontal plane between a fixed reference axis and the horizontal projection of the aircraft's forward axis. It is the angle through which the aircraft must first be rotated followed by the pitch and roll rotations to bring it to its present orientation. See Euler angles.

Index

Printed in the United States
by Baker & Taylor Publisher Services